Pipeline Operations

Level Two

Trainee Guide
Third Edition

PEARSON

Boston Columbus Indianapolis New York San Francisco Amsterdam
Cape Town Dubai London Madrid Milan Munich Paris Montreal Toronto Delhi
Mexico City Sao Paulo Sydney Hong Kong Seoul Singapore Taipei Tokyo

NCCER

President: Don Whyte
Vice President: Steve Greene
Chief Operations Officer: Katrina Kersch
Director of Product Development: Daniele Dixon
Pipeline Mechanical Project Managers: Patty Bird, Elizabeth Schlaupitz, Justin Street, Chris Wilson
Senior Development Manager: Mark Thomas
Senior Production Manager: Tim Davis
Quality Assurance Coordinator: Karyn Payne
Desktop Publishing Coordinator: James McKay
Permissions Specialist: Adrienne Payne
Production Specialist: Adrienne Payne
Editors: Graham Hack, Karyn Payne, Debie Hicks, Jackie Strawbridge, Adrienne Payne

Writing and development services provided by Topaz Publications, Liverpool, NY

Lead Writers/Project Managers: Don Congdon, Veronica Westfall
Desktop Publisher: Joanne Hart
Art Director: Alison Richmond
Permissions Editors: Andrea LaBarge, Briana Morgan
Writers: Don Congdon, Tony Clay, Thomas Burke

Pearson

Director of Alliance/Partnership Management: Andrew Taylor
Editorial Assistant: Collin Lamothe
Program Manager: Alexandrina B. Wolf
Digital Studio Project Managers: Heather Darby, Tanika Henderson, Jose Carchi
Director of Marketing: Leigh Ann Simms
Senior Marketing Manager: Brian Hoehl
Composition: NCCER
Printer/Binder: LSC Communications
Cover Printer: LSC Communications
Text Fonts: Palatino and Univers

Credits and acknowledgments for content borrowed from other sources and reproduced, with permission, in this textbook appear at the end of each module.

ScoutAutomatedPrintCode

PEARSON

ISBN-13: 978-0-13-472539-0
ISBN-10: 0-13-472539-5

Preface

To the Trainee

NCCER is pleased to release the 3rd edition of Pipeline Training and Qualification with updates that will engage you and give you the best training in the industry. The curriculum has been redesigned to better align with the learning objectives, and also includes new end-of-section review questions to compliment the Module Review. Changes to the NCCER's covered task (CT) modules include updates to the latest technology and standards, as well as alignment to the American Petroleum Institute's *API Recommended Practice for Pipeline Operator Qualification (OQ) 1161*, 3rd edition (2014). Each CT module contains task-specific Abnormal Operating Conditions (AOCs) to which the trainee must be prepared to respond. General AOCs for both Field (Liquid and Gas) and Control Center settings are also available as single modules.

CT modules are available as traditional printed levels and as individual module eBooks. Modules are completed by successfully passing the proctored Module Examination, and completing the Performance Profile under the supervision of an NCCER-certified evaluator. These completions are tracked within the NCCER Registry for training delivered through an Accredited Training Sponsor. Covered task module completions in the Registry are also documented in ISNetworld.

The modules are designed to be used for Operational Qualification, and they are each a vital part of NCCER's pipeline workforce development curricula. They provide in-depth training on the fundamental knowledge and skills required for pipeline work, as well as technical training that increases the depth and breadth of a pipeline worker's capability within several specialty areas.

We invite you to visit the NCCER website at **www.nccer.org** for information on the latest product releases and training, as well as online versions of the *Cornerstone* magazine and Pearson's NCCER product catalog.

Your feedback is welcome. You may email your comments to **curriculum@nccer.org** or send general comments and inquiries to **info@nccer.org**.

NCCER Standardized Curricula

NCCER is a not-for-profit 501(c)(3) education foundation established in 1996 by the world's largest and most progressive construction companies and national construction associations. It was founded to address the severe workforce shortage facing the industry and to develop a standardized training process and curricula. Today, NCCER is supported by hundreds of leading construction and maintenance companies, manufacturers, and national associations. The NCCER Standardized Curricula was developed by NCCER in partnership with Pearson, the world's largest educational publisher.

Some features of the NCCER Standardized Curricula are as follows:

- An industry-proven record of success
- Curricula developed by the industry for the industry
- National standardization providing portability of learned job skills and educational credits
- Compliance with the Office of Apprenticeship requirements for related classroom training (*CFR 29:29*)
- Well-illustrated, up-to-date, and practical information

NCCER also maintains the NCCER Registry, which provides transcripts, certificates, and wallet cards to individuals who have successfully completed a level of training within a craft in NCCER's Curricula. *Training programs must be delivered by an NCCER Accredited Training Sponsor in order to receive these credentials.*

Additional Pipeline Resources

The following additional Pipeline Resources are available from NCCER's website at **www.nccer.org/pipeline**:

1. A complete copy of NCCER's Covered Task (CT) list
2. Access to:
   a. An online bookstore to order print and digital versions of NCCER's pipeline curricula.
   b. Instructor resources (Lesson Plans and PowerPoints)
   c. Exams for Operator Qualification
   d. A site to custom build curriculum with pipeline and/or other NCCER-developed modules

Pipeline Career Pathway Training

Entry-level Pipeline Craft Trainee → Pipeline Fundamentals → Level 1

The newly established pipeline career pathway represented on this page provides the recognition pipeline professionals deserve through standardized training and industry-recognized credentials.

Levels of training have been established by combining training modules into specific focus areas within the industry. This allows trainees to progress through a standardized program and earn a professional credential.

In addition, Levels Two and Three are comprised of covered task training modules that can lead to operator qualifications. *Pipeline Corrosion Control is the only exception which actually provides covered task training starting in Level One.*

The blue boxes denote the professional credential that can be earned with successful completion of each program.

A trained workforce is a safe and qualified workforce!

Pipeline Fundamentals

Core Curriculum

Introduction to the Pipeline Industry

and

a.) Field Abnormal Operating Conditions
Module: AOCFG-17

or

b.) Control Center Abnormal Operating Conditions
Module: AOCCC_17

Level 1

Field and Control Center Operations

Pipeline Maintenance and Mechanical

Electrical and Instrumentation

Pipeline Corrosion Control (CT)

Level 2: Craft-Specific Training with Operator Qualification* → **Level 3:** Craft-Specific Training with Operator Qualification* → **Trained and Qualified Pipeline Professional Credential**

Level 2: Craft-Specific Training with Operator Qualification*	Level 3: Craft-Specific Training with Operator Qualification*	Trained and Qualified Pipeline Professional Credential
Pipeline Operations (Control Center/ Gas/Liquid) CT		Pipeline Operations Technician
Pipeline Mechanical CT	Pipeline Mechanical CT	Pipeline Mechanical Technician
Pipeline Maintenance CT	Pipeline Maintenance CT	Pipeline Maintenance Technician
Pipeline Electrical and Instrumentation CT		Pipeline Electrical and Instrumentation Technician
Pipeline Corrosion Control CT		Pipeline Corrosion Technician

CT *Each of these titles have corresponding operator qualifications. The Covered Task list is available online at www.nccer.org/pipeline-program*

Special Features

In an effort to provide a comprehensive and user-friendly training resource, this curriculum showcases several informative features. Whether you are a visual or hands-on learner, these features are intended to enhance your knowledge of the construction industry as you progress in your training. Some of the features you may find in the curriculum are explained below.

Introduction

This introductory page, found at the beginning of each module, lists the module Objectives, Performance Tasks and Trade Terms. The Objectives list the knowledge you will acquire after successfully completing the module. The Performance Tasks give you an opportunity to apply your knowledge to real-world tasks. The Trade Terms are industry specific vocabulary that you will learn as you study this module.

00104-15
INTRODUCTION TO POWER TOOLS

Objectives

When you have completed this module, you will be able to do the following:

1. Identify and explain how to use various types of power drills and impact wrenches.
   a. Identify and explain how to use common power drills and bits.
   b. Identify and explain how to use a hammer drill.
   c. Identify and explain how to use pneumatic drills and impact wrenches.
2. Identify and explain how to use various types of power saws.
   a. Identify and explain how to use a circular saw.
   b. Identify and explain how to use saber and reciprocating saws.
   c. Identify and explain how to use a portable band saw.
   d. Identify and explain how to use miter and cutoff saws.
3. Identify and explain how to use various grinders and grinder attachments.
   a. Identify and explain how to use various types of grinders.
   b. Identify and explain how to use various grinder accessories and attachments.
4. Identify and explain how to use miscellaneous power tools.
   a. Identify and explain how to use pneumatic and powder-actuated fastening tools.
   b. Identify and explain how to use pavement breakers.
   c. Identify and explain the uses of hydraulic jacks.

Performance Tasks

Under supervision of your instructor, you should be able to do the following:

1. Safely and properly demonstrate the use of three of the following tools:
   - Electric drill
   - Hammer drill or rotary hammer
   - Circular saw
   - Reciprocating saw
   - Portable band saw
   - Miter or cutoff saw
   - Portable or bench grinder
   - Pneumatic nail gun
   - Pavement breaker

Trade Terms

Abrasive	Chuck key	Ground fault circuit interrupter (GFCI)	Revolutions per minute (rpm)
Alternating current (AC)	Countersink	Ground fault protection	Ring test
Arbor	Direct current (DC)	Kerf	Shank
Auger bit	Forstner bit	Masonry bit	Trigger lock
Carbide	Grit	Reciprocating	
Chuck			

Industry Recognized Credentials

If you are training through an NCCER-accredited sponsor, you may be eligible for credentials from NCCER's Registry. The ID number for this module is 00104-15. Note that this module may have been used in other NCCER curricula and may apply to other level completions. Contact NCCER's Registry at 888.622.3720 or go to **www.nccer.org** for more information.

Trade Features

Trade features present technical tips and professional practices based on real-life scenarios similar to those you might encounter on the job site.

Bowline Trivia

Some people use this saying to help them remember how to tie a bowline: "The rabbit comes out of his hole, around a tree, and back into the hole."

Figures and Tables

Photographs, drawings, diagrams, and tables are used throughout each module to illustrate important concepts and provide clarity for complex instructions. Text references to figures and tables are emphasized with *italic* type.

Notes, Cautions, and Warnings

Safety features are set off from the main text in highlighted boxes and categorized according to the potential danger involved. Notes simply provide additional information. Cautions flag a hazardous issue that could cause damage to materials or equipment. Warnings stress a potentially dangerous situation that could result in injury or death to workers.

NOTE

Nameplates must be posted on each material handling device. The nameplate must indicate

CAUTION

It is essential to note the revision designation on a construction drawing and to use only the latest

WARNING!

Saw teeth are very sharp. Use gloves and do not handle the saw teeth with bare hands. When cutting with a saw, ensure that your fingers remain clear of the teeth at all times.

Case History

Case History features emphasize the importance of safety by citing examples of the costly (and often devastating) consequences of ignoring best practices or OSHA regulations.

Case History

Requesting an Outage

An electrical contractor requested an outage when asked to install two bolt-in, 240V breakers in panels in a data processing room. It was denied due to the 24/7 worldwide information processing hosted by the facility. The contractor agreed to proceed only if the client would sign a letter agreeing not to hold them responsible if an event occurred that damaged computers or resulted in loss of data. No member of upper management would accept liability for this possibility, and the outage was scheduled.

The Bottom Line: If you can communicate the liability associated with an electrical event, you can influence management's decision to work energized.

Going Green

Going Green features present steps being taken within the construction industry to protect the environment and save energy, emphasizing choices that can be made on the job to preserve the health of the planet.

GOING GREEN

Reducing Your Carbon Footprint

Many companies are taking part in the paperless movement. They reduce their environmental impact by reducing the amount of paper they use. Using email helps to reduce the amount of paper used,

Did You Know

Did You Know features introduce historical tidbits or interesting and sometimes surprising facts about the trade.

Did You Know?

Safety First

Safety training is required for all activities. Never operate tools, machinery, or equipment without prior training. Always refer to the manufacturer's instructions.

Step-by-Step Instructions

Step-by-step instructions are used throughout to guide you through technical procedures and tasks from start to finish. These steps show you how to perform a task safely and efficiently.

Perform the following steps to erect this system area scaffold:

Step 1 Gather and inspect all scaffold equipment for the scaffold arrangement.

Step 2 Place appropriate mudsills in their approximate locations.

Step 3 Attach the screw jacks to the mudsills.

Trade Terms

Each module presents a list of Trade Terms that are discussed within the text and defined in the Glossary at the end of the module. These terms are presented in the text with **bold, blue** type upon their first occurrence. To make searches for key information easier, a comprehensive Glossary of Trade Terms from all modules is located at the back of this book.

During a rigging operation, the **load** being lifted or moved must be connected to the apparatus, such as a crane, that will provide the power for movement. The connector—the link between the load and the apparatus—is often a sling made of synthetic, chain, or **wire rope** materials. This section focuses on three types of slings:

Section Review

Each section of the module wraps up with a list of Additional Resources for further study and Section Review questions designed to test your knowledge of the Objectives for that section.

Additional Resources

Materials Handling Handbook, The American Society of Mechanical Engineers (ASME) and The International Material Management Society (IMMS), Raymond A. Kulwiec, Editor-in-Chief. 1985. New York, NY: Wiley-Interscience.

Manufacturing Facilities Design & Material Handling, Matthew P. Stevens, Fred E. Meyers. 2013. West Lafayette, IN: Purdue University Press.

1.0.0 Section Review

1. For material handling tasks, it is just as important to be mentally fit as it is to be _____.
   a. physically fit
   b. physically aggressive
   c. closely supervised
   d. over 200 pounds

2. Which of the following is a type of knot that is often used to join the ends of two ropes in non-critical, low-strain applications?
   a. Bowline
   b. Clove hitch
   c. Half hitch
   d. Square knot

Review Questions

The end-of-module Review Questions can be used to measure and reinforce your knowledge of the module's content.

Review Questions

1. Identification tags for slings must include the _____.
   a. type of protective pads to use
   b. type of damage sustained during use
   c. color of the tattle-tail
   d. manufacturer's name or trademark

2. The type of wire rope core that is susceptible to heat damage at relatively low temperatures is the _____.
   a. fiber core
   b. strand core
   c. independent wire rope core
   d. metallic link supporting core

3. Synthetic slings must be inspected _____.
   a. once every month
   b. visually at the start of each work week
   c. before every use
   d. once wear or damage becomes apparent

4. An alloy steel chain sling must be removed from service if there is evidence that _____.
   a. the sling has been used in different hitch configurations
   b. replacement links have been used to repair the chain
   c. the sling has been used for more than one year
   d. strands in the supporting core have weakened

5. A piece of rigging hardware used to couple the end of a wire rope to eye fittings, hooks, or other connections is a(n) _____.
   a. eyebolt
   b. hitch
   c. shackle
   d. U-bolt

6. A lifting clamp is most likely to be used to move loads such as _____.
   a. steel plates
   b. piping bundles
   c. concrete blocks
   d. plastic tubing

7. Chain hoists are able to lift heavy loads by utilizing a _____.
   a. rope and pulley system
   b. rigger's strength
   c. stationary counterweight
   d. gear system

8. Before attempting to lift a load with a chain hoist, make sure that the _____.
   a. hoist is secured to a come-along
   b. load is properly balanced
   c. tag lines are properly anchored
   d. tackle is connected to its power source

9. A hitch configuration that allows slings to be connected to the same load without using a spreader beam is a _____.
   a. double-wrap hitch
   b. choker hitch
   c. bridle hitch
   d. basket hitch

10. To make the emergency stop signal that is used by riggers, extend both arms _____.
    a. horizontally with palms down and quickly move both arms back and forth
    b. directly in front and then move both arms up and down repeatedly
    c. vertically above the head and wave both arms back and forth
    d. horizontally with clenched fists and move both arms up and down

00106-15 Introduction to Basic Rigging Module Six 31

NCCER Standardized Curricula

NCCER's training programs comprise more than 80 construction, maintenance, pipeline, and utility areas and include skills assessments, safety training, and management education.

Boilermaking
Cabinetmaking
Carpentry
Concrete Finishing
Construction Craft Laborer
Construction Technology
Core Curriculum: Introductory Craft Skills
Drywall
Electrical
Electronic Systems Technician
Heating, Ventilating, and Air Conditioning
Heavy Equipment Operations
Highway/Heavy Construction
Hydroblasting
Industrial Coating and Lining Application Specialist
Industrial Maintenance Electrical and Instrumentation Technician
Industrial Maintenance Mechanic
Instrumentation
Insulating
Ironworking
Masonry
Millwright
Mobile Crane Operations
Painting
Painting, Industrial
Pipefitting
Pipelayer
Plumbing
Reinforcing Ironwork
Rigging
Scaffolding
Sheet Metal
Signal Person
Site Layout
Sprinkler Fitting
Tower Crane Operator
Welding

Maritime

Maritime Industry Fundamentals
Maritime Pipefitting
Maritime Structural Fitter

Green/Sustainable Construction

Building Auditor
Fundamentals of Weatherization
Introduction to Weatherization
Sustainable Construction Supervisor
Weatherization Crew Chief
Weatherization Technician
Your Role in the Green Environment

Energy

Alternative Energy
Introduction to the Power Industry
Introduction to Solar Photovoltaics
Introduction to Wind Energy
Power Industry Fundamentals
Power Generation Maintenance Electrician
Power Generation I&C Maintenance Technician
Power Generation Maintenance Mechanic
Power Line Worker
Power Line Worker: Distribution
Power Line Worker: Substation
Power Line Worker: Transmission
Solar Photovoltaic Systems Installer
Wind Turbine Maintenance Technician

Pipeline

Pipeline Fundamentals
Pipeline Corrosion Control
Pipeline Electrical and Instrumentation
Pipeline Maintenance
Pipeline Mechanical
Pipeline Operations (Control Center, Gas, Liquid)

Safety

Field Safety
Safety Orientation
Safety Technology

Supplemental Titles

Applied Construction Math
Tools for Success

Management

Construction Workforce Development Professional
Fundamentals of Crew Leadership
Mentoring for Craft Professionals
Project Management
Project Supervision

Spanish Titles

Acabado de concreto: nivel uno (*Concrete Finishing Level One*)
Aislamiento: nivel uno (*Insulating Level One*)
Albañilería: nivel uno (*Masonry Level One*)
Andamios (*Scaffolding*)
Carpintería: Formas para carpintería, nivel tres (*Carpentry: Carpentry Forms, Level Three*)
Currículo básico: habilidades introductorias del oficio (*Core Curriculum: Introductory Craft Skills*)
Electricidad: nivel uno (*Electrical Level One*)
Herrería: nivel uno (*Ironworking Level One*)
Herrería de refuerzo: nivel uno (*Reinforcing Ironwork Level One*)
Instalación de rociadores: nivel uno (*Sprinkler Fitting Level One*)
Instalación de tuberías: nivel uno (*Pipefitting Level One*)
Instrumentación: nivel uno, nivel dos, nivel tres, nivel cuatro (*Instrumentation Levels One through Four*)
Orientación de seguridad (*Safety Orientation*)
Paneles de yeso: nivel uno (*Drywall Level One*)
Seguridad de campo (*Field Safety*)

Acknowledgments

This curriculum was revised as a result of the farsightedness and leadership of the following sponsors:

American Petroleum Institute
Central NDT
Construction Craft Academy
CP Masters
EMS USA, Inc.
Enbridge Pipelines, Inc.
Enhance Co.
Enterprise
Flint Energy Services, Inc.
Jomax Construction
Kinder Morgan, Inc.
L.E. Bell Construction Co. Inc.
Marathon Pipe Line
Midwestern Contractors
OneOK
Plains All-American Pipeline
Praxair Services
Protech EIS
Safety Advantage LLC
Southland Safety
Sunoco Logistics, L.P.
Sunoco Pipeline
Thorco Holdings, LLC
URS/Flint Energy

This curriculum would not exist were it not for the dedication and unselfish energy of those volunteers who served on the Authoring Team. A sincere thanks is extended to the following:

Gary Affolter
Carl Allebach
Travis Ashby
Steve Barker
Art Birx
James Carroll
Leslie Carroll
Gordon Carson
Bill Emery
Paul Fontenot II
Paula Gates-Rippe
Trobi Hall
Sid Harrell
David Hoffman
Kimberly-Joy Harris
Les Harrison
Lisa Hartnup
Benjamin Johnson
Dan Lake
Tom Magness
Wallace McCracken
Tim Mills
Steve Montoy
Michael Nordstrom
Tom Pickthall
Donnie Pittsinger
William Rado
David Reeder
Tom Reeder
Randall Rogers
Dan Rohret
Dan Ruane
DeWayne Schapansky
Mark Seitz
Blaine Shellhammer
Lynn Simpson
Cameron Strother
Charles Whiten
Rick Wilenzick
Michael Williams
Erin Worthington
Donavon Wright
Will Wright

NCCER Partners

American Council for Construction Education
American Fire Sprinkler Association
Associated Builders and Contractors, Inc.
Associated General Contractors of America
Association for Career and Technical Education
Association for Skilled and Technical Sciences
Construction Industry Institute
Construction Users Roundtable
Design Build Institute of America
GSSC – Gulf States Shipbuilders Consortium
ISN
Manufacturing Institute
Mason Contractors Association of America
Merit Contractors Association of Canada
NACE International
National Association of Women in Construction
National Insulation Association
National Technical Honor Society
National Utility Contractors Association
NAWIC Education Foundation
North American Crane Bureau
North American Technician Excellence
Pearson
Prov
SkillsUSA®
Steel Erectors Association of America
U.S. Army Corps of Engineers
University of Florida, M. E. Rinker Sr., School of Construction Management
Women Construction Owners & Executives, USA

NCCER Business Partners

Contents

CT63_1-17

Start-Up of a Liquid Pipeline (Field)

Introduces the methods and procedures used by field operations personnel to start up a liquid pipeline. It describes the required knowledge and skills, as well as abnormal operating conditions associated with starting up a liquid pipeline. (Module ID CT63_1-17; 5 Hours)

CT63_2-17

Shutdown of a Liquid Pipeline (Field)

Introduces the methods and procedures related to shutting down a liquid pipeline. It describes the knowledge, skills, and abnormal operating conditions associated with shutting down a liquid pipeline. (Module ID CT63_2-17; 5 Hours)

CT63_3-17

Monitor Pressures, Flows, Communications, and Line Integrity and Maintain Them Within Allowable Limits on a Liquid Pipeline System (Field)

Introduces the trainees to the procedures for monitoring pipeline parameters, safety devices, alarms, and purging operations. The module also introduces the trainees to the procedures for monitoring remote locations and weather conditions. (Module ID CT63_3-17; 5 Hours)

CT63_4-17

Locally Operate Valves on a Liquid Pipeline System

Introduces the trainees to the methods and procedures related to locally operating valves on a liquid pipeline. Trainees will also review the knowledge, skills, and abnormal operating conditions associated with locally operating valves on a liquid pipeline. (Module ID CT63_4-17; 5 Hours)

CT64_1-17

Start-Up of a Liquid Pipeline (Control Center)

Introduces the methods and procedures used by control center operations personnel to start up a liquid pipeline. It identifies key liquid pipeline components and describes their function and operation. This module explains the required knowledge and skills, as well as abnormal operating conditions associated with starting up a liquid pipeline. (Module ID CT64_1-17; 5 Hours)

CT64_2-17

Shutdown of a Liquid Pipeline (Control Center)

Introduces the methods and procedures used by control center operations personnel to shut down a liquid pipeline. It identifies key liquid pipeline components and describes their function and operation. This module explains the required knowledge and skills, as well as abnormal operating conditions associated with shutting down a liquid pipeline. (Module ID CT64_2-17; 5 Hours)

CT64_3-17

Monitor Pressures, Flows, Communication, and Line Integrity and Maintain Them Within Allowable Limits on a Liquid Pipeline (Control Center)

Introduces the technologies that pipeline controllers use to monitor the pipeline and maintain safe and efficient operation. This module explains the general procedures involved in monitoring pipeline conditions and maintaining them within normal operating limits. It describes potential abnormal operating conditions that can occur during these procedures. (Module ID CT64_3-17; 5 Hours)

(continued)

CT64_4-17

Remotely Operate Valves on a Liquid Pipeline System

Introduces valves used to control product as it flows through the different components of a pipeline. It describes different types of valves and explains how they operate. Pumps are also introduced as the pipeline's power source, and supervisory control is discussed as the means of coordinating pipeline operations. This module presents the general procedures used to operate valves remotely, and describes abnormal operating conditions that may occur during valve operation. (Module ID CT64_4-17; 5 Hours)

CT65_1-17

Start-up of a Pipeline (Gas)

Describes how to join pipe segments using flange fittings, as well as how to store and protect components prior to joining. (Module ID CT65_1-17; 15 Hours)

CT65_2-17

Shutdown of a Pipeline (Gas)

Introduces the trainees to the methods and procedures related to shutting down a pipeline. Trainees will also review the knowledge, skills, and abnormal operating conditions associated with shutting down a pipeline. (Module ID CT65_2-17; 5 Hours)

CT65_3-17

Monitor Pressures, Flows, Communications, And Line Integrity And Maintain Them Within Allowable Limits (Gas)

Introduces the trainees to the methods and procedures related to monitoring pipeline parameters. Trainees will also review the knowledge, skills, and abnormal operating conditions associated with monitoring pipeline parameters. (Module ID CT65_3-17; 5 Hours)

CT65_4-17

Manually or Remotely Open or Close Valves or Other Equipment

Describes the procedures for manually or remotely opening and closing valves, starting and stopping compressors, and monitoring the compressor drivers. It also reviews the knowledge, skills, and abnormal operating conditions associated with valve operation. (Module ID CT65_4-17; 15 Hours)

AOCCC-17

Abnormal Operating Conditions – Control Center

Overview

Pipeline operators and contractors have programs, procedures, training, safety equipment, and warning devices for practically every facet of pipeline operations. Situations that are outside of normal operating conditions may arise at any time, and control center personnel must know what to do to help keep pipeline operations safe for everyone. In this module, personnel will learn how to recognize, properly react to, and properly report AOCs that may occur during pipeline operations.

Trainees with successful module completions may be eligible for credentialing through NCCER's National Registry. To learn more, go to **www.nccer.org** or contact us at **1.888.622.3720**. Our website has information on the latest product releases and training, as well as online versions of our *Cornerstone* magazine and Pearson's product catalog.

Your feedback is welcome. You may email your comments to **curriculum@nccer.org,** send general comments and inquiries to **info@nccer.org**, or fill in the User Update form at the back of this module.

This information is general in nature and intended for training purposes only. Actual performance of activities described in this manual requires compliance with all applicable operating, service, maintenance, and safety procedures under the direction of qualified personnel. References in this manual to patented or proprietary devices do not constitute a recommendation of their use.

AOCCC-17
Abnormal Operating Conditions – Control Center

Objectives

When you have completed this module, you will be able to do the following:

1. Explain how to recognize abnormal operating conditions in a control center for both gas and liquid pipelines.
   a. Explain how to recognize abnormal facility conditions and pipeline system damage.
   b. Explain how to recognize the activation of a safety device.
   c. Explain how to recognize a communications failure and a control system failure.
   d. Explain how to recognize power interruptions.
   e. Explain how to recognize a fire, explosion, and natural disaster occurring with a pipeline.
   f. Explain how to recognize hazardous liquid or gas encounters.
   g. Explain how to recognize an unexplained flow rate/pressure change.
2. Describe the proper reaction to abnormal operating conditions in a control center for both gas and liquid pipelines.
   a. Describe the proper reaction to abnormal facility conditions and pipeline system damage.
   b. Describe the proper reaction to the activation of a safety device.
   c. Describe the proper reaction to a communications failure and a control system failure.
   d. Describe the proper reaction to power interruptions.
   e. Describe the proper reaction to a fire, explosion, and natural disaster occurring with a pipeline.
   f. Describe the proper reaction to hazardous liquid or gas encounters.
   g. Describe the proper reaction to an unexplained flow rate/pressure change.

Performance Tasks

This is a knowledge-based module; there are no performance tasks.

Trade Terms

Abnormal operating condition (AOC)
American National Standards Institute (ANSI)
Breakout tank
Computational pipeline monitoring (CPM)
Disbondment
Encroachment
Highly volatile liquid (HVL)
Lower explosive limit (LEL)
Maximum allowable operating pressure (MAOP)
Maximum operating pressure (MOP)
Naturally occurring radioactive materials (NORMs)
One-Call notification system
Pipeline and Hazardous Materials Safety Administration (PHMSA)
Pipeline facility
Pipeline operator
Redundant system
Resistance temperature device (RTD)
Right-of-way (ROW)
Self-contained breathing apparatus (SCBA)
Supervisory control and data acquisition (SCADA)
Surge pressure
Uninterruptible power system (UPS)

Industry-Recognized Credentials

If you're training through an NCCER-accredited sponsor, you may be eligible for credentials from NCCER's Registry. The ID number for this module is AOCCC-17. Note that this module may have been used in other NCCER curricula and may apply to other level completions. Contact NCCER's Registry at 888.622.3720 or go to **www.nccer.org** for more information.

Contents

Topics to be presented in this module include:

Figures and Tables

Section One

1.0.0 Recognizing Abnormal Operating Conditions

Objective

Explain how to recognize abnormal operating conditions in a control center for both gas and liquid pipelines.

a. Explain how to recognize abnormal facility conditions and pipeline system damage.
b. Explain how to recognize the activation of a safety device.
c. Explain how to recognize a communications failure and a control system failure.
d. Explain how to recognize power interruptions.
e. Explain how to recognize a fire, explosion, and natural disaster occurring with a pipeline.
f. Explain how to recognize hazardous liquid or gas encounters.
g. Explain how to recognize an unexplained flow rate / pressure change.

Trade Terms

Abnormal operating condition (AOC): A condition within a pipeline system, identified by the operator, that may indicate (1) a malfunction of a component or (2) a deviation from normal operations that may indicate a condition exceeding design limits or result in a hazard to persons, property, or the environment.

American National Standards Institute (ANSI): An organization that adopts standards for a variety of products, services, and processes in the United States to ensure consistency.

Breakout tank: A tank used to (a) relieve surges in a hazardous liquid pipeline system or (b) receive and store hazardous liquid transported by a pipeline for reinjection and continued transportation by pipeline.

Computational pipeline monitoring (CPM): Technology that provides monitoring tools used by control center personnel to detect temperature, flow, and pressure irregularities in a pipeline.

Disbondment: Separation of the protective coating from the pipeline.

Encroachment: Activity carried out by and/or equipment belonging to outside parties on the pipeline right-of-way. A particular encroachment might be authorized or not, so each situation requires specific handling.

Highly volatile liquid (HVL): a hazardous liquid which will form a vapor cloud when released to the atmosphere and which has a vapor pressure exceeding 276 kPa (40 psia) at 37.8°C (100°F).

Lower explosive limit (LEL): The lowest concentration of a gas or a vapor in the atmosphere capable of producing fire in the presence of an ignition source. Different products have different LELs.

Maximum allowable operating pressure (MAOP): The maximum pressure at which a pipeline or segment of a pipeline may be operated (gas).

Maximum operating pressure (MOP): The maximum pressure at which a pipeline or segment of a pipeline may be normally operated (liquid).

One-Call notification system: A centralized communication system established by utilities, government agencies, and operators of underground facilities. This system provides one telephone number for contractors and the general public to call to notify of their intent to use equipment on or near a pipeline. It is also in place for any individual to call if pipeline damage is discovered. The One-Call system is generally set up on a state-by-state basis.

Pipeline and Hazardous Materials Safety Administration (PHMSA): An agency of the US Department of Transportation responsible for developing and enforcing regulations for pipeline operations in the United States.

Pipeline facility: New and existing pipe; rights-of-way; and any equipment, facility, or building used in the transportation of hazardous liquids or carbon dioxide, or new and existing pipelines; rights-of-way; and any equipment, facility, or building used in the transportation of gas or in the treatment of gas during the course of transportation.

Pipeline operator: The company that operates the pipeline, as opposed to an individual person who works for the company that operates the pipeline.

Redundant system: A system that duplicates the operations of a system so that, in the event of a failure, a backup system is available.

Resistance temperature device (RTD): A device used on pumps that reads the temperature of the pump case and the product in the pipeline. It will sound an alarm when the temperature reaches a critical level and shut down the pump automatically if needed.

Right-of-way (ROW): The area along the pipeline which has been legally permitted to be used by the pipeline operator to construct, operate, and maintain the pipeline. The ROW extends the entire length of the pipeline.

Supervisory control and data acquisition (SCADA): A computer-based system or systems used by a controller in a control room that collects and displays information about a pipeline facility and may have the ability to send commands back to the pipeline facility.

Surge pressure: Pressure produced by a change in velocity of the moving stream that results from shutting down a pump station or pumping unit, closure of a valve, or any other blockage of the moving stream.

Uninterruptible power system (UPS): A system that provides continuity of power when the normal power supply is interrupted. A UPS provides AC power to the system it serves.

An **abnormal operating condition (AOC)** is defined under federal regulations as a condition identified by the **pipeline operator** that may indicate a malfunction of a component or a deviation from normal operations that may indicate a condition exceeding design limits or result in a hazard to persons, property, or the environment. AOCs are defined as such by federal pipeline safety law 49 *United States Code (U.S.C.)* 60102 and pipeline safety regulations 49 *Code of Federal Regulations (CFR)* 192 (for transportation of gas) and 49 *CFR* 195 (for transportation of liquids). The federal agency that regulates pipeline maintenance is the **Pipeline and Hazardous Materials Safety Administration (PHMSA)** in Washington, DC.

In the control room, control center personnel may become aware of AOCs from reports directly from field personnel, by reports from third parties, or by alerts from computer systems in the control room. Some pipeline equipment is fitted with safety devices that automatically alert control room systems to problems. The control center console (see *Figure 1*) consists of computer systems and monitors that allow control center personnel to observe pipeline operations and take actions as needed.

Types of AOCs, along with examples and proper reactions, are listed in *Table 1*. Note that the reactions are general.

1.1.0 Recognizing Abnormal Pipeline Facility Conditions

Any problem with the physical structure of the pipeline is defined as an abnormal **pipeline facility** condition. A pipeline facility includes the pipeline itself; **right-of-way (ROW)**; and any equipment, facility, or building used in the transportation of a liquid or gas. Examples of components of a pipeline facility include compressors for gas pipelines and pump stations for liquid pipelines (this module refers to both as units). The terms *pipeline* or *pipeline system* mean all parts of a pipeline facility through which a hazardous liquid or gas moves in transportation, including, but not limited to, line pipe, valves, and other equipment connected to line pipe; pumping or compressor units and the fabricated assemblies associated with them; metering and delivery stations; and regulator stations. The following are examples of abnormal pipeline facility conditions that can be reported to the control center:

- Exposed pipeline that should be covered
- Damaged pipeline
- Missing or damaged pipeline supports
- Corrosion
- Coating separating from the pipeline (**disbondment**)
- Missing signs or markers
- Pressure reduction for various reasons
- Reports of unauthorized activity and/or equipment on the pipeline right-of-way (unauthorized **encroachment**)
- Sunken floating tank roof

65102-13_F01.EPS

Figure 1 Control center console.

Table 1A Abnormal Operating Conditions: Examples and General Reactions (1 of 2)

Abnormal Operating Condition	Examples	Appropriate Reaction
Abnormal pipeline facility condition	• Exposed pipe • Evidence of corrosion • Incorrect ANSI pressure rating of a component or material type • Disbondment (coatings separating from the pipe) • Unauthorized activity and/or equipment on the right-of-way (unauthorized encroachment) • Incorrect data or records caused by equipment malfunction • Mechanical/hydraulic system failures • Fire and explosion • Pipeline system damage, such as: – Missing or damaged pipeline supports – Damaged or inoperable pipeline components (pipe, valves, pumps, compressors, meters, tank gauges) due to external force – Gouges, dents, pitting, wrinkles, buckles, or cracks – Cracked or corroded flange or bolts – Sunken floating tank roof – Damaged or malfunctioning pontoon – Defective tank roof seal – Damaged tank shell – Damaged pipe due to natural disaster	Follow standard operating procedures and make notifications per company policy. Isolate and/or shut down the pipeline segment or facility if necessary. This may require coordination with field personnel. Complete the required documentation.
Activation of a safety device	• Emergency or abnormal shutdown alarm • High-pressure shutdown alarm • Activation of a pressure relief valve • Activation of a smoke or fire detector • Activation of a hazardous atmosphere alarm • Activation of an overfill protection device • High sump-level shutdown alarm • High tank-level shutdown alarm • High pump-case pressure alarm • Seal failure alarm	Follow standard operating procedures and make notifications per company policy. Isolate and/or shut down the pipeline segment or facility if necessary. This may require coordination with field personnel. Complete the required documentation.
Communications failures	Personnel communication tools: • Lost use of radio communications • Lost use of landline telephone • Lost use of cell phone System communications: • Facility outages • SCADA communication outage • System programming errors/incorrect command logic • Loss of satellite communications • Loss of microwave signals • Stale data (communications going from the control center to the field, but not being received by the control center from the field)	Switch to alternate equipment or backup systems. Follow standard operating procedures and make notifications per company policy. Complete the required documentation.

- Incorrect data or records caused by equipment malfunction
- Incorrect **American National Standards Institute (ANSI)** pressure rating of a component or material type
- Malfunctions due to human error

Control center personnel may be notified of an abnormal facility condition by field personnel. A report may also come in to the control room as

Table 1B Abnormal Operating Conditions: Examples and General Reactions (2 of 2)

Abnormal Operating Condition	Examples	Appropriate Reaction
Power interruptions	• Transformer failure • Downed power lines • Broken leads or wires • Incorrect wire termination • Station or terminal power problems (due to motor overload or a tripped breaker, for example)	Follow standard operating procedures. Switch to an alternate control system or power source if necessary. Notify the power company of the outage. Make additional notifications per company policy. Complete the required documentation.
Fire, explosion, or natural disaster	• Electrical fire • Ignition of a hazardous vapor • Unintentional detonation • Weather conditions, such as an ice storm • Fire, explosion, or pipeline movement caused by natural disaster, such as: – Lightning strike – Flooding – Landslide – Earthquake – Tornado	Follow company emergency action plans. Call local emergency personnel if appropriate. Isolate and/or shut down the pipeline segment or facility if necessary. Limit access to the area as much as possible. Make notifications per company policy. Complete the required documentation.
Unexpected release of a hazardous gas or liquid	Reports of: • Vapor clouds or frost balls • Smell of hydrocarbon release • Dead vegetation/contaminated soil • Dead wildlife • Rainbow sheen on water Hazardous products include: • Hydrogen sulfide • Carbon dioxide • Anhydrous ammonia • Propane • Natural gas • Refined fuels/products Note that the controlled transfer, sampling, venting, draining, and handling of hazardous products are not considered AOCs.	Follow company emergency action plans. Call local emergency personnel if appropriate. Isolate and/or shut down the pipeline segment or facility if necessary. Make notifications per company policy. Complete the required documentation.
Unexplained flow rate / pressure change	Any change from the expected flow or pressure, such as: • Increased pressure • Decreased pressure • No pressure • High flow limit • Low flow limit More often than not, irregularities are detected through a computational pipeline monitoring (CPM) system	Follow standard operating procedures and make notifications per company policy. Isolate and/or shut down the pipeline segment or facility if necessary. Complete the required documentation.

a result of a call made to the **One-Call notification system** or from a person calling the pipeline operator directly. When a report comes in, personnel must obtain complete information so that they can quickly find the problem and address it. Many pipeline operators use a form to capture the following information:

- Caller's name and telephone number
- Date and time of the call
- Was anyone injured? Was 911 notified by the caller?
- Type of product involved, if relevant and if known
- Location of the problem, including latitude and longitude, if possible
- Nearby landmarks such as highways, streams, mile markers, or any other structure
- The nature of the damage and any conditions observed at the site

One of the most important things to remember as a control center operator is to always follow company policy and procedures with any abnormal operating condition.

1.1.1 Pipeline System Damage

The control center may receive the indication of pipeline damage on the control center console as an unexplained change in operating conditions. Pipeline damage or exposed pipe may also be discovered by field personnel or a third party who then reports the damage to the control center. Third-party examples include the aerial patroller of the pipeline ROW who notifies the pipeline operator, or an individual landowner who calls the One-Call notification system.

One way that control center personnel notice damage is by monitoring the pressure and flow changes on the control center console. Control center personnel use monitoring tools, referred to as **computational pipeline monitoring (CPM)** systems, to detect pressure, flow, and temperature irregularities in a pipeline. These irregularities may indicate a release. Most CPM systems provide an alarm and display other data related to pressures, flows, and temperatures that personnel use to detect problems.

Physical signs of damage to the pipeline and its components may include gouges, cracks, pitting, dents, buckles, and inoperable valves, pumps, or tank gauges. Flanges or bolts may be cracked or show significant corrosion. Tank shells or pipeline supports may show damage. A sunken roof or a damaged pontoon also indicates damage. Less obvious signs of damage could include a rainbow sheen on the water near submerged pipelines, which may indicate a release. The release of a **highly volatile liquid (HVL)** may be discovered by observing a frost ball or vapor cloud along the ROW of a liquid pipeline. While the control center personnel are not able to see physical damage to a pipeline, unless viewed through a camera system, they often receive notice of these types of damage indications.

1.1.2 Damaged Pipeline Supports

Control center personnel may receive reports about damage to pipeline supports (see *Figure 2*) from maintenance or inspection personnel who are traversing the pipeline. Damage can include visible cracking, sagging, or buckling of a support, or bolts or other components of the support that have broken or become loose. This type of damage can be a serious problem and can require a shutdown. Sometimes pipeline operations can continue while repairs are made.

65102-13_F02.EPS

Figure 2 Pipeline on pipe supports.

1.1.3 Malfunctions Due to Human Error

Malfunctions are not always due to mechanical or other system failures; they can occur simply due to human error. Control center personnel should be able to recognize the following possible malfunctions as a result of human error:

- Deviation from the design limits
- Valves that are not sequenced correctly
- Cavern upset
- Units or valves that are not aligned properly
- Units that are set incorrectly, which may cause overpressure in the line
- Overflow of a tank

1.2.0 Recognizing the Activation of a Safety Device

Safety devices are part of the pipeline system designed to sense and alert personnel of any potential danger and/or hazardous condition. Hazardous conditions can include changes in pressure that might exceed pipeline design limits as well as the release of harmful gases. Safety devices include, but are not limited to, pressure relief systems, pressure limiting devices, high pump-case-pressure alarms, seal failure alarms, overfill protection devices, tank-level alarms, main-line pressure shutdown alarms, and hazardous atmosphere alarms.

Generally when a safety device activates, a sensor on the device sends a signal to the control center. An alarm may sound in the control room, and the signal appears as an alarm on the control center console. All safety alarms must be verified and the problem corrected before clearing the alarm.

1.2.1 Pressure Limiting Devices

Pressure-limiting devices include control valves. These valves are monitored and controlled by the control center. If the pressure in the pipeline is different than it should be, the control valves are adjusted as needed.

1.2.2 Seal Failure Alarms

Seal failure alarms are activated when a leak is detected in the seal on a pump. There are different types of seals that can activate seal failure alarms when not working properly. One common type of seal is a tandem seal. Tandem seals are a combination of two seals with a barrier liquid between the seals in a chamber that is not pressurized. When the barrier fluid rises to a certain level from process liquid entering the chamber the alarm is activated.

Another type of seal commonly used is a double seal. Double seals consist of two seals that are mounted back-to-back. Usually a seal failure results in product leaking into the atmosphere. This could cause a hazardous atmosphere alarm to activate.

1.2.3 Overfill Protection Devices

Overfill protection devices activate if product reaches a certain level in the tank. As a result of activation, the system may shut off product flow, restrict product flow, or signal an alarm. Two types of overfill protection device are a high and a high-high level alarm that detects the level of product when it reaches a predetermined level. The alarm may also activate a pump shutdown or valve to divert the product flow to prevent a release. Another device related to the overfill protection device is the low-low level alarm. This alarm activates when the product level reaches a predetermined low level. The pump does not shut down with a low-low alarm activation, but monitoring may be required to ensure the levels do not become dangerously low and damage the pump.

1.2.4 Pressure Switches

Pressure switches activate when electrical contacts within the switch close when the pressure reaches a set level. The switch may be designed to make contact on pressure rise or on pressure fall. Pressure switches can have either automatic or manual resets. The automatic resets will reset when the pressure returns to normal operating range. Manual resets need to be reset by personnel.

1.2.5 Activation of a Pressure Relief System

The activation of relief devices along the pipeline system is a time-sensitive occurrence. Relief valves can discharge into tanks or vessels that are designed and constructed in compliance with American Petroleum Institute (API) *Standard 650* or *API 2510* for pressure vessels. Relief valves can also open into the atmosphere. Relief valves (see *Figure 3*) can be installed on a pipeline or **breakout tank**. The purpose of a pressure relief device is to protect the pipeline or tank from overpressure. Breakout tanks can also be used to temporarily absorb **surge pressure** from the main-line and manifold pressure-relief valves to lower pressure on the line. Pressure-relief devices must be provided for each low-pressure and high-pressure breakout tank. There are different specifications for low-pressure and high-pressure tanks.

Some valves can be manually operated by field personnel, and some are remotely operated by control center personnel. When a relief valve is activated, an alarm may sound.

1.2.6 High Pump-Case Temperature/Pressure Alarm

Main-line pumps are fitted with a **resistance temperature device (RTD)** that give temperature readings for the pump case and product. These devices are preset to signal an alarm in the control center

65102-13_F03.EPS

Figure 3 Relief valve.

when the temperature reaches a critical level, and will shut the pump down automatically if some action cannot be taken to lower the temperature. When a pump-case temperature reaches such a critical level and automatically shuts down, it is considered an AOC. *Figure 4* shows RTDs on a pump case.

1.3.0 Recognizing a Communications Failure

Communications refers to the tools that personnel use to communicate with each other and with systems. Communications failures include the failure of communications among personnel, or the failure of communications between the control center and equipment on the pipeline.

1.3.1 Failure of Personnel Communication Tools

Pipeline personnel use landline telephones, cell phones, and radios to communicate with one another on the job. A breakdown in personnel communications equipment may be easy to recognize. Personnel may not hear messages or be able to make phone calls. Personnel must keep their equipment in good working order and test their equipment regularly to ensure that it is working properly.

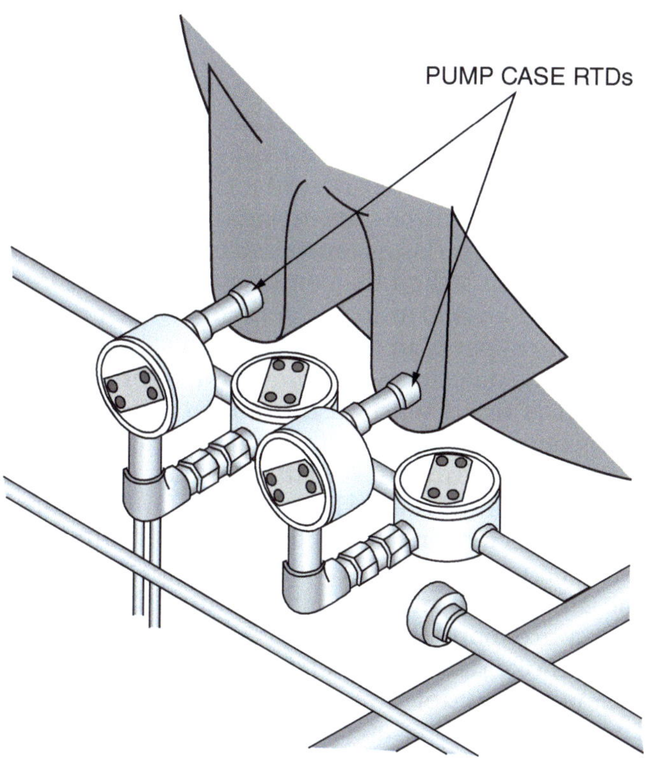

Figure 4 Resistance temperature devices (RTDs) on a pump case.

1.3.2 System Communications Failures

System communications refers to the computer equipment located in the control center and on some pipeline equipment. Personnel in the control center use the **supervisory control and data acquisition (SCADA)** system to manage pipeline operations. The system receives signals from and sends signals to pipeline equipment that may be many miles away. The SCADA system allows control center personnel to monitor pressure, volume, and temperature and keep products flowing safely and on schedule. If the system fails, control center personnel lose the ability to communicate with equipment on the pipeline. Some systems are set up to switch to a **redundant system**—a backup that comes on line if the main system fails.

Some failures of communications systems show up as warnings or messages on the control center console. Another potential sign of a communications failure is stale data showing on the computer monitor. A system failure may include any of the following events:

- Random facility outages
- Partial or complete facility outages
- Loss of the SCADA system
- Loss of satellite communications
- Loss of microwave signals
- Loss of a signal point within the system
- Stale data

If a communication failure occurs, the control center computer signals control center personnel of a problem so that an investigation can begin and the problem can be isolated, if needed.

If an individual remote terminal unit (RTU) loses communication with the control center, control center personnel must follow appropriate company procedures. In an emergency shutdown situation, if communication is lost to a main-line station that must be shut down, control center personnel must use other means to shut down the station. One method is to allow the suction pressure to decrease below the low-suction set point and call the appropriate field personnel.

The discovery that a large group of RTUs is not responding reveals a serious problem. Some SCADA systems are set up to switch to a backup communication system so that an attempt to control the situation can begin as soon as possible. Some companies are equipped with two backup computer systems in case the SCADA fails. Control center personnel should request a switch to one of the backup systems if available.

Control center personnel must consult with appropriate personnel to try to estimate how long the communications system will be in failure. The console monitors will reveal how many systems are affected. If control center personnel are advised that the system will be in failure for a specified amount of time, they will initiate appropriate procedures.

1.3.3 Control System Failures

Mechanical and hydraulic failures, also known as control system failures, are related to system communications failures. For example, the SCADA system sends a signal to a piece of equipment, but the equipment does not respond to the signal. In some cases, the problem is a result of a programming or command logic error. In other cases, equipment may not work because of a mechanical, electrical, or hydraulic problem. Examples include frozen valves, electrical component burnouts, and compressor breakdowns. A valve may have a mechanical failure that prevents it from being operated remotely by the control system. Sometimes valve actuators, electrical components, or connections burn out. This prevents the equipment from responding to a command by the system to start up or shut down.

1.4.0 Recognizing Power Interruptions

Electric power is essential to pipeline operations. Nearly every part of the pipeline depends on electricity in some way. A loss of power impacts the abilities of the control center or facility operations to safely control and monitor the pipeline facility. Power interruptions that result from bad weather or a problem at the local utility company, such as transformer failure or downed power lines, can be common.

Some pipeline operators have an **uninterruptible power system (UPS)** that turns on if electrical power is off for more than a few seconds. UPS systems supply emergency alternating current (AC) power to pipeline components to keep the pipeline operating until power is restored. Instead of electricity, these systems run on engines fueled with gasoline, diesel, or natural gas that are designed to start up if power is off for more than a few seconds. When power fails, an AC power failure alarm may be sent to the control center. To ensure that the backup system has turned on, control center personnel check to see whether they are receiving updated information from the pipeline.

If power is lost, electronic indicators may be lost. The SCADA system may switch off and pumps or compressors may shut down. In addition, lights, heat, and air conditioning may fail. An area-wide power outage may cause a loss of telephone communications. Local power outages may shut down only a portion of the pipeline system. Power outages can also cause pumps or compressors at the receiving terminal to stop operating, causing a liquid or gas shipment to overflow in tanks or be unable to be shipped further.

Some equipment is protected from power failure by battery systems. These systems run on direct current (DC). The following types of equipment can operate on DC power:

- Fire detection equipment and alarms
- Halon dispenser systems
- Fire suppression systems
- Emergency lighting systems
- SCADA
- Alarm panels
- Control modules
- Process control panels
- Specialized compressor units

1.5.0 Recognizing Fire, Explosion, or Natural Disaster

Fire, explosion, or natural disaster can occur at any time. Some events may be anticipated, but many are not. Control center personnel must be trained to recognize these events.

1.5.1 Fire

Detection devices installed along the pipeline and in pipeline facility buildings alert personnel in case of a fire. These devices send an alarm when their sensors detect heat or flame. In most cases, control center personnel will only see or hear the alarm, not the fire itself. Detection devices include thermal detectors, flame detectors, and smoke detectors.

A thermal detector is a heat-sensitive unit. It is fitted with electrical contacts that stay open as long as conditions are normal. When the unit senses rapidly rising temperatures, the contacts start to close. When the contacts connect, the unit may send an alarm to the control center and the fire control panel in the workstation.

A flame detector (see *Figure 5*) is a light-sensitive unit. It is fitted with an ultraviolet detector that turns on an alarm when the unit senses light from a flame. The unit is sensitive enough to detect a flame even in the earliest stages of a fire. Some flame detectors are connected to a firefighting system that may dump fire suppression material into the area. Personnel can also dump this material over a fire manually by using pull switches.

A smoke detector can be a light-sensitive unit, an ionization detector, or a combination of both types. Ionization detectors are more sensitive to fires in the early smoldering stage, while light detectors are more sensitive to the flaming stage of fires than ionization detectors.

65102-13_F05.EPS

Figure 5 Flame detector.

An ionization smoke detector produces ionization in air; when there is smoke, a difference is detected. If any smoke particles pass into the chamber of the detector, the ions attach to the particles, and become less able to carry the current. An electronic circuit detects the current drop, and sounds the alarm. The light detector operates by detecting light beams. In the absence of smoke, the light passes in front of the detector in a straight line. When smoke enters the chamber across the path of the light beam, light is scattered by smoke particles, directing it at the sensor that triggers the alarm.

Monitoring devices that detect a lower explosive limit (LEL) produce alarms to indicate that there is a risk of explosion. The LEL is the lowest concentration of a gas or a vapor in the atmosphere capable of producing fire in the presence of an ignition source. Different products have different LELs. The monitoring device must be set to detect the particular LEL in an enclosed area. Hazardous atmosphere detectors also need to be used in an enclosed area where an LEL has gone off before anyone re-enters the area.

1.5.2 Explosion

Pipeline operators install a number of safety devices to warn personnel about conditions that could cause explosions. However, operators cannot control every circumstance, such as intentional damage, that could lead to an explosion on a pipeline. An explosion on a pipeline registers in the control center as an unexplained change in pressure. An explosion may occur in any enclosed space where hazardous vapors can accumulate to dangerous levels. Examples of such spaces include pipeline facility buildings.

If natural gas leaks into an enclosed space, the odorant added to the gas sometimes alerts people to the problem. However, gas can sometimes accumulate undetected. For example, a building may be vacant or the odors may be too far away from people inside the building. In addition, products may not have an odorant added during transport. Be aware that a spark from electrical equipment can cause the gas to explode.

Whenever a call comes into the control center in which the caller describes a smell related to a pipeline, always make note of it for reporting purposes.

WARNING!

Personnel should never rely on their sense of smell to detect a natural gas leak. Products do not always have a detectable odor.

1.5.3 Natural Disaster

Natural disasters include earthquakes, hurricanes, tornadoes, landslides, wildfires, or flooding. These events may cause soil movement, which may result in forces sufficient enough to cause pipeline failure or product release. Even heavy rains resulting in saturated soil conditions may cause soil movement in some areas.

Pipeline operators monitor for weather that can develop into a natural disaster. Control center personnel may monitor for earthquakes with seismic indicators. In addition, they may obtain real-time information on the internet. Up-to-date earthquake information is available from the National Earthquake Information Center, which is located in Denver, Colorado. Personnel may track weather alerts and warnings through the National Weather Service website. In case of a power failure, personnel can continue to receive information through the National Weather Service's National Weather Radio or other electronic devices. A list of weather and natural disaster information sources is included in *Appendix A*.

1.6.0 Recognizing the Unexpected Release of Hazardous Gas or Liquid

Control center personnel will usually receive reports about an unexpected release from field personnel. The field personnel may notice the physical signs of a release such as a pool of liquid on the ground, a frost ball, vapor cloud or mist in the air, dead or discolored vegetation, or a rainbow sheen on water. Other signs include dead fish in waterways or other dead wildlife. Know company policies and procedures for each type of release and be prepared to act quickly and decisively to protect people, the pipeline, and the environment.

1.7.0 Recognizing Unexplained Flow Rate / Pressure Changes

Flow rate and pressure are important indicators of both normal and abnormal operating conditions. Minor changes in pressure may occur normally on the pipeline. However, significant unexplained pressure changes are considered AOCs. Unexplained pressure changes include absences of, increases in, and decreases in pressure. The maximum pressure a gas pipeline or segment of a gas pipeline may be safely operated is called the **maximum allowable operating pressure (MAOP)** and is defined in 49 *CFR* 192. MAOP is based on the materials transported and the design of the system. The maximum internal pressure permitted during the operation of a hazardous liquid or carbon dioxide pipeline is called the **maximum operating pressure (MOP)** and is defined in 49 *CFR* 195. MOP must be documented if it reaches 110 percent or higher.

Pressure-indicating devices located near mainline valves, compressor stations, and custody transfer stations monitor pressure and send readings to the control center. When an unexpected change in pressure occurs, an alarm may alert control center personnel. Sudden pressure increases or decreases might indicate a problem such as the failure of a valve or a transmitter. Pressures may also increase when inspection or monitoring devices called pigs are launched into the pipeline or if field personnel are manually operating valves. Personnel must be familiar enough with the pipeline system, equipment, product characteristics, and environmental factors to quickly identify any important changes in pressure and flow.

Additional Resources

It is assumed that you will learn to perform these procedures under the direct supervision of your crew leader, foreman, or instructor. For independent study, refer to the recommended Additional Resources at the back of this module.

1.0.0 Section Review

1. Which of the following is an AOC?
   a. Aboveground pipe on pipe supports
   b. Damaged pipeline
   c. Valves operating properly
   d. Authorized encroachment

2. Safety devices include _____.
   a. hazardous atmosphere alarms
   b. tank shells
   c. SCADA
   d. compressors

3. SCADA serves as a _____.
   a. regulatory agency
   b. notification system for all utilities
   c. device worn to provide breathable air where the atmosphere is hazardous
   d. command system that assists control center personnel

4. An uninterruptible power system (UPS) supplies _____.
   a. temperature readings
   b. emergency AC power
   c. emergency DC power
   d. pressure readings

5. A flame detector is designed to activate when it detects _____.
   a. an increase in temperature
   b. light
   c. smoke
   d. pipeline pressure

6. Dead vegetation or dead wildlife may be a sign of an unexpected release of a hazardous gas or liquid.
   a. True
   b. False

7. The maximum operating pressure of a gas pipeline is called _____.
   a. MOP
   b. MAOP
   c. surge pressure
   d. compression

Section Two

2.0.0 Reacting to Abnormal Operating Conditions

Objective

Describe the proper reaction to abnormal operating conditions in a control center for both gas and liquid pipelines.

a. Describe the proper reaction to abnormal facility conditions and pipeline system damage.
b. Describe the proper reaction to the activation of a safety device.
c. Describe the proper reaction to communications failure and a control system failure.
d. Describe the proper reaction to power interruptions.
e. Describe the proper reaction to a fire, explosion, and natural disaster occurring with a pipeline.
f. Describe the proper reaction to hazardous liquid or gas encounters.
g. Describe the proper reaction to an unexplained flow rate / pressure change.

Trade Terms

Naturally occurring radioactive materials (NORMs): Materials with radioactive elements that can be extracted during oil and gas extraction from the earth and can get into product streams and become transported in the pipeline.

Self-contained breathing apparatus (SCBA): A device worn to provide breathable air for an individual in an area where the atmosphere is hazardous.

Time is of the essence when AOCs occur, and control center personnel are expected to take decisive action. Personnel must always follow established company-specific policies and procedures. Personnel must make use of their training, skills, and good judgment to maintain safe operation until solutions are found.

2.1.0 Reacting to Abnormal Pipeline Facility Conditions

Some abnormal pipeline facility conditions may require immediate action. Others may require scheduling repairs at a later date. In most cases, if control center personnel are notified of an abnormal pipeline facility condition, they must notify the appropriate personnel so that repairs can be made safely and in a timely manner. If notified of a condition by a third-party call, keep in mind that a caller's description may not be entirely accurate. The problem may be other than what was described. If a report is confirmed and the condition does not require shutdown, make appropriate notifications as required by company policy. Reports of exposed pipe and damaged pipeline supports are examples of conditions that may require immediate action.

2.1.1 Pipeline System Damage

If a report of exposed pipe comes in to the control center, control center personnel should check the data received from the monitors and SCADA and send field personnel to investigate according to company policy. It is possible that communications were destroyed at the time the damage occurred and thus no alarm was signaled. The pipeline may need to be shut down until the investigation is complete, according to company policies and procedures. The decision to do so should be based on the assessment provided by trained and qualified personnel. The sooner qualified personnel can arrive on the scene, the more quickly a practical assessment of conditions can be obtained and control of the situation can begin.

Some incidents may be serious enough to require an emergency shutdown of the pipeline. One example of a serious incident is the loss of control of a vehicle on the ROW, extensively gouging or denting a large section of pipeline or damaging a valve. Another common example is third-party damage to pipeline from excavation. Whenever pipeline damage results in a release and/or explosion, control center personnel should receive a report from field personnel. In some cases, control center personnel may be able to drain a portion of the pipeline into a delivery system to lessen the amount of product that could be released from the pipeline.

Pipeline shutdowns do not have to occur only because of obvious damage to the pipeline. A blocked valve that causes increased pressure in the line can be reason for a shutdown. However, if pressures remain within the normal operating parameters, a shutdown may not be necessary, but the control center personnel must report the status change to the appropriate personnel.

Once the pipeline has resumed normal operations, the pipeline operator will conduct an internal investigation. This investigation focuses on evaluating control systems and mechanical equipment, as well as personnel performance. Personnel actions must be evaluated to determine whether company policies and procedures were followed. The work must also be evaluated to determine the effectiveness of existing company procedures. If the operator suspects that the damage was intentional, the appropriate personnel may contact regulatory and law enforcement agencies that may then conduct a criminal investigation.

2.1.2 Damaged Pipeline Supports

The proper reaction to the discovery of damaged pipeline supports varies depending on the severity of the damage. If the damage to the support is critical or indicative of a possible release, control center personnel may receive a report from field personnel. Notification is especially important if the pipe is in danger of excessive movement. Follow company policy precisely to investigate the condition and dispatch qualified personnel to the scene.

2.2.0 Reacting to the Activation of a Safety Device

Control center personnel must respond to safety device alarms quickly and calmly. They must be able to dispatch the appropriate personnel and inform them of the potential for a hazardous environment. Control center personnel may need to shut down the pipeline, if necessary. Field personnel may need to reset alarms. Company procedures must always be followed.

It is important to ensure that an alarm is not a false alarm. In case of a false alarm, the control center should notify affected personnel and have the alarm equipment checked. Generally, no further action is required when an alarm is false.

For example, if responding to a seal failure alarm, personnel should investigate the validity of the alarm, check the station sump level, and notify the appropriate personnel. This may not necessarily result in an AOC, but it may require control center personnel to document the alarm and notify the appropriate personnel to check and possibly replace the seal. Control center personnel should never attempt to restart a pump before investigating the cause of the alarm and responding to the alarm according to company policy.

During an emergency, personnel must focus on three key areas: ensuring personnel safety, reducing the damage to equipment, and protecting the environment. The general rule is to put people first. The goal is also to prevent small problems from turning into larger incidents. All reactions must be made safely.

WARNING!

For safety reasons, only authorized personnel who have had the proper training may use PPE.

2.2.1 Activation of a Pressure Relief System

Each relief event should be recognized and must be documented by control center personnel. If a relief device is activated, control center personnel must immediately minimize the situation and verify the device is operating correctly. If the alarm is signaled and the pressure remains high, personnel must take immediate action to reduce pressure. A pressure reduction is usually achieved by shutting down the upstream pumping or compressor station or the pipeline system.

2.2.2 High Pump-Case Temperature/Pressure Alarm

Some main-line pumps are set to shut down on low pressure. Pipeline operators fit pumps with differential pressure switches and temperature switches to protect the equipment from damage caused by low pressure, no pressure, or no flow; to avoid exceeding maximum operating pressure (MOP)/maximum allowable operating pressure (MAOP), which means no more than 110 percent MOP/MAOP; and to ensure that the temperatures do not increase in the pipeline facility. Pumps also have high discharge-pressure alarms that will shut down the pump when the discharge pressure exceeds a preset limit. Many times, this indicates a downstream line obstruction that is not allowing the product to flow properly from the pump.

2.3.0 Reacting to a Communications Failure

The proper reaction to a communications failure depends on whether the failure involves a communications tool that personnel use or the failure of system communications.

2.3.1 *Failure of Personnel Communication Tools*

When communication tools fail, personnel should first check to see if the problem is a simple one that can be quickly and easily fixed. For example, some telephones have a volume control that may accidentally be turned off. The telephone may seem to be dead, but a simple adjustment to the volume corrects the problem. Other simple solutions include replacing a battery or recharging a unit.

Some communications problems are more serious. Weather or problems at the local utility company can knock out communications completely. Regardless of the outage, the best preparation is to ensure personnel are trained to follow these steps:

Step 1 Switch to alternate equipment if necessary.

Step 2 Notify the appropriate personnel per company policy.

Step 3 Complete any required documentation.

2.3.2 *System Communications Failures*

Control center personnel must recognize an abnormal condition, investigate immediately, and react according to company policy. The condition must be verified, and the cause(s) should be determined. If the failure involves a major pipeline component such as a remote gate valve or block valve, control center personnel must try to solve the problem as quickly as possible to prevent an incident. In some instances, control center personnel might be able to bypass the problem component until field personnel can perform appropriate maintenance. In other cases, they may be required to shut down.

If one of the pumps at a station that is running suddenly indicates a stop status with no apparent changes to the station pressures, control center personnel should notify appropriate field personnel, and investigate as required per company policy. If a control center loses its monitoring capabilities for a remotely unmanned station, but SCADA is showing normal operating pressures and flow rates upstream and downstream of the station, control center personnel should not make multiple unit switches upstream or downstream of the facility before it is confirmed that a problem actually exists. Appropriate reactions include restoring communications with an alternate system, notifying repair personnel, and having the pipeline facility manned.

The control center must notify field personnel and/or systems support about system communications failures so that they can isolate, investigate, and address the problem as needed. If field personnel can operate the component manually, the problem can be alleviated until repairs can be made. If the component cannot be manually operated, the pipeline may have to be shut down. Control center personnel must monitor pressure remotely or shut down. If pressure is remaining within limits or the component is not one that will affect the operation directly, appropriate personnel must decide if the pipeline can safely remain in operation. If the pipeline remains in operation, the condition must be monitored closely.

After normal conditions have been restored, control center personnel must continue to monitor the condition to make certain that the procedures have been successful. If SCADA communication is lost, control center personnel must follow their company's policies and procedures, usually covered in the business continuity plan. These procedures may include switching to a backup system that may be in another location.

2.4.0 Reacting to Power Interruptions

The proper reaction to a power failure depends on which part(s) and how much of the pipeline is affected. Regardless of how the outage affects the pipeline facility, the appropriate personnel should always notify the power company of the outage as soon as possible. If only a single component or station is involved, control center and/or field personnel may be able to adjust line pressure and flow to keep the pipeline in operation. For example, if a power interruption occurs while a valve is in operation, field personnel may have to operate the valve manually. However, if a station is needed to keep pressure within nnormal operating limits, control center personnel may have to isolate a pipeline segment. If a shutdown is not required, the gas or liquid shipment will continue to arrive, and the schedule must be followed and the tasks completed manually.

Power interruptions can happen in the control center itself. If backup power isn't available, the control center personnel should contact field personnel to go to their control stations to maintain pipeline operations.

Each company has specific procedures for responding properly to a power interruption. Always follow company policy and make appropriate notifications per company policy.

2.5.0 Reacting to a Fire, Explosion, or Natural Disaster

The pipeline operator must have procedures in place so that personnel know what to do in case of a fire, explosion, or natural disaster that affects a pipeline. Always react according to company policy, which includes the policies regarding shutting down the pipeline.

When any of these events occur, the safety of personnel is always top priority. Personnel must ensure that any injured individuals receive medical attention as soon as possible. Consider any alarm as an actual alarm and react appropriately and decisively until it has been verified as false. Control center personnel must investigate the cause of any alarms or other notifications and confirm the nature of the problem in order to react appropriately. Personnel must make appropriate notifications and complete the required documentation.

In emergency situations, including evacuation if needed, follow the company emergency procedures manual as appropriate. Personnel must be adequately trained and familiar with emergency procedures in advance.

2.5.1 Emergency Shutdowns

Some events occurring on the pipeline that present immediate hazards require an emergency shutdown. Such an event might be caused by human error, mechanical failure, electrical failure, or an accident on the pipeline. It might also be brought about by threats from weather or civil disobedience. Regardless of the reason, when an emergency shutdown is required the line must be shut down quickly and safely. Each company has a standard operating procedure for emergency shutdowns.

In the case of a fire, explosion, or release, time for discussion may be very limited. Once the need is confirmed, decisive action must be taken. The pipeline must be shut down by control center personnel who will shut down the appropriate pumps. Control center personnel may isolate or authorize the isolation of the system by closing block valves as necessary.

In emergency conditions such as a possible release, the pipeline operator has full responsibility and authority to make shutdown decisions and take appropriate action. These actions are governed by the company release policy above all other considerations.

2.5.2 Evacuations

The safety of pipeline personnel and the public is the first priority when any AOC occurs. If there is any threat of injury or death to any persons, an evacuation of pipeline personnel must occur. Events that could result in the need for an evacuation are fire, explosion, release of hydrocarbon or other hazardous vapors, seismic activity, or other natural disasters.

Companies have standard procedures in place to guide evacuation of personnel. By law, emergency response plans must be in place and evaluated annually. Facility layouts with evacuation maps are required to be posted in every area or department. Safe areas are designated for all personnel. When the command to evacuate is given, trained personnel may act as safety coordinators. Their job is to conduct a roll call to confirm that all personnel are safe. To assist in this process, personnel must report to their designated safe areas and identify themselves to the designated safety coordinator. If any personnel cannot be accounted for, authorized personnel must be informed so that missing individuals can be located and rescued, if necessary.

2.6.0 Reacting to the Unexpected Release of Hazardous Gas or Liquid

When a hazardous gas or liquid is released from a pipeline unexpectedly, personal safety should be the first consideration. Beyond that, the proper reaction to the unexpected release of a product depends on the characteristics of the product and the location of the release. Some gases and liquids can explode under certain conditions, and some are toxic to people and the environment. To identify the hazards associated with the chemical being released, consult the material safety data sheets (MSDSs) / safety data sheets (SDSs).

Once notified by field personnel of a release, the source of the release must be determined quickly. Control center personnel may need to reduce pressure and shut down the pipeline. If power cannot safely be cut off at the switchgear, control center personnel must call the power company and have power cut off from outside the pipeline facility. Releases in outdoor areas include the additional concern for the safety of the general public, as well as the safety of the environment. In all cases, follow company policies and procedures, make the appropriate notifications, and complete the required documentation.

Personnel must take special precautions when dealing with the release of products that can become toxic. Examples of toxic products include carbon dioxide (CO_2), hydrogen sulfide (H_2S), propane, anhydrous ammonia, and natural gas.

Carbon dioxide is a colorless, odorless, gaseous compound used as a propellant in many domestic applications, such as fire extinguishing equipment, and is a by-product of combustion. Because it is heavier than air, it can collect in low-lying areas where it may displace oxygen; exposure can result in suffocation and death. A leak detector can detect a hazardous atmosphere, but without it, personnel can be overcome without realizing the problem. Some pipeline operations pump carbon dioxide commercially, and in such operations the hazards are increased. To enter an area where a carbon dioxide release is suspected, personnel must use the proper gas detection meter, calibrated for the detection of carbon dioxide, and wear the appropriate PPE, such as a **self-contained breathing apparatus (SCBA)** (see *Figure 6*).

65102-13_F06.EPS

Figure 6 Self-contained breathing apparatus.

Hydrogen sulfide (H_2S) is a colorless gas that often smells like rotten eggs. However, in fairly low concentrations, it can deaden the sense of smell. Exposure to a higher concentration can cause death. In addition, it is flammable and may cause a flash fire. When a gas detection unit shows that H_2S is present, personnel must wear the appropriate PPE or immediately leave the area.

Propane is a common natural gas that is colorless and heavier than air, so it can collect in low-lying areas where it may displace oxygen. Exposure can cause death. At high concentrations, it is flammable.

Anhydrous ammonia is a liquefied gas that has a sharp odor. Exposure in high concentration can cause death. Even exposure in low concentrations can cause severe reactions from mild irritation to severe burns. The reaction depends on the length of exposure and concentration amounts. If anhydrous ammonia is released, call emergency personnel.

Naturally occurring radioactive materials (NORMs) may be present when oil and gas is extracted from the earth. These NORMs are then introduced into the product streams and transported in the pipeline. Natural gas liquids (NGLs), along with processing associated with these products, may trap NORMs in the form of sludge, scale, and particulates. Filters, valves, scrapers, and even internal surfaces of pipelines may contain high levels of NORMs. Precautions should be taken to protect personnel by monitoring for NORMs in accordance with all local, state, federal, and company requirements. If high levels are detected, personnel should distance and/or shield themselves from the source and notify the appropriate personnel.

WARNING!

Exposure to hydrogen sulfide (H_2S), carbon dioxide (CO_2), and propane can be fatal. H_2S is flammable and may cause a flash fire. When a gas detection unit shows that H_2S is present, wear the appropriate PPE or immediately leave the area. CO_2 and propane are colorless, odorless gases that are heavier than air, so they can collect in low-lying areas where they may displace oxygen.

2.7.0 Reacting to Unexplained Flow Rate/Pressure Changes

Control center personnel must respond to a flow rate/pressure change alarm immediately to prevent serious damage to the pipeline. The sooner a response is initiated, the more quickly the problem will be solved and the more likely a serious condition will be averted. Control center personnel should check pressure and flow rates upstream and downstream for comparison. If the time frame involved in the change is very short, a release might be indicated and the pipeline shut down per company policy.

2.7.1 Unexplained Flow Rate Changes

Control center personnel must determine if there has been an intentional change in pumping status or compressor status on the pipeline that was not scheduled. A valve might be the cause of the problem. If the main-line flow has dropped, a pump unit or compressor unit may have shut down upstream, and control center personnel can adjust pumping or compressor units upstream in an attempt to compensate. It is the responsibility of control center personnel to initiate corrective action.

If a pipeline block valve has been closed without authorization or if personnel notice a high flow rate at the origin station, personnel should follow established troubleshooting procedures, perform a shutdown as necessary, and make all proper notifications.

If a pipeline is running, and the origin meter flow rate increases without justification, possible causes for this include SCADA malfunction, a pipeline rupture downstream of the origin meter, or maintenance personnel interfering with data transmission prior to notification. Again, control center personnel should follow established troubleshooting procedures, perform a shutdown as necessary, and make all proper notifications.

If there is not a release, it may be possible to safely work around the AOC. Control center personnel should try to adjust the pipeline pressure and flow to achieve a steady-state operation. During line upsets or valve swings, control center personnel must avoid frequent or unnecessary line upsets because data indicating a release is very difficult to identify during these conditions. If the pressure and flow cannot be corrected within a reasonable length of time, control center personnel should shut the pipeline down as quickly as possible without jeopardizing safety. Company policy determines exact procedures that must be followed. Appropriate personnel should be notified of the shutdown as soon as possible.

2.7.2 Unexplained Pressure Changes

Unexplained pressure changes can include an absence of pressure, a sudden increase in pressure, or a sudden decrease in pressure. Deviations in pressure may indicate an unusual condition on the pipeline. Occurrences such as unintended valve closures may cause AOCs that could endanger the pipeline. The condition must be recognized, and the reaction must be quick and decisive.

If a high-pressure alarm is received from a compressor or pumping unit, personnel should investigate the alarm and determine the appropriate action. This could include shutting down the unit or line as needed.

If the condition is a valve closure or a sudden decrease in discharge pressure followed by a sudden increase in discharge pressure, control center personnel must troubleshoot to determine the cause of the closure. If there is not a high-pressure condition downstream and a valve closure is identified between the two stations, then a command to open the valve must be sent by control center personnel.

If the valve cannot be opened automatically or if the condition occurs at a location with personnel available, opening the valve manually may be possible. If the valve does not open, the pipeline may have to be shut down due to increasing pressure. Company policy and procedure must be followed.

If control center personnel conclude that there may be time to make adjustments before the pipeline reaches MOP/MAOP, immediate action must be taken. Personnel must notify all operations upstream to reduce or stop pipeline injections. Personnel may reduce strategic pumping units upstream or initiate other means to relieve the pressure. Appropriate personnel may be consulted if there is time. If the condition can be corrected by such efforts, the pipeline operation may be continued. Control center personnel must monitor the pressure readings closely until satisfied that the pipeline has stabilized.

If there is no time to accomplish the adjustments or if it appears that the efforts are having no effect, control center personnel must perform a shutdown according to company policy and procedure.

At a steady-state, multipump station pipeline, if there is an increased pressure reduction between two stations, a flow increase upstream of the pressure loss, and all analog data indicates the high probability of a release, control center personnel should shut down the line with the lowest possible pressure on the segment with the suspected release. If personnel notice a mainline block valve status has changed from opened to closed with an increase in discharge pressure at the upstream pump station, personnel should shut down the pipeline until the correct valve position is determined.

Additional Resources

It is assumed that you will learn to perform these procedures under the direct supervision of your crew leader, foreman, or instructor. For independent study, refer to the recommended Additional Resources at the back of this module.

2.0.0 Section Review

1. If damaged pipeline is reported to the control center, the proper reaction always includes shutting down the pipeline.
   a. True
   b. False

2. If a safety device activates, personnel should focus *first* on _____.
   a. protecting the pipeline
   b. ensuring the safety of personnel
   c. notifying PHMSA
   d. verifying that the alarm is not false

3. Field personnel may need to operate equipment in the field manually if there is a systems failure.
   a. True
   b. False

4. When a power outage occurs, the appropriate pipeline personnel should contact _____.
   a. the One-Call notification system
   b. the power company
   c. OSHA
   d. local authorities

5. When reacting to a fire, explosion, or natural disaster, the first priority is always to protect _____.
   a. the pipeline facility
   b. the environment
   c. people
   d. the pipeline product

6. One reason that the unexpected release of carbon dioxide can be dangerous is because it is _____.
   a. odorless
   b. a compound
   c. lighter than air
   d. white in color

7. A pipeline rupture downstream of the origin meter could be the cause of _____.
   a. a power outage
   b. an unexplained communication failure
   c. an unexplained flow rate change
   d. damaged pipe

SUMMARY

Pipeline operators have developed effective routines and policies that enable personnel to respond quickly, safely, and effectively to AOCs. Usually an AOC is a routine matter that can be handled safely by pipeline personnel. Sometimes, however, an AOC can result in an emergency situation that may involve working with outside agencies.

In this module, you learned about the different types of AOCs that can occur on the pipeline and the proper reactions to each type of AOC. During your career, you may have to respond to some or all of the AOCs outlined in this module. When responding to these AOCs, follow company policies and procedures. Protecting people, property, and the environment should be of the utmost concern.

Review Questions

1. By definition, an AOC is a condition that may exceed _____.
   a. maximum operating pressure
   b. design limits
   c. safety laws
   d. size limits

2. By definition, a pipeline facility includes the right-of-way.
   a. True
   b. False

3. Which of the following is an example of pipeline system damage?
   a. Gouges or dents in the pipeline
   b. Transformer failure
   c. Decreased pressure
   d. Activation of a pressure-relief valve

4. Pressure-relief systems, tank-level alarms, and main-line pressure shutdown alarms are examples of _____.
   a. prevention devices
   b. warning devices
   c. hazardous atmosphere alarms
   d. fire alarms

5. The loss of SCADA is an example of which type of AOC?
   a. Hydraulic failure
   b. Mechanical failure
   c. Communications failure
   d. Power failure

6. To keep the pipeline running in case of a power failure, some companies use an _____.
   a. operations override power source
   b. uninterruptible power system
   c. interface power system
   d. indirect power system

7. The failure of a valve or a transmitter may be indicated by _____.
   a. ongoing pressure increases or decreases
   b. sudden pressure increases or decreases
   c. high-level pressure increases or decreases
   d. maximum pressure increases or decreases

8. When pipeline support damage is discovered, notification is especially important if the pipe _____.
   a. is in danger of excessive movement
   b. has changed color
   c. is dented
   d. is corroded

9. If a relief device alarm is activated, control center personnel must immediately _____.
   a. shut down or call 911
   b. reduce pressure or shut down
   c. contact field personnel
   d. complete the required reports

10. At a station that is running, if a pump suddenly indicates a stop status with no apparent changes to the station pressures, control center personnel should _____.
    a. notify appropriate field personnel
    b. reduce pressure
    c. shut down the SCADA system
    d. activate safety devices

11. If SCADA fails, it may be necessary for equipment to be operated manually by _____.
    a. PHMSA
    b. field personnel
    c. control center personnel
    d. local authorities

12. When a power interruption occurs, personnel may be able to keep the pipeline in operation by adjusting _____.
    a. MOP/MAOP
    b. line pressure and pipe location
    c. line pressure and flow
    d. product temperature

13. Regardless of the nature of an emergency, if there is a threat of injury or death, pipeline personnel must be _____.
    a. contacted
    b. evacuated
    c. warned
    d. involved

14. The top priority is always the safety of _____.
    a. personnel
    b. the pipeline
    c. the environment
    d. equipment
15. Emergency shutdowns are generally required when _____.
    a. equipment is reported on the ROW
    b. natural disasters are anticipated
    c. there is an immediate hazard to persons
    d. pipeline damage is reported by third parties
16. Product releases in outdoor areas involve an extra concern for _____.
    a. safety of third-party equipment
    b. safety of pipeline equipment
    c. safety of the pipeline
    d. safety of the environment
17. Control center personnel should rely solely on reports of unexpected releases by field personnel or third parties to determine the proper reaction.
    a. True
    b. False
18. AOCs are not necessarily considered safety-related conditions.
    a. True
    b. False
19. All of the following are examples of toxic products, *except* _____.
    a. CO_2
    b. anhydrous ammonia
    c. propane
    d. H_2O
20. What type of device is preset to signal an alarm in the control center when the temperature reaches a critical level?
    a. RTD
    b. seal failure alarm
    c. LEL
    d. high-pressure alarm

Trade Terms Quiz

Fill in the blank with the correct term that you learned from your study of this module.

1. A computer-based system or systems used by a controller in a control room that collects and displays information about a pipeline facility and may have the ability to send commands back to the pipeline facility is called ________.
2. The ________ is the area along the pipeline which has been legally permitted to be used by the pipeline operator to construct, operate, and maintain the pipeline.
3. Activity carried out by and/or equipment belonging to outside parties on the pipeline ROW is referred to as ________.
4. A(n) ________ is a device worn by an individual to provide breathable air in an atmosphere that is hazardous.
5. ________ is the separation of the protective coating from the pipeline.
6. A(n) ________ is a backup system that provides AC power to the systems it serves.
7. The maximum pressure at which a gas pipeline or segment of a gas pipeline may be normally operated is the ________.
8. A(n) ________ is used to relieve surges in a hazardous liquid pipeline system or receive and store hazardous liquid transported by a pipeline for reinjection and continues transportation by pipeline.
9. A device used on pumps that reads the temperature of the pump case and the product in the pipeline that will sound an alarm when the temperature reaches a critical level and shut down the pump automatically if needed is a(n) ________.
10. A(n) ________ duplicates the operations of a system so that a complete backup system is available in the event of systems failure.
11. A(n) ________ is identified by the operator and may indicate (1) a malfunction of a component or (2) a deviation from normal operations that may indicate a condition exceeding design limits or result in a hazard to persons, property, or the environment.
12. The pressure produced by a change in the velocity of the moving stream that can result from the closure of a valve is called ________.
13. The maximum pressure at which a liquid pipeline or segment of a liquid pipeline may be operated is the ________.
14. The ________ is a centralized communication system established by utilities, government agencies, and operators of underground facilities that provides one telephone number for contractors and the general public to call to notify of their intent to use equipment on or near a pipeline.
15. A hazardous liquid that has a vapor pressure exceeding 40 pounds per square inch absolute at 100°F is classified as a(n) ________.
16. The government agency that is responsible for developing and enforcing regulations for pipeline operations in the United States is the ________.
17. The ________ is the company that operates the pipeline.

18. Materials with radioactive elements that can be extracted during oil and gas extraction from the earth and can get into product streams and become transported in the pipeline are called ________.

19. An organization that adopts standards for a variety of products, services, and processes in the United States to ensure consistency is the ________.

20. The lowest concentration of a gas or a vapor in the atmosphere capable of producing fire in the presence of an ignition source is the ________.

21. Technology that provides monitoring tools used by control center personnel to detect temperature, flow, and pressure irregularities in a pipeline is ________.

22. New and existing pipe on the pipeline itself, the right-of-way (ROW), and any equipment, facility, or building used in the transportation of a liquid or gas are defined as part of a(n) ________.

Trade Terms

Abnormal operating condition (AOC)
American National Standards Institute (ANSI)
Breakout tank
Computational pipeline monitoring (CPM)
Disbondment
Encroachment
Highly volatile liquid (HVL)
Lower explosive limit (LEL)
Maximum allowable operating pressure (MAOP)
Maximum operating pressure (MOP)
Naturally occurring radioactive materials (NORMs)
One-Call notification system
Pipeline and Hazardous Materials Safety Administration (PHMSA)
Pipeline facility
Pipeline operator
Redundant system
Resistance temperature device (RTD)
Right-of-way (ROW)
Self-contained breathing apparatus (SCBA)
Supervisory control and data acquisition (SCADA)
Surge pressure
Uninterruptible power system (UPS)

Trade Terms Introduced in This Module

Abnormal operating condition (AOC): A condition identified by the operator that may indicate (1) a malfunction of a component or (2) a deviation from normal operations that may indicate a condition exceeding design limits or result in a hazard to persons, property, or the environment.

American National Standards Institute (ANSI): An organization that adopts standards for a variety of products, services, and processes in the United States to ensure consistency.

Breakout tank: A tank used to (a) relieve surges in a hazardous liquid pipeline system or (b) receive and store hazardous liquid transported by a pipeline for reinjection and continued transportation by pipeline.

Computational pipeline monitoring (CPM): Technology that provides monitoring tools used by control center personnel to detect temperature, flow, and pressure irregularities in a pipeline.

Disbondment: Separation of the protective coating from the pipeline.

Encroachment: Activity carried out by and/or equipment belonging to outside parties on the pipeline right-of-way (ROW). A particular encroachment might be authorized or not, so each situation requires specific handling.

Highly volatile liquid (HVL): a hazardous liquid which will form a vapor cloud when released to the atmosphere and which has a vapor pressure exceeding 276 kPa (40 psia) at 37.8°C (100°F).

Lower explosive limit (LEL): The lowest concentration of a gas or a vapor in the atmosphere capable of producing fire in the presence of an ignition source. Different products have different LELs.

Maximum allowable operating pressure (MAOP): The maximum pressure at which a pipeline or segment of a pipeline may be operated under this part (liquid).

Maximum operating pressure (MOP): The maximum pressure at which a pipeline or segment of a pipeline may be normally operated under this part (gas).

Naturally occurring radioactive materials (NORMs): Materials with radioactive elements that can be extracted during oil and gas extraction from the earth and can get into product streams and become transported in the pipeline.

One-Call notification system: A centralized communication system established by utilities, government agencies, and operators of underground facilities. This system provides one telephone number for contractors and the general public to call to notify of their intent to use equipment on or near a pipeline. It is also in place for any individual to call if pipeline damage is discovered. The One-Call system is generally set up on a state-by-state basis.

Pipeline and Hazardous Materials Safety Administration (PHMSA): An agency of the US Department of Transportation responsible for developing and enforcing regulations for pipeline operations in the United States.

Pipeline facility: New and existing pipe; rights-of-way; and any equipment, facility, or building used in the transportation of hazardous liquids or carbon dioxide, or new and existing pipelines; rights-of-way; and any equipment, facility, or building used in the transportation of gas or in the treatment of gas during the course of transportation.

Pipeline operator: The company that operates the pipeline, as opposed to an individual person who works for the company that operates the pipeline.

Redundant system: A system that duplicates the operations of an entire system so that, in the event of a failure, a backup system is available.

Resistance temperature device (RTD): A device used on pumps that reads the temperature of the pump case and the product in the pipeline. It will sound an alarm when the temperature reaches a critical level and shut down the pump automatically if needed.

Right-of-way (ROW): The area along the pipeline which has been legally permitted to be used by the pipeline operator to construct, operate, and maintain the pipeline. The ROW extends the entire length of the pipeline.

Self-contained breathing apparatus (SCBA): A device worn to provide breathable air for an individual in an area where the atmosphere is hazardous.

Supervisory control and data acquisition (SCADA): A computer-based system or systems used by a controller in a control room that collects and displays information about a pipeline facility and may have the ability to send commands back to the pipeline facility.

Surge pressure: Pressure produced by a change in the velocity of the moving stream that results from shutting down a pump station or pumping unit, closure of a valve, or any other change in the velocity of the moving stream.

Uninterruptible power system (UPS): A system that provides continuity of power when the normal power supply is interrupted. A UPS provides AC power to the system it serves.

Appendix A

Weather and Natural Disaster Information Sources

Emergency Managers Weather Information. US Department of Commerce, National Oceanic and Atmospheric Administration, National Weather Service, 1325 East-West Highway, Silver Spring, MD 20910. **www.nws.noaa.gov**

National Earthquake Information Center, US Geological Survey. Box 25046, DFC, MS 967 Denver, CO 80225. **www.earthquake.usgs.gov**

National Hurricane Center Tropical Prediction Center. 11691 SW 17th Street, Miami, FL 33165-2149. **www.nhc.noaa.gov**

National Weather Service and National Weather Service Weather Radio, National Oceanic and Atmospheric Administration (NOAA). 1325 East-West Highway, Silver Spring, MD 20910. **www.weather.gov**

Organizations of Interest to the Pipeline Industry

American Gas Association (AGA). 400 N. Capitol Street, NW, Washington, DC 20001. **www.aga.org**

American Petroleum Institute (API). 1220 L Street, NW, Washington, DC 20005. **www.api.org**

Interstate Natural Gas Association of America (INGAA). 20 F Street, NW, Suite 450, Washington, DC 20001. **www.ingaa.org**

National Institute for Occupational Safety and Health (NIOSH). 395 E Street, SW, Suite 9200, Patriots Plaza Building, Washington, DC 20201. **www.cdc.gov/niosh**

National Response Center (NRC). c/o United States Coast Guard (CG-5335 - Stop 7581), 2100 2nd Street, SW, Washington, DC 20593-0001. **www.nrc.uscg.mil**

Occupational Safety and Health Administration (OSHA). 200 Constitution Avenue, NW, Washington, DC 20210. **www.osha.gov**

Pipeline Hazardous Material Safety Administration (PHMSA). East Building, 2nd Floor, 1200 New Jersey Ave., SE, Washington, DC 20590. **www.phmsa.dot.gov**

Additional Resources

The following recommended resources can provide additional helpful information related to the requirements for this covered task.

Code of Federal Regulations 49, Parts 192 and 195.

OSHA *Occupational Safety and Health Standards* 1910, Subpart L, Standard 1910.157.

Figure Credits

Courtesy of Sunoco Logistics, Module Opener, Figure 1

U.S. Geological Survey/photo by Don Becker, Figure 2

PORV image courtesy of Farris Engineering, Figure 4

Spectrex Inc., Figure 5

Courtesy of MSA, Figure 6

Section Review Answer Key

Answer	Section Reference	Objective
Section One		
1. b	1.1.0	1a
2. a	1.2.0	1b
3. d	1.3.2	1d
4. b	1.4.0	1d
5. b	1.5.1	1e
6. a	1.6.0	1f
7. b	1.7.0	1g
Section Two		
1. b	2.1.1	2a
2. b	2.2.0	2b
3. a	2.3.2	2d
4. b	2.4.0	2e
5. c	2.5.0	2f
6. a	2.6.0	2h
7. c	2.7.1	2i

NCCER CURRICULA — USER UPDATE

NCCER makes every effort to keep its textbooks up-to-date and free of technical errors. We appreciate your help in this process. If you find an error, a typographical mistake, or an inaccuracy in NCCER's curricula, please fill out this form (or a photocopy), or complete the online form at **www.nccer.org/olf**. Be sure to include the exact module ID number, page number, a detailed description, and your recommended correction. Your input will be brought to the attention of the Authoring Team. Thank you for your assistance.

Instructors – If you have an idea for improving this textbook, or have found that additional materials were necessary to teach this module effectively, please let us know so that we may present your suggestions to the Authoring Team.

NCCER Product Development and Revision

13614 Progress Blvd., Alachua, FL 32615

Email: curriculum@nccer.org

Online: www.nccer.org/olf

❑ Trainee Guide ❑ Lesson Plans ❑ Exam ❑ PowerPoints Other ____________

Craft / Level: Copyright Date:

Module ID Number / Title:

Section Number(s):

Description:

Recommended Correction:

Your Name:

Address:

Email: Phone:

AOCFG-17

Abnormal Operating Conditions – Field and Gas

Overview

Pipeline operators and contractors have programs, procedures, training, safety equipment, and warning devices for practically every facet of pipeline operations. Situations that are outside of normal operating conditions may arise at any time, and field personnel must know what to do to help keep pipeline operations safe for everyone. In some cases, an abnormal operating condition (AOC) is a routine matter that can be quickly and safely addressed by personnel. In other cases, an AOC can result in an emergency situation. In this module, personnel will learn how to recognize, properly react to, and properly report AOCs that may occur during pipeline operations.

Trainees with successful module completions may be eligible for credentialing through the NCCER Registry. To learn more, go to **www.nccer.org** or contact us at **1.888.622.3720.** Our website has information on the latest product releases and training, as well as online versions of our *Cornerstone* magazine and Pearson's product catalog.

Your feedback is welcome. You may email your comments to **curriculum@nccer.org**, send general comments and inquiries to **info@nccer.org**, or fill in the User Update form at the back of this module.

This information is general in nature and intended for training purposes only. Actual performance of activities described in this manual requires compliance with all applicable operating, service, maintenance, and safety procedures under the direction of qualified personnel. References in this manual to patented or proprietary devices do not constitute a recommendation of their use.

AOCFG-17
Abnormal Operating Conditions – Field and Gas

Objectives

When you have completed this module, you will be able to do the following:

1. Explain how to recognize abnormal operating conditions for both gas and liquid pipelines.
   a. Explain how to recognize abnormal pipeline facility conditions.
   b. Explain how to recognize the activation of a safety device.
   c. Explain how to recognize a communications failure and a control system failure.
   d. Explain how to recognize power interruptions.
   e. Explain how to recognize a fire, explosion, and natural disaster occurring with the pipeline.
   f. Explain how to recognize the unexpected release of hazardous liquid or gas.
   g. Explain how to recognize unexplained pressure/flow rate changes.
2. Describe the proper reaction to abnormal operating conditions for both gas and liquid pipelines.
   a. Describe the proper reaction to abnormal pipeline facility conditions.
   b. Describe the proper reaction to the activation of a safety device.
   c. Describe the proper reaction to a communications failure and a control system failure.
   d. Describe the proper reaction to power interruptions.
   e. Describe the proper reaction to a fire, explosion, and natural disaster occurring with the pipeline.
   f. Describe the proper reaction to the unexpected release of hazardous liquid or gas.
   g. Describe the proper reaction to unexplained pressure/flow rate changes.

Performance Tasks

This is a knowledge-based module; there are no performance tasks.

Trade Terms

Aboveground span
Abnormal operating condition (AOC)
American National Standards Institute (ANSI)
Breakout tank
Disbondment
Encroachment
Flame ionization detection unit
Highly volatile liquid (HVL)
Incipient fire
Intrinsically safe/explosionproof
Lower explosive limit (LEL)
Maximum allowable operating pressure (MAOP)
Maximum operating pressure (MOP)
Naturally occurring radioactive materials (NORMs)
One-Call notification system
Pipeline and Hazardous Materials Safety Administration (PHMSA)
Pipeline facility
Pipeline operator
Redundant system
Right-of-way (ROW)
Self-contained breathing apparatus (SCBA)
Supervisory control and data acquisition (SCADA)
Surge pressure
Uninterruptible power system (UPS)

Industry-Recognized Credentials

If you're training through an NCCER-accredited sponsor, you may be eligible for credentials from NCCER's Registry. The ID number for this module is AOCFG-17. Note that this module may have been used in other NCCER curricula and may apply to other level completions. Contact NCCER's Registry at 888.622.3720 or go to **www.nccer.org** for more information.

Contents

Topics to be presented in this module include:

Figures and Tables

Section One

1.0.0 Recognizing Abnormal Operating Conditions

Objective

Explain how to recognize abnormal operating conditions for both gas and liquid pipelines.

a. Explain how to recognize abnormal pipeline facility conditions.
b. Explain how to recognize the activation of a safety device.
c. Explain how to recognize a communications failure and a control system failure.
d. Explain how to recognize power interruptions.
e. Explain how to recognize a fire, explosion, and natural disaster occuring with the pipeline.
f. Explain how to recognize the unexpected release of hazardous liquid or gas.
g. Explain how to recognize unexplained pressure/flow rate changes.

Trade Terms

Aboveground span: A length of pipeline that is entirely aboveground and supported by pipe supports.

Abnormal operating condition (AOC): A condition identified by the operator that may indicate (a) a malfunction of a component or (b) a deviation from normal operations that may indicate a condition exceeding design limits or result in a hazard to persons, property, or the environment.

American National Standards Institute (ANSI): An organization that adopts standards for a variety of products, services, and processes in the United States to ensure consistency.

Breakout tank: A tank used to (a) relieve surges in a hazardous liquid pipeline system or (b) receive and store hazardous liquid transported by a pipeline for reinjection and continued transportation by pipeline.

Disbondment: Separation of the protective coating from the pipeline.

Encroachment: Activity carried out by, and/or equipment belonging to outside parties on the pipeline right-of-way (ROW). A particular encroachment might be authorized or not, so each situation requires specific handling.

Flame ionization detection unit: A device that detects some hazardous gases in the atmosphere.

Highly volatile liquid (HVL): A hazardous liquid which will form a vapor cloud when released to the atmosphere and which has a vapor pressure exceeding 276 kPa (40 psia) at 37.8°C (100°F).

Intrinsically safe/explosion-proof: Describes electrical equipment that is designed specifically for use in explosive or flammable locations so that a spark cannot ignite the atmosphere.

Lower explosive limit (LEL): The lowest concentration of a gas or a vapor in the atmosphere capable of producing fire in the presence of an ignition source. Different products have different LELs.

Maximum allowable operating pressure (MAOP): The maximum pressure at which a pipeline or segment of a pipeline may be operated (gas).

Maximum operating pressure (MOP): The maximum pressure at which a pipeline or segment of a pipeline may be normally operated (liquids).

One-Call notification system: A centralized communication system established by utilities, government agencies, and operators of underground facilities. This system provides one telephone number for contractors and the general public to call to notify of their intent to use equipment on or near a pipeline. It is also in place for any individual to call if pipeline damage is discovered. The One-Call system is generally set up on a state-by-state basis.

Pipeline and Hazardous Materials Safety Administration (PHMSA): An agency of the US Department of Transportation responsible for developing and enforcing regulations for pipeline operations in the United States.

Pipeline facility: New and existing pipe, rights-of-way, and any equipment, facility, or building used in the transportation of hazardous liquids or carbon dioxide, or new and existing pipelines, rights-of-way, and any equipment, facility, or building used in the transportation of gas or in the treatment of gas during the course of transportation.

Pipeline operator: The company that operates the pipeline, as opposed to an individual person who works for the company that operates the pipeline.

Redundant system: A system that duplicates the operations of a system so that, in the event of a failure, a backup system is available.

Right-of-way (ROW): The area along the pipeline which has been legally permitted to be used by the pipeline operator to construct, operate, and maintain the pipeline. The ROW extends the entire length of the pipeline.

Supervisory control and data acquisition (SCADA): A computer-based system or systems used by a controller in a control room that collects and displays information about a pipeline facility and may have the ability to send commands back to the pipeline facility.

Surge pressure: Pressure produced by a change in velocity of the moving stream that results from shutting down a pump station or pumping unit, closure of a valve, or any other blockage of the moving stream.

Uninterruptible power system (UPS): A system that provides continuity of power when the normal power supply is interrupted. A UPS provides AC power to the system it serves.

An abnormal operating condition (AOC) is defined under federal regulations as a condition identified by the pipeline operator that may indicate (1) a malfunction of a component or (2) a deviation from normal operations that may indicate a condition exceeding design limits or result in a hazard to persons, property, or the environment. AOCs are defined as such by federal pipeline safety law 49 *United States Code* (*U.S.C.*) 60102 and pipeline safety regulations 49 *Code of Federal Regulations (CFR)* 192 (for transportation of gas) and 49 *CFR* 195 (for transportation of liquids). The federal agency that regulates pipeline maintenance is the Pipeline and Hazardous Materials Safety Administration (PHMSA) in Washington, DC.

Personnel in the field are responsible for being aware of pipeline surroundings at all times so they can recognize AOCs when they occur. Types of AOCs, along with examples and proper reactions, are listed in *Table 1*. Note that the reactions are general.

Some pipeline equipment is fitted with warning devices to alert personnel to problems. Sometimes field workers will be notified of AOCs by control center personnel. In other cases, field workers must use their skills and training to observe the physical signs of AOCs. Field personnel play a critical role as the "eyes, ears, and nose" of the pipeline operator. They must be able to recognize the signs of AOCs and report them appropriately. They must stay in close communication with control center personnel who may need to take the proper action remotely. In the most serious cases, the control center may need to shut down the pipeline.

One of the most important responsibilities of field personnel is to be able to recognize the unexpected release of hazardous gases or liquids. In some cases, early recognition can prevent releases from turning into fires and explosions. Dead vegetation, dead wildlife (*Figure 1*), and an oily sheen on water are just a few of the signs that a release has occurred. *Table 2* illustrates the ways that field personnel must use their senses to be aware of the signs of releases.

1.1.0 Recognizing Abnormal Pipeline Facility Conditions

Any problem with the physical structure of the pipeline is defined as an abnormal pipeline facility condition. A pipeline facility is defined as new and existing pipe, right-of-way (ROW), and any equipment, facility, or building used in the transportation of hazardous liquids or carbon dioxide, or new and existing pipelines, rights-of-way, and any equipment, facility, or building used in the transportation of gas or in the treatment of gas during the course of transportation. Examples of components of a pipeline facility include compressors (for gas pipelines) and pump stations (for liquid pipelines).

The following are examples of abnormal pipeline facility conditions:

- Exposed pipeline that should be covered
- Pipeline system damage
- Missing or damaged pipeline supports
- Corrosion
- Coating separating from the pipeline (disbondment)
- Unauthorized activity and/or equipment on the pipeline right-of-way (unauthorized encroachment)
- Missing signs or markers
- Sunken floating tank roof
- Incorrect data or records caused by equipment malfunction
- Incorrect American National Standards Institute (ANSI) pressure rating of a component or material type
- Damage caused by fire or explosion
- Malfunctions due to human error

Work crews, line patrols, or other field personnel are often the first to discover abnormal pipeline facility conditions. However, a report may come in as a result of a call made to the One-Call notification system or from a person calling the pipeline operator directly. When a report comes in, personnel must obtain complete information so that they can quickly find the problem and address it. Many pipeline operators use a form to capture the following information:

- Caller's name and telephone number
- Date and time of the call

Table 1A Abnormal Operating Conditions: Examples and General Reactions (1 of 2)

Abnormal Operating Condition	Examples	Appropriate Reaction
Abnormal pipeline facility condition	• Exposed pipe • Evidence of corrosion • Incorrect ANSI pressure rating of a component or material type • Disbondment (coatings separating from the pipe) • Missing signs or markers • Unauthorized activity and/or equipment on the right-of-way (unauthorized encroachment) • Incorrect data or records caused by equipment malfunction • Mechanical/hydraulic system failures • Fire and explosion • Pipeline system damage, such as: – Missing or damaged pipeline supports – Damaged or inoperable pipeline components (pipe, valves, pumps, compressors, meters, tank gauges) due to external force – Gouges, dents, pitting, wrinkles, buckles, or cracks – Cracked or corroded flange or bolts – Sunken floating tank roof – Damaged or malfunctioning pontoon – Defective tank roof seal – Damaged tank shell – Damaged pipe due to natural disaster	Follow standard operating procedures and make notifications per company policy. Isolate and/or shut down the pipeline segment or facility if necessary. Complete repairs as necessary. Complete the required reporting.
Activation of a safety device	• Emergency or abnormal shutdown alarm • High-pressure shutdown alarm • Activation of a pressure relief valve • Activation of a smoke or fire detector • Activation of a hazardous atmosphere alarm • Activation of an overfill protection device • High sump-level shutdown alarm • High tank-level shutdown alarm • High pump-case pressure alarm • Seal failure alarm	Follow standard operating procedures and make notifications per company policy. Isolate and/or shut down the pipeline segment or facility if necessary. Complete the required reporting.
Communications failures	Personnel communication tools: • Lost use of radio communications • Lost use of landline telephone • Lost use of cell phone System communications: • Facility outages • SCADA communication outage • System programming errors/incorrect command logic • Loss of satellite communications • Loss of microwave signals • Stale data (communications going from the control center to the field, but not being received by the control center from the field)	Switch to alternate equipment or backup systems. Follow standard operating procedures and make notifications per company policy. Complete the required reporting.

Table 1B Abnormal Operating Conditions: Examples and General Reactions (2 of 2)

Power interruptions	• Transformer failure • Downed power lines • Broken leads or wires • Incorrect wire termination • Station or terminal power problems (due to motor overload or a tripped breaker, for example) • Weather conditions, such as an ice storm	Follow standard operating procedures. Switch to an alternate control system or power source if necessary. Notify the power company of the outage. Make additional notifications per company policy. Complete the required reporting.
Fire, explosion, or natural disaster	• Electrical fire • Ignition of a hazardous vapor • Unintentional detonation • Fire, explosion, or pipeline movement caused by natural disaster, such as: – Lightning strike – Flooding – Landslide – Earthquake –Tornado	Follow company emergency action plans. Call 911 if appropriate. Isolate and/or shut down the pipeline segment or facility if necessary. Limit access to the area as much as possible. Make notifications per company policy. Complete the required reporting.
Unexpected release of a hazardous gas or liquid	Signs include: • Vapor clouds or frost balls • Smell of hydrocarbon release • Dead vegetation/contaminated soil • Dead wildlife • Rainbow sheen on water Hazardous products include: • Hydrogen sulfide • Carbon dioxide • Propane • Natural gas • Refined fuels/products Note that the controlled transfer, sampling, venting, draining, and handling of hazardous products are not considered AOCs.	Follow company emergency action plans. Call 911 if appropriate. Isolate and/or shut down the pipeline segment or facility if necessary. Make notifications per company policy. Complete the required reporting.
Unexplained flow rate/ pressure change	Any change from the expected flow or pressure, such as: • Increased pressure • Decreased pressure • No pressure • High-pressure limit • Low-pressure limit • High flow • Low flow • No flow	Follow standard operating procedures and make notifications per company policy. Isolate and/or shut down the pipeline segment or facility if necessary. Complete the required reporting.

- Location of the problem (obtain the latitude/longitude of the location from a global positioning system [GPS] device or a phone, if possible)
- Nearby landmarks such as highways, streams, mile markers, or any other structure that can help repair crews pinpoint the location
- The nature of the damage and any conditions observed at the site

1.1.1 Pipeline System Damage

Accidents that can damage the pipeline and its components occasionally happen. Personnel working on the right-of-way may strike the pipeline without realizing it. An excavator operator may accidentally hit the line. Although the locations of pipelines are clearly marked with signs or markers, construction work near the pipeline may hide markings. Ships' anchors may hit offshore

67107-13_F01.EPS

Figure 1 Dead wildlife may be a sign of an unexpected release.

pipelines. In some parts of the country, hunters have accidentally shot pipelines. Intentional damage due to vandalism, terrorist activity, or other criminal activity may also occur.

Pipeline damage may be discovered by field personnel, control center personnel, or a third party. Field personnel may notice physical signs of damage while out in the field. The control center may receive the indication of pipeline damage on the control center console as an unexplained change in operating conditions.

Physical signs of damage to the pipeline and its components may include gouges, cracks, pitting, dents, buckles, and inoperable valves, pumps, or tank gauges. Flanges or bolts may be cracked or show significant corrosion. Tank shells or pipeline supports may show damage. A sunken roof or a damaged pontoon can also indicate damage. Less obvious signs of damage could include a rainbow sheen on the water near submerged pipelines, which may indicate a release. The release of a highly volatile liquid (HVL) may be discovered by observing a frost ball or vapor cloud along the ROW of a liquid pipeline.

Table 2 Recognizing Warning Signs of Hazardous Gases or Liquids

Sense	Examples of Warning Signs
Sight	• Dead, discolored vegetation, often near healthy vegetation • Dead wildlife • Dry or frozen spots in soil during warm weather • Oily sheen on water/"rainbow" sheen • Vapor fog • Blowing dirt around the pipeline • Bubbles in standing water
Sound	Hissing, blowing, or roaring sound
Smell	Gaseous, skunk-like, or "rotten egg" odor is sometimes recognizable. Never rely on sense of smell only; a release can be completely odorless.

1.1.2 Damaged Pipeline Supports

Damage to pipeline supports can be a serious problem and can lead to a shutdown. In other cases, pipeline operations may continue while repairs are made. Personnel may recognize pipeline support damage while performing maintenance or inspection tasks in the field. Examples of damage include evidence of corrosion, visible cracking, sagging, or buckling of a support, or bolts or other components of the support that have broken or become loose. Damage to supports may be caused by corrosion, settling of soil, erosion due to wind or water, landslides, washouts, or earthquakes.

Any aboveground span must adhere to specific regulations. These spans refer to lengths of pipeline that are entirely aboveground and supported by pipe supports (see *Figure 2*). 49 *CFR* Part 195.110 requires that pipeline systems be designed to handle anticipated external loads from such conditions as earthquakes, vibration, thermal expansion, and contraction. In addition, the pipe and other components must be supported in such a way that the support does not cause excess stress to the pipe wall. 49 *CFR* Part 195.254 states each component of the pipeline that is aboveground must be protected from the forces exerted by the anticipated loads.

1.2.0 Recognizing the Activation of a Safety Device

Safety devices are a part of the pipeline system and are designed to sense and alert personnel to potential danger and/or hazardous conditions.

67107-13_F02.EPS

Figure 2 Aboveground span of pipeline on pipe supports.

Hazardous conditions can include changes in pressure that may exceed pipeline design limits as well as the release of harmful gases. The purpose of a safety device is to prevent injury or accidents. Safety devices include but are not limited to pressure relief systems, pressure limiting devices, high pump-case-pressure alarms, seal failure alarms, overfill protection devices, tank-level alarms, main-line pressure shutdown alarms, and hazardous atmosphere alarms.

When a safety device activates, it may sound an alarm. The device may ring loudly, beep continuously, or flash a light. A sensor on the device may also send a signal to the control center, where it appears as an alert on the control center console. Never ignore a safety alarm and never turn off a safety alarm until the problem has been checked and cleared by authorized personnel. Notify the control center when an alarm sounds, to maintain communication about appropriate steps to take.

1.2.1 Activation of a Pressure Relief System

The activation of relief devices along the pipeline system is a time-sensitive occurrence. Relief valves reduce pressure by discharging product. Depending on the product, it may be discharged into the atmosphere, tanks, flares, or sumps. Relief valves (see *Figure 3*) can be installed on pipeline or **breakout tanks**. The purpose of a pressure relief device is to protect the pipeline or pipeline facility from overpressure. Pressure relief valves are usually self-actuating at preset limits determined by the operator. However, according to DOT regulations, the pressure in a pipeline cannot exceed 110 percent of the operating pressure, so all preset limits must be within acceptable parameters.

1.2.2 Unintended Valve Closure or Shutdown

The unintended closure of a main-line block valve, remote block valve, manual block valve, pump-station suction or discharge valve, and certain terminal valves (manifold/tank) may cause a pressure change which could activate a pressure-relief system. Breakout tanks can be used to temporarily absorb **surge pressure** from the main-line and manifold pressure-relief valves to lower pressure on the line. According to 49 *CFR* 195, relief venting must be provided for each breakout tank. Pressure relief devices must be provided for each low-pressure and high-pressure breakout tank. There are different specifications for low-pressure and high-pressure tanks.

The unintended closure of the valves could also result in a shutdown of the pipeline. In these cases, personnel should inspect the equipment to determine why closure occurred, and reset, open, or close the valve when it is safe to do so. Such occurrences may not cause an AOC, but if the event causes the pipeline to exceed design limits in any way, its classification changes to abnormal.

1.3.0 Recognizing a Communications Failure

Communications refers to both the tools that personnel use to communicate with each other and with communication systems. Communications failures include the failure of communications among personnel, or the failure of communications between the control center and equipment on the pipeline.

1.3.1 Failure of Personnel Communication Tools

Pipeline personnel use landline telephones, cell phones, and radios to communicate with one another on the job. If using a cell phone near the pipeline facility, personnel must make sure it is **intrinsically safe/explosion-proof**. Personnel must keep their equipment in good working order and test their equipment regularly to ensure that it is working properly.

A breakdown in personnel communications equipment is easy to recognize. Personnel cannot hear messages or make phone calls. Some equipment may beep or flash a liquid crystal display to alert users to a problem. These tools are usually reliable, but breakdowns can occur. If radio communication goes down, personnel must address the problem immediately. A communications breakdown can also occur due to weather-related conditions and location.

67107-13_F03.EPS

Figure 3 Relief valve.

1.3.2 System Communications Failures

System communications refers to the computer equipment located in the control center and on some pipeline equipment. Personnel in the control center use the **supervisory control and data acquisition (SCADA)** system to manage pipeline operations. They receive signals from and send signals to pipeline equipment that may be many miles away. The SCADA system allows them to monitor pressure, volume, and temperature and keep products flowing safely and on schedule. If the system fails, control center personnel lose the ability to communicate with equipment on the pipeline. Some systems are set up to switch to a **redundant system**—a backup that comes on line if the main system fails.

Some failures of communications systems show up as warnings or messages on the control center console. A system failure may include any of the following events:

- Random facility outages
- Partial or complete facility outages
- Loss of the SCADA system
- Loss of satellite communications
- Loss of microwave signals
- Stale data
- Loss of signal point within the system

Mechanical and hydraulic failures are related to system failures. For example, the SCADA system sends a signal to a piece of equipment, but the equipment does not respond to the signal. In some cases, the problem is a result of a programming or command logic error. In other cases, equipment may not work because of a mechanical, electrical, or hydraulic problem. Examples of monitoring device failures include, but are not limited to, frozen valves, electrical component burnouts, and compressor breakdowns. A valve may have a mechanical failure that prevents it from being operated remotely by the control system. Sometimes electrical components or connections burn out, preventing the equipment from responding to a command by the system to start up or shut down.

1.4.0 Recognizing Power Interruptions

Electric power is essential to pipeline operations. Nearly every part of the pipeline depends on electricity in some way. A loss of power impacts the abilities of the control center or pipeline facility operations to safely control and monitor the pipeline facility. Power interruptions that result from bad weather or a problem at the local utility company, such as transformer failure or downed power lines, are not uncommon.

Some pipeline operators have an **uninterruptible power system (UPS)** that turns on if electrical power is off for more than a few seconds. UPS systems supply emergency alternating current (AC) power to pipeline components to keep the pipeline operating until power is restored. Instead of electricity, these systems run on engines fueled with gasoline, diesel, or natural gas. When power fails, an AC power failure alarm may be sent to the control center. To ensure that the backup system has turned on, control center personnel check to see whether they are receiving updated information from the pipeline.

If power is lost, electronic indicators will turn off. Some compressors and remote control valves may not operate. The SCADA system may switch off and pumps may shut down. In addition, lights, heat, and air conditioning may fail. An area-wide power outage may cause a loss of telephone communications. Local power outages may shut down only a portion of the pipeline system. Power outages can also cause pumps at the receiving terminal to stop operating, causing a gas or liquid shipment to overflow in tanks or be unable to be shipped further without pumps.

Some equipment is protected from power failure by batteries that supply temporary low voltage power. These batteries run on direct current (DC).

The following types of equipment can operate on DC power:

- Fire detection equipment and alarms
- Fire suppression systems
- Emergency lighting systems
- SCADA
- Alarm panels
- Control modules
- Process control panels
- Specialized compressor units

> **NOTE** Equipment plugged into electric sockets uses AC power. AC power reverses its direction at regular intervals. Batteries supply DC power. DC power flows in one direction.

1.5.0 Recognizing Fire, Explosion, or Natural Disaster

Fire, explosion, or natural disaster can occur at any time. Some events may be anticipated, but many cannot. Field personnel must be trained to recognize these events.

1.5.1 Fire

Personnel must take steps to minimize the likelihood of fire and explosions in a pipeline facility. Two examples include using intrinsically safe/explosion-proof electronic equipment and checking atmospheric conditions before starting any hot work. Always follow company procedures related to the prevention of fire and explosion. Detection devices installed along the pipeline and in pipeline facility buildings alert personnel in case of a fire. These devices send an alarm when their sensors detect heat or flame. Such equipment includes thermal detectors, flame detectors, and smoke detectors.

> **NOTE** Before performing any work that may produce an ignition source, verify that a hazard analysis has been performed. Examples of such work include welding or cutting, opening electrical boxes, or operating non-explosionproof tools in hazardous classified areas.

A thermal detector is a heat-sensitive unit. It is fitted with electrical contacts that stay open as long as conditions are normal. When the unit senses rapidly rising temperatures, the contacts start to close. When the contacts connect, the unit may send an alarm, which may be audible, to the control center and the fire control panel in the workstation. If a thermal detector activates, personnel should evacuate the area and, investigate the cause of the alarm (if appropriately trained according to company procedures).

A flame detector (*Figure 4*) is a light-sensitive unit. It is fitted with an ultraviolet detector that turns on an alarm when the unit senses light from a flame. The unit is sensitive enough to detect a flame even in the earliest stages of a fire. Some flame detectors are connected to a firefighting system that may dump fire suppression material into the area. Personnel can also dump this material over a fire manually by using pull switches.

A smoke detector can be a light-sensitive unit or an ionization detector, or it can be a combination of both types. Ionization detectors are more sensitive to fires in the early smoldering stage. Light detectors are more sensitive to the flaming stage of fires.

An ionization smoke detector detects ionization in air caused by smoke. If any smoke particles pass into the chamber of the detector, the ions will attach to the particles, and so become less able to carry the current. An electronic circuit detects the current drop and sounds the alarm. The light detector operates by detecting light beams. In the absence of smoke, the light passes in front of the detector in a straight line. When smoke enters the chamber across the path of the light beam, light is scattered by smoke particles, directing it at the sensor that triggers the alarm.

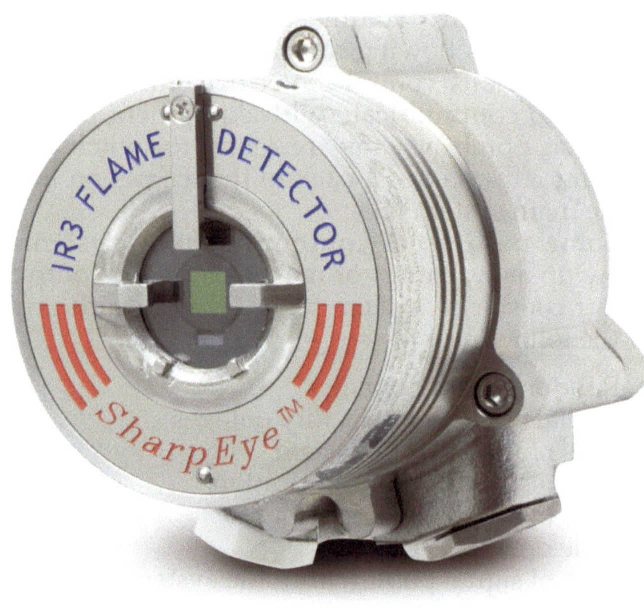

67107-13_F04.EPS

Figure 4 Flame detector.

CAUTION

Before using any fire suppression material, personnel must know how to handle it. Be sure to read the material safety data sheet (MSDS)/ safety data sheet (SDS) for information about how to use the material safely.

Monitoring devices that detect a lower explosive limit (LEL) produce alarms to indicate that there is a risk of explosion. The LEL is the lowest concentration of a gas or a vapor in the atmosphere capable of producing fire in the presence of an ignition source. Different products have different LELs. The monitoring device must be set to detect the particular LEL in an enclosed area. If an LEL alarm activates while welding is taking place, the first action is to stop the welding immediately. Always check atmospheric conditions before reentering an area after an LEL alarm has activated.

1.5.2 Explosion

Pipeline operators install a number of warning devices to warn personnel about conditions that could cause explosions. However, operators cannot control every circumstance, such as intentional damage, that could lead to an explosion on the pipeline. An explosion on a pipeline registers in the control center as an unexplained change in pressure. An explosion may occur in any enclosed space where flammable vapors can accumulate to dangerous levels. Examples of such spaces include deep pipeline trenches and pipeline facilities.

A natural gas explosion may occur when natural gas leaks into an enclosed space. The odorant added to the gas sometimes alerts people to the problem. However, gas can sometimes accumulate undetected. For example, the building may be vacant or the odors may be too far away from people inside the building. In addition, products may not have an odorant added during transport. Be aware that a spark from electrical equipment can cause the gas to explode.

WARNING!

Personnel should never rely on their sense of smell to detect a natural gas leak. Products do not always have a detectable odor.

For an explosion to take place in an enclosed space, there must be a source of ignition and fuel. The vapors creating the fuel for an explosion can be of various types. Each type of flammable vapor or gas has a different lower and upper explosive limit (LEL/UEL). For example, when natural gas escapes into the air, its molecules generally dissipate or scatter because natural gas is lighter than air. However, in enclosed areas natural gas can accumulate, and if there is an ignition source an explosion may occur.

1.5.3 Natural Disasters

Natural disasters include earthquakes, hurricanes, tornadoes, landslides, wildfires, and flooding. These events may cause soil movement, which may result in forces sufficient to cause pipeline failure or product release. Even heavy rains resulting in saturated soil conditions may cause soil movement in some areas.

Pipeline operators generally have designated personnel to monitor weather conditions that can develop into a natural disaster. Control center personnel monitor for earthquakes with seismic indicators. In addition, they may obtain real-time information on the internet. Up-to-date earthquake information is available from the National Earthquake Information Center, which is located in Denver, Colorado. Personnel may track weather alerts and warnings through the National Weather Service website. In case of a power failure, personnel can continue to receive information through the National Weather Service's National Weather Radio or other electronic devices. A list of weather and natural disaster information sources is included in *Appendix A*.

Operators should also have field personnel designated to patrol the ROW to monitor for soil movement and to ensure the accessibility of pipeline components, such as valves, that are needed in case of emergency response actions such as shutdown, isolation, and containment in case of a pipeline failure.

1.6.0 Recognizing the Unexpected Release of Hazardous Gas or Liquid

Pipeline releases are often noticed when someone sees the physical signs. Depending on the product, the release may result in a pool of liquid on the ground, a frost ball, vapor cloud, mist in the air, dead or discolored vegetation, or a rainbow sheen on water. Other signs include dead fish in waterways or other dead wildlife. Gas leaks may also cause a whistling or hissing sound. Natural gas leaks may be identified by a loud, roaring noise. Note that the controlled transfer, sampling, venting, draining, and handling of hazardous products are not AOCs.

Personnel may use a combustible gas indicator or a flame ionization detection unit to detect hazardous gases such as hydrogen sulfide (H_2S), propane, natural gas, and in some cases, carbon dioxide (CO_2). An unexpected release may lead to a hazardous atmosphere, an explosion, or other serious condition. Additional PPE may be required with an unexpected release. Always follow company procedures when dealing with any release.

Personnel must not rely on their sense of smell to detect a hazardous gas for the following reasons:

- Some hazardous gases are odorless. Products may not have an odorant added to alert workers to the presence of gas.
- A low concentration of some hazardous gases can deaden the sense of smell.
- A person's sense of smell may be deadened by a medical problem.

1.7.0 Recognizing Unexplained Flow Rate / Pressure Changes

Flow rate and pressure are important indicators of both normal and abnormal operating conditions. Minor changes in pressure may occur normally in the pipeline. However, significant deviations from the expected flow or pressure are considered AOCs. Unexplained pressure changes include absences of, increases in, and decreases in pressure. The maximum internal pressure permitted during the operation of a gas pipeline is called the **maximum allowable operating pressure (MAOP)** and is defined in 49 *CFR* 192. The maximum internal pressure permitted during the operation of a liquid pipeline is called the **maximum operating pressure (MOP)** and is defined in 49 *CFR* 195.

Pressure-indicating devices located near mainline valves, compressor stations, and custody transfer stations monitor pressure and send readings to the control center. When an unexpected change in pressure occurs, an alarm may alert control center personnel, who then notify field personnel. Sudden pressure increases or decreases might indicate a problem such as the failure of a valve or a transmitter. Personnel must be familiar enough with the pipeline system, equipment, product characteristics, and environmental factors to quickly identify any important changes in pressure and flow.

Internal Combustion Engines

Pipeline facilities/systems that process or store flammable liquids or gases are particularly vulnerable to the risk of fire and explosion. But did you know that the internal combustion engines that are typically used in pipeline facilities can ignite flammable liquids and gases? Internal combustion engines include vehicles of any kind (trucks, cranes, excavators, fire engines, and ambulances) as well as stationary or portable engines such as compressors, generators, pumps, and welders.

Be sure that you know and understand your company's specific safety policies about internal combustion engines to ensure your own safety, the safety of other personnel, and the safety of the entire pipeline facility. To prevent internal combustion engines from becoming an ignition source, pipeline operators are responsible for implementing safety measures such as:

- Ensuring that materials and equipment are stored and used in accord with applicable Occupational Safety and Health Administration (OSHA) standards
- Ensuring that permanently mounted internal combustion engines are not stored where flammable vapors or gases could exist
- Establishing and marking safe boundaries and traffic routes for vehicles and portable equipment through hazardous areas
- Installing flammable gas and vapor detectors in areas where hazards could exist
- Providing training to personnel about the signs and hazards of flammable vapors and gas, the safe handling of internal combustion engines, and all relevant emergency procedures.

Additional Resources

It is assumed that you will learn to perform these procedures under the direct supervision of your crew leader, foreman, or instructor. For independent study, refer to the recommended Additional Resources at the back of this module.

1.0.0 Section Review

1. Which of the following is an abnormal pipeline facility condition?
   a. Aboveground pipe on pipe supports
   b. Missing signs or markers
   c. Valves operating properly
   d. Authorized encroachment

2. Safety devices include _____.
   a. hazardous atmosphere alarms
   b. tank shells
   c. SCADA
   d. compressors

3. SCADA serves as a _____.
   a. regulatory agency
   b. notification system for all utilities
   c. device worn to provide breathable air where the atmosphere is hazardous
   d. command system that assists control center personnel

4. An uninterruptible power system (UPS) supplies _____.
   a. temperature readings
   b. emergency backup power
   c. nuclear power
   d. pressure readings

5. A thermal detector, unlike a flame detector, senses _____.
   a. rapidly rising temperatures
   b. ultraviolet light
   c. flames
   d. pipeline pressure

6. Which of the following is a reason that personnel should *not* rely on their sense of smell to detect a hazardous gas release?
   a. Hazardous gases always have an odor.
   b. Low concentrations of hazardous gases heighten the sense of smell.
   c. A person's sense of smell may be deadened by a medical problem.
   d. Personnel can hear a gas release.

7. The maximum operating pressure of a gas pipeline is called _____.
   a. MOP
   b. MAOP
   c. surge pressure
   d. compression

Section Two

2.0.0 Reacting to Abnormal Operating Conditions

Objective

Describe the proper reaction to abnormal operating conditions for both gas and liquid pipelines.

a. Describe the proper reaction to abnormal pipeline facility conditions.
b. Describe the proper reaction to the activation of a safety device.
c. Describe the proper reaction to a communications failure and a control system failure.
d. Describe the proper reaction to power interruptions.
e. Describe the proper reaction to a fire, explosion, and natural disaster occuring with the pipeline.
f. Describe the proper reaction to the unexpected release of hazardous liquid or gas.
g. Describe the proper reaction to unexplained pressure / flow rate changes.

Trade Terms

Incipient fire: A fire in its beginning stage.

Naturally occurring radioactive materials (NORMs): Materials with radioactive elements that can be extracted during oil and gas extraction from the earth and can get into product streams and become transported in the pipeline.

Self-contained breathing apparatus (SCBA): A device worn to provide breathable air for an individual in an area where the atmosphere is hazardous.

Time is of the essence when AOCs occur and field personnel are expected to take decisive action. Personnel must always follow established company-specific policies and procedures. As regulation 49 *CFR* 195.402 states, each pipeline operator must ensure its personnel maintain a thorough knowledge of its operations, maintenance, and emergency manuals. Personnel must make use of their training, skills, and good judgment to maintain safe operation until solutions are found.

2.1.0 Reacting to Abnormal Facility Conditions

Some abnormal pipeline facility conditions may require immediate action. Others may require scheduling repairs for a later date. If field personnel recognize an abnormal pipeline facility condition, they must notify the appropriate personnel so that repairs can be made safely and in a timely manner. If notified of a condition by a third-party call, it is important that personnel record information received from the third party. They must also keep in mind that a caller's description may not be entirely accurate. The problem may be different than described. In rare cases, a call might not be valid. After a report is confirmed, field personnel should generally follow these steps:

Step 1 Investigate and evaluate the report to determine the extent of the damage, if any.

Step 2 Carry out company safety procedures to protect people, property, and the environment if necessary.

Step 3 Notify the appropriate personnel so that proper notifications or repairs can be made.

2.1.1 Pipeline System Damage

When field personnel notice or accidently cause pipeline damage, they investigate the extent of the damage, note the location on the pipeline, and isolate the damage, as applicable. They must also contact the control center, according to company procedures. If an immediate hazard exists, contact 911 emergency personnel.

NOTE

Before excavation or other digging near a pipeline begins, responsible personnel must contact the One-Call notification system.

Some incidents may be serious enough to require an emergency shutdown of the pipeline. One example of a serious incident is someone losing control of a vehicle on the ROW, extensively gouging or denting a large section of pipeline or damaging a valve. Whenever pipeline damage results in a release and/or explosion, field personnel must notify the control center as soon as possible. In some cases, control center personnel may be able to drain a portion of the pipeline into a delivery system to reduce the amount of product released from the pipeline.

Once the pipeline has resumed normal operations, the pipeline operator conducts an internal investigation. This investigation focuses on evaluating control systems and mechanical equipment as well as personnel performance.

Each company has specific procedures for reacting to pipeline damage. These procedures explain how and when to contact the control center, emergency personnel, and local, state, and federal law enforcement agencies. Always follow company policy, make appropriate notifications per company policy, and complete the required reporting.

2.1.2 Damaged Pipeline Supports

The proper reaction to the discovery of damaged pipeline supports varies depending on the severity of the damage. If the damage to the support is critical or indicative of a possible release, the individual making the discovery should contact the control center or other appropriate personnel immediately. Notification is especially important if the pipe is in danger of excessive movement. The appropriate personnel must then follow company policy precisely to investigate the condition and dispatch qualified personnel to the scene. Until the arrival of qualified personnel, all other personnel should stay a safe distance away from the site, evacuate the area if necessary, and make the appropriate notifications as required by company policy.

In some cases, the pipeline operator may need to isolate the pipeline segment. Whether or not the segment needs to be isolated, a damage report must be submitted so that the repair work can be scheduled. The decision to continue or halt operations is a judgment call made by management and qualified inspection personnel. If the pipeline segment must be isolated, affected personnel must be notified.

Field personnel must know and follow their company's procedures. Those procedures are especially important when a condition requires a decision about whether to isolate a segment of the pipeline facility. Make notifications and complete the reporting as required by company policy.

2.2.0 Reacting to the Activation of a Safety Device

Personnel must respond to safety device alarms quickly and calmly. To react to an alarm, personnel should follow these steps:

Step 1 Notify affected personnel.

Step 2 Carry out company warning procedures to protect people, property, and the environment.

Step 3 Troubleshoot the problem and make necessary repairs if qualified to do so.

Step 4 Reset and test the warning device alarm.

Step 5 Make the appropriate notifications and complete the required reporting.

It is important to ensure that the alarm is not a false alarm. In case of a false alarm, the control center should notify affected personnel and have the alarm equipment checked. Generally, no further action is required when an alarm is false.

When a hazardous atmosphere alarm sounds, the proper reaction is somewhat different. In this case, personnel wearing the appropriate personal protective equipment (PPE) must follow these steps:

Step 1 Evacuate the area and don the appropriate PPE if available.

Step 2 Make the appropriate notifications per company policy.

During an emergency, personnel must focus on three key areas: ensuring personnel safety, reducing the damage to equipment, and protecting the environment. The general rule is to put people first. The goal is also to prevent small problems from turning into larger incidents. All reactions must be made safely. Personnel investigating the alarm must wear the appropriate PPE. If told by authorized personnel to evacuate an area, they must leave immediately.

If reacting to a seal failure alarm, personnel should investigate the alarm, check the station sump level if it is equipped with one, and notify the appropriate personnel.

If a pressure-relief system activates, an audible alarm may activate. Immediate action must be taken to reduce pressure. A pressure reduction is usually achieved by shutting down the upstream pumping station, closing a block valve, or shutting down the pipeline system. Field personnel may need to inspect the relief system, test it, and recalibrate it as necessary.

2.3.0 Reacting to a Communications Failure

The proper reaction to a communications failure depends on whether the failure involves a communications tool that personnel use or the failure of system communications.

2.3.1 Failure of Personnel Communication Tools

When communication tools fail, personnel should first check to see if the problem is a simple one that can be quickly and easily fixed. Immediately correct any communications failure, if possible. For example, some telephones have a volume control that may accidentally be turned off. The telephone may seem to be dead, but a simple adjustment to the volume corrects the problem. Other simple solutions include replacing a battery or recharging a unit.

Some communications problems are more serious. Weather or problems at the local utility company can knock out communications completely. Regardless of the outage, the best preparation is to ensure personnel are trained to follow these steps:

Step 1 Switch to alternate equipment if necessary.

Step 2 Notify the appropriate personnel per company policy.

Step 3 Complete any required reporting or documentation.

If communications are interrupted for long periods, it may not be possible to ask for instructions from a remote location. Training before an emergency situation occurs is the best way to ensure that field personnel know how to react properly. However, field personnel must do everything that they can to re-establish communications as soon as possible.

2.3.2 System Communications Failures

The control center will generally notify field personnel about system communications failures so that they can isolate, investigate, and react according to company policy. The proper reactions for field personnel may include some or all of the following activities:

- Switch to alternate or backup systems
- Isolate a pipeline segment if necessary
- Determine the scope of the problem
- Take corrective actions to restore the system
- Estimate how long the main system will be down
- Notify affected personnel about how long repairs will take
- Complete the required reporting or documentation

If the problem is with the SCADA system, or equipment cannot otherwise be operated from the control center, field personnel may have to operate equipment manually. The pipeline may be able to continue to operate while repairs are made. Operations are dependent on company procedures.

2.4.0 Reacting to Power Interruptions

The proper reaction to a power failure depends on which part(s) and how much of the pipeline is affected. Regardless of how the outage affects the pipeline facility, the appropriate personnel should always notify the power company of the outage as soon as possible. If only a single component or station is involved, control center and/or field personnel may be able to adjust line pressure and flow to keep the pipeline in operation. For example, if a power interruption occurs while a valve is in operation, field personnel may have to operate the valve manually. However, if a station is needed to keep pressure within normal operating limits, for example, control center personnel may have to isolate a pipeline segment. If a shutdown is not required, the gas or liquid shipment will continue to arrive, and the schedule must be followed and the tasks completed manually.

Each company has specific procedures for responding to a power interruption. Always follow company policy, make appropriate notifications per company policy, and complete the required documentation.

2.5.0 Reacting to a Fire, Explosion, or Natural Disaster

The pipeline operator must have a training program in place so that personnel know what to do in case of a fire, explosion, or natural disaster that affects a pipeline. Many companies conduct regular drills so that personnel can practice their responses to such events. Drills help ensure that personnel are prepared to handle emergencies safely and calmly. Each operator has its own procedures for emergencies. In general, trained personnel should know the following information:

- The location of fire suppression equipment and how to operate it; certain fire suppression equipment and PPE may require additional training
- The process for shutting down equipment manually if communication with the control center is lost
- The company's evacuation plan
- The procedures for contacting local emergency response agencies, if necessary

Depending on the situation, pipeline personnel may be able to safely handle an emergency without calling for outside help. Follow company procedures for contacting local emergency response agencies.

2.5.1 Fire

When a fire occurs, the safety of personnel is always top priority, so personnel should evacuate the area. Protection of property is always second.

Personnel must ensure that any injured personnel receive medical attention as soon as possible. Consider any alarm as an actual alarm and react appropriately and decisively.

Personnel must also eliminate the source of the fire, but only as authorized. For example, only qualified personnel are authorized to turn off valves. Company policy dictates procedures for emergency actions and must be followed. Procedures are site-specific and may allow for some flexibility in responding. However, personnel should not need to consult a manual in case of an alarm. They must be adequately trained and familiar with emergency procedures in advance. It may even be necessary to shut down the pipeline if required to control the emergency. Always notify the appropriate personnel and emergency agencies and complete any required documentation or reporting.

According to 49 *CFR* 192 and 195, operators shall establish and conduct a continuing training program to instruct emergency personnel in the proper use of firefighting procedures and equipment, in the identification of hazardous conditions likely to cause emergencies, in the steps necessary to control accidental releases, and in the characteristics of hazardous liquids or carbon dioxide being transported.

The proper reaction differs according to the type of fire in progress. If the fire is small or just beginning (**incipient fire**), personnel should use the available fire extinguishing equipment to control the situation. According to *OSHA 1910.157*, companies must provide designated personnel with training on fire extinguishing equipment as well as the hazards involved with incipient firefighting.

Use the proper type of fire extinguisher for each class of fire. It is important to know the extinguisher's fire class rating before using it. The fire class rating is printed at the bottom of the label on a fire extinguisher. There are four fire class ratings (see *Table 3*). Most extinguishers can be used for multiple types of fires.

If the fire is larger than the immediate workforce can contain, personnel should try to mitigate the situation, notify appropriate personnel and emergency agencies, and complete the required documentation and/or reporting. For example, during an electrical storm, if lightning strikes a full tank at a facility and the roof seal catches on fire, personnel should follow company policy and contact the fire department to contain the fire.

Once the fire is out, the air is cleared, and any injured personnel have received medical attention, the following actions are taken:

Table 3 Fire Class Ratings

Fire Class	Materials and Proper Fire Extinguisher
A	Involves ordinary combustibles, such as wood or paper. This type of fire is suppressed by cooling the fire. Class A extinguishers contain water. A Class A fire extinguisher can never be used on a Class B, C, or D fire.
B	Involves grease, flammable liquids, and gases. Class B extinguishers contain carbon dioxide or another material that smothers the fire by removing oxygen.
C	Involves electrical equipment. Class C extinguishers are designed to protect the firefighter from electric shock. This type of extinguisher smothers the fire.
D	Involves metals. Class D extinguishers contain a powder that either forms a crust around the burning metal or gives off gases that prevent oxygen from reaching the fire. Some metals may keep burning even after they have been coated with powder from a Class D extinguisher. The best way to fight these fires is to keep using the extinguisher so that the fire does not spread.

- Alarms are reset and tested.
- Equipment damage is assessed and reported to the appropriate personnel.
- Work orders are issued and required permits are obtained for repairs.
- Personnel performance is evaluated and corrective action is determined if necessary.
- Injuries or deaths are reported to the appropriate regulatory agencies if required.

The evaluation of personnel performance during a fire may reveal problems in personnel response. Management generally recommends retraining so that personnel can improve their responses to a fire.

A false alarm or an expected alarm due to maintenance activities requires no further action unless the control center determines that the system needs to be investigated. If an alarm has been verified as a false alarm, the appropriate personnel follow these steps:

Step 1 Notify affected personnel.

Step 2 Test alarms to ensure that they are operating properly.

Step 3 Reset alarms or ensure that they have automatically reset.

2.5.2 Explosion

Each operator has specific procedures for reacting to an explosion. In addition, the correct order of the steps to take may vary from operator to operator. A response to an explosion must be appropriate, decisive, and calm. If the main-line valves do not close automatically, designated personnel must initiate actions to ensure that the valves are closed. In general, the proper reactions to an explosion are:

- Evacuate and notify appropriate personnel
- Evaluate and assess the situation.
- Carry out company safety procedures to protect people, property, and the environment.
- Call local emergency response agencies if necessary.
- Confirm the location of all pipeline personnel near the explosion location.
- Confirm the location of any members of the public near the explosion location.
- Call local, state, or federal law enforcement if the cause indicates criminal or terrorist activity.
- Make the appropriate notifications and complete the required documentation and/or reporting.
- Conduct an incident investigation.

2.5.3 Natural Disaster

Much of the pipeline's ability to withstand extreme weather that can result in natural disaster is already present in its design and installation.

Each company has specific procedures for dealing with natural disasters. These procedures reflect local conditions. For example, hurricanes occur more frequently in Florida or the Gulf states. Therefore, companies located in these areas include hurricane response in their procedures.

The following are examples of proper reactions to weather-related problems:

- Monitor weather conditions such as tropical storms that may become hurricanes
- Carry out company safety procedures for protecting people, property, and the environment
- Adjust work schedules as needed to respond to emergency conditions
- Follow company procedures for responding to fire, explosion, or damage resulting from natural disasters
- Set up a command post if conditions warrant such a step
- Work with local emergency response agencies if required
- Conduct investigations when appropriate

2.5.4 Emergency Shutdowns

Events occurring on the pipeline that present immediate hazard to persons or property require an emergency shutdown. An event might be caused by human error, mechanical failure, electrical failure, or an accident on the pipeline. It might also be brought about by threats from weather or civil disobedience. Regardless of the reason, the line must be shut down quickly and safely. Each company has a standard operating procedure for emergency shutdowns.

In the case of a fire, explosion, or release, time for discussion may be very limited. Once the need is confirmed, decisive action must be taken. The pipeline must be isolated by control center personnel by safely shutting down the originating station and continuing downstream in a safe manner until all pumps are offline. Control center personnel will then close block valves to complete the shutdown and notify other personnel as needed. In case of a pump shutdown from low pressure, there may be some time to discuss the decision with appropriate management if the line has enough flexibility to withstand the stress.

Keep in mind that threats from natural disaster or civil disobedience may come without warning. If field personnel witness these kinds of threats to the pipeline, they should contact their supervisor or control center immediately. Affected facilities may need to be isolated. Once facilities are isolated, pipeline shutdown may follow.

In emergency conditions such as a possible release, the pipeline operator has full responsibility and authority to make shutdown decisions and take appropriate action. These actions are governed by the company release policy above all other considerations.

2.5.5 Evacuations

The safety of pipeline personnel and the public is the first priority when any AOC occurs. If there is any threat of injury or death to any persons, an evacuation of pipeline personnel must occur. Events that could result in the need for an evacuation are fire, explosion, release of hydrocarbon or other hazardous vapors, seismic activity, or other natural disasters.

Companies have standard procedures in place to guide the evacuation of personnel. By law, emergency response plans must be in place and evaluated annually. Facility layouts with evacuation maps are required to be posted in every area or department. Safe areas are designated for all personnel. When the command to evacuate is given, trained personnel may act as safety coordinators. Their job is to conduct a roll call to confirm that all personnel are safe. To assist in this process, personnel must report to their designated safe areas and identify themselves to the designated safety coordinator. If any personnel cannot be accounted for, authorized personnel must locate and rescue them if necessary.

Some emergencies may require an evacuation of the public located near the pipeline. The pipeline operator may work with the local authorities to provide details about the incident, determine whether a general evacuation is required, provide contact information for people residing or working near the pipeline, and provide personnel to help staff a command center if requested.

> **NOTE**
> Local emergency planning committees (LEPCs) exist in every jurisdiction in the United States. Law enforcement agencies, firefighting companies, local government agencies, hospitals, companies, and citizen's groups make up the memberships of LEPCs. Committee members meet on a regular basis to plan for emergencies. Planning for a general evacuation is just one example of the projects handled by an LEPC.

2.6.0 Reacting to the Unexpected Release of Hazardous Gas or Liquid

When a hazardous gas or liquid is released from a pipeline unexpectedly, personal safety should be the first consideration. Beyond that, the proper reaction to the unexpected release of a product depends on the characteristics of the product and the location of the release. Some gases and liquids can explode under certain conditions, and some are toxic to people and the environment.

> **NOTE**
> To identify the hazards associated with the chemical being released, consult the material safety data sheets (MSDS)/safety data sheets (SDS).

Regardless of location or the type of product, when a product is released personnel *must* take the following steps:

Step 1 Evacuate the area.

Step 2 Identify the source of the release and eliminate potential ignition sources.

Step 3 Vent the area if possible.

Step 4 Prevent anyone who is not wearing the appropriate PPE from entering the area (personnel must be trained to use PPE).

Step 5 Call for emergency rescue and medical help, if required.

Step 6 Contact the control center; appropriate personnel may need to isolate the pipeline.

Step 7 Have the power company turn off the power, if appropriate.

> **WARNING**
> When witnessing a release and there is a downed person in the area of the release, call 911 and evacuate the area. Do not attempt to rescue personnel.

> **NOTE**
> OSHA stipulates that employers must ensure that contract personnel are trained in the work practices necessary to perform their job safely.

Of equal importance are actions personnel must avoid. When reacting to a release, personnel *must not*:

- Open switchgear doors.
- Operate electrical switches or equipment in the facility, including lights, doors, louvers, or vents.
- Operate personal electrical equipment (including cell or landline telephones) unless the equipment is classified as intrinsically safe/explosion-proof.

Intrinsically safe/explosion-proof electrical equipment is designed so that a spark cannot ignite the atmosphere. It is made to meet the following specifications:

- Outside air cannot penetrate the casing.
- The casing must be strong enough to contain an explosion inside the equipment without igniting the surrounding atmosphere.
- The interior must be filled with nonflammable powder or inert gas to prevent ignition.

This equipment is determined to be safe by the National Electrical Manufacturers Association (NEMA), and is certified according to the standards of the *National Electric Code®*. It must also meet the requirements of Underwriters Laboratories (UL). Equipment that is intrinsically safe/explosion-proof should always be labeled as such.

WARNING

Electronic equipment must not be used in the presence of hazardous gases/vapors unless it is classified as intrinsically safe/explosion-proof. Failure to use intrinsically safe/explosion-proof equipment in a hazardous atmosphere may result in an explosion.

2.6.1 Releases in Enclosed Areas

If a product release occurs in an enclosed area, the source of the release must be determined quickly. It is also important that personnel move to a safe location as quickly as possible and then notify the appropriate personnel, who can operate equipment remotely. If trained, personnel may close the cutoff valve to mitigate the situation.

2.6.2 Releases in Outdoor Areas

Any release must be approached by personnel from the upwind side. Releases in outdoor areas include the additional concern for the safety of the general public, as well as the safety of the environment. Personnel should evacuate the area, go upwind, and take additional appropriate action per company policy. If an odor of hydrocarbon is detected even a few hundred feet from the source, contact control center personnel to shut down the pipeline. Report the incident and any action taken, and complete the required documentation.

2.6.3 Toxic Products

Personnel must take special precautions when dealing with the release of products that can become toxic. Examples include carbon dioxide (CO_2), hydrogen sulfide (H_2S), propane, anhydrous ammonia, and natural gas.

Carbon dioxide is a colorless, odorless, gaseous compound used as a propellant in many domestic applications, such as fire extinguishing equipment, and is a byproduct of combustion. Because it is heavier than air, it can collect in low-lying areas where it may displace oxygen; exposure can result in suffocation and death. A leak detector can detect a hazardous atmosphere, but without it, personnel can be overcome without realizing the problem. Some pipeline operations pump carbon dioxide commercially, and in such operations, the hazards are increased. To enter an area where a carbon dioxide release is suspected, personnel must use the proper gas detection meter, calibrated for the detection of carbon dioxide, and wear the appropriate PPE, such as **self-contained breathing apparatus (SCBA)** (see *Figure 5*).

67107-13_F05.EPS

Figure 5 Self-contained breathing apparatus.

WARNING

Carbon dioxide (CO_2) is a colorless, odorless gas that is heavier than air, so it can collect in low-lying areas where it may displace oxygen. Exposure can be fatal.

Hydrogen sulfide (H_2S) is a colorless gas that smells like rotten eggs. However, in fairly low concentrations, it can deaden the sense of smell. Exposure to high concentrations can be fatal.

WARNING

Exposure to high concentrations of hydrogen sulfide (H_2S) can be fatal. Even in low concentrations, it can cause headaches or eye irritation. In addition, it is flammable and may cause a flash fire. When a gas detection unit shows that H_2S is present, wear the appropriate PPE or immediately leave the area.

Anhydrous ammonia is a liquefied gas that has a sharp odor. Exposure in high concentration can cause death. Even exposure in low concentrations can cause severe reactions from mild irritation to severe burns. The reaction depends on the length of exposure and concentration amounts. If anhydrous ammonia is released, call emergency personnel.

> **WARNING**
>
> Exposure to high concentrations of anhydrous ammonia can cause death. Even in low concentrations, it can cause severe reactions. If detected, wear the appropriate PPE or immediately leave the area. If a release occurs, call emergency personnel.

Propane is a common natural gas that is colorless and heavier than air, so it can collect in low-lying areas where it may displace oxygen. Exposure can cause death. At high concentrations, it is flammable.

> **WARNING**
>
> Exposure to propane gas can cause death. At high concentrations, it is flammable. If detected, wear the appropriate PPE or immediately leave the area.

Naturally occurring radioactive materials (NORMs) may be present when oil and gas is extracted from the earth. These NORMs are then introduced into the product streams and transported in the pipeline. Natural gas liquids (NGLs), along with the processing associated with these products, may trap NORMs in the form of sludge, scale, and particulates. Filters, valves, scrapers, and even internal surfaces of pipelines may contain high levels of NORMs. Precautions should be taken to protect personnel by monitoring for NORMs in accordance with all local, state, federal, and company requirements. If high levels are detected, personnel should distance and/or shield themselves from the source and notify the appropriate personnel.

Only appropriately trained personnel may enter an area where a hazardous gas has been detected. These personnel must be trained and qualified to use SCBA. Other personnel must stay a safe distance away until the deadly atmosphere is cleared. Warning signs must be posted at access points so that personnel do not enter until the area has been inspected and tested.

2.7.0 Reacting to Unexplained Flow Rate/ Pressure Changes

If a change in pressure or flow rate occurs that is different from what is expected, alarms may be initiated to alert personnel to the problem. Personnel must act quickly to address the problem. Qualified personnel will investigate and evaluate the problem. Some problems may require field personnel to manually adjust components, such as valves and transmitters, or make repairs. More serious problems may result in the need to isolate a segment of the pipeline.

Additional Resources

It is assumed that you will learn to perform these procedures under the direct supervision of your crew leader, foreman, or instructor. For independent study, refer to the recommended Additional Resources at the back of this module.

2.0.0 Section Review

1. If a damaged pipeline support is discovered and the damage appears to be critical, the individual who made the discovery should _____.
   a. begin repairs immediately
   b. stay close to the site
   c. report it to the appropriate personnel
   d. assume it is safe

2. If a hazardous atmosphere alarm sounds, field personnel should first _____.
   a. call 911
   b. evacuate the area
   c. notify control center personnel
   d. verify that the alarm is not false

3. Field personnel may need to operate equipment in the field manually if there is a system failure.
   a. True
   b. False

4. When a power outage occurs, the appropriate pipeline personnel should contact _____.
   a. the One-Call notification system
   b. the power company
   c. OSHA
   d. local authorities

5. Fire extinguishing equipment can be used on an incipient fire.
   a. True
   b. False

6. Electronic equipment must not be used in the presence of hazardous gases unless it is classified as _____.
   a. OSHA approved
   b. intrinsically safe/explosion-proof
   c. NRC approved
   d. mechanically safe/disruption proof

7. When an unexplained flow rate or pressure change occurs, field personnel may need to _____.
   a. notify the One-Call system
   b. contact the power company
   c. make repairs
   d. remove pipe supports

Summary

Pipeline operators have developed effective procedures and policies that enable personnel to respond quickly, safely, and effectively to AOCs. Usually an AOC is a routine matter that can be handled safely by pipeline personnel. Sometimes, however, an AOC can result in an emergency situation that may involve working with outside agencies.

In this module, you learned about the different types of AOCs that can occur on the pipeline and the proper reactions to each type of AOC. During your career, you may have to respond to some or all of the AOCs outlined in this module. When responding to these AOCs, you must follow your company's policies and procedures. Protecting people, property, and the environment should be your utmost concern.

Review Questions

1. By definition, an AOC is a condition that may exceed _____.
   a. company operating budget
   b. design limits
   c. safety laws
   d. size limits
2. The federal agency that regulates pipeline maintenance is _____.
   a. OSHA
   b. GAO
   c. EPA
   d. PHMSA
3. By definition, a pipeline facility includes the right-of-way.
   a. True
   b. False
4. Which of the following is *not* always an AOC?
   a. Exposed pipe
   b. Reduced pressure
   c. Lost radio communication
   d. Unexpected release of a hazardous gas
5. Pressure relief systems, tank-level alarms, and main-line pressure shutdown alarms are examples of _____.
   a. prevention devices
   b. safety devices
   c. hazardous atmosphere alarms
   d. fire alarms
6. The loss of SCADA is an example of which type of AOC?
   a. Hydraulic failure
   b. Mechanical failure
   c. Communications failure
   d. Power failure
7. In the event of a power failure, pipeline companies may be able to keep the pipeline running by using an _____.
   a. operations override power source
   b. uninterruptible power system
   c. interface power system
   d. indirect power system
8. In the control center, one indication of a pipeline explosion could be a(n) _____.
   a. change in product temperature
   b. valve malfunction
   c. power outage
   d. unexplained change in pressure
9. Which of the following is *not* true of hazardous gas?
   a. Odorants are often added.
   b. High concentrations can deaden a person's sense of smell.
   c. Some gases are odorless.
   d. Low concentrations can deaden a person's sense of smell.
10. Which of the following best defines an unexplained change in flow rate or pressure?
    a. Reduced pressure only
    b. No flow only
    c. High flow rate
    d. Any change from the intended flow or pressure
11. When pipeline support damage is discovered, notification is especially important if the pipe _____.
    a. is in danger of excessive movement
    b. has changed color
    c. is metal
    d. is corroded
12. When a safety device sounds an alarm, field personnel must _____.
    a. turn it off
    b. never ignore it
    c. reset it
    d. contact the local emergency agency
13. If personnel communication tools fail, personnel should first _____.
    a. check for problems that have simple solutions
    b. contact the control center
    c. call 911
    d. complete any required reporting

14. If SCADA fails, equipment may have to operated manually by _____.
    a. PHMSA
    b. field personnel
    c. control center personnel
    d. local authorities

15. When a power interruption occurs, personnel may be able to keep the pipeline in operation by adjusting the _____.
    a. MAOP
    b. MOP
    c. line pressure and flow
    d. product temperature

16. A false alarm may trigger an investigation of the system, which is determined by _____.
    a. the control center
    b. field personnel
    c. the job site supervisor
    d. SCADA

17. Before using a fire extinguisher, it is important to know its _____.
    a. fire class rating
    b. OSHA rating
    c. combustibility
    d. capacity

18. The pipeline operator may work with local authorities to facilitate an evacuation of the area surrounding a pipeline.
    a. True
    b. False

19. An area where a hazardous gas has been detected may only be entered by personnel trained in rescue operations and qualified to use a _____.
    a. combustible gas indicator
    b. flame ionization indicator
    c. self-contained breathing apparatus
    d. fire suppression kit

20. All of the following are examples of a pipeline facility *except* _____.
    a. compressors
    b. pump stations
    c. control center
    d. rights-of-way

Trade Terms Quiz

Fill in the blank with the correct term that you learned from your study of this module.

1. The computer-based system(s) used by a controller in a control room that collects information about a pipeline facility and may have the ability to send commands back to the pipeline facility is called________.
2. The ________ is the area along the pipeline which has been legally permitted to be used by the pipeline to construct, operate, and maintain the pipeline.
3. A fire in its beginning stage is called a(n) ________.
4. ________ is the separation of the protective coating from the pipeline.
5. A(n) ________ is a system that provides continuity of power when the normal power supply is interrupted.
6. The maximum pressure at which a pipeline or segment of a pipeline may be operated under this part (gas) is called ________.
7. A(n) ________ is used to relieve surges in a hazardous liquid pipeline system or to receive and store hazardous liquid transported by a pipeline for injection and continued transportation by pipeline.
8. The lowest concentration of a gas or a vapor in the atmosphere capable of producing fire in the presence of an ignition source is the ________.
9. A ________ duplicates the operations of a system so that, in the event of a failure, a backup system is available.
10. A(n) ________ indicates a malfunction of a component or a deviation from normal operations that may indicate a condition exceeding design limits or result in a hazard to persons, property, or the environment.
11. A device that detects some hazardous gases in the atmosphere is a(n) ________.
12. The pressure produced by a change in the velocity of the moving stream that results from shutting down a pump station or pumping unit, closure of a valve, or any other blockage of the moving stream is called ________.
13. ________ is the maximum pressure at which a pipeline or segment of a pipeline may be normally operated under this part (liquids).
14. ________ is a centralized communication system established by utilities, government agencies, and operators of underground facilities.
15. A hazardous liquid which will form a vapor cloud when released to the atmosphere and which has a vapor pressure exceeding 276 kPa (40 psia) at 37.8°C (100°F) is classified as a(n) ________.
16. The government agency that is responsible for developing and enforcing regulations for pipeline operations in the United States is the ________.
17. Electrical equipment that is ________ is designed specifically for use in explosive or flammable locations so that a spark cannot ignite the atmosphere.
18. An organization that adopts standards for a variety of products, services, and processes in the United States to ensure consistency is the ________.
19. A ________ is a device worn by an individual to provide breathable air in an atmosphere that is hazardous.
20. The ________ is the company that operates the pipeline, as opposed to an individual person who works for the company that operates the pipeline.

21. New and existing pipe on the pipeline itself, the right-of-way (ROW), and any equipment, facility, or building used in the transportation of a liquid or gas are defined as part of a(n) _______.
22. Activity carried out by, and/or equipment belonging to outside parties on the pipeline ROW is referred to as _______.
23. A length of pipeline that is entirely aboveground and supported by pipe supports is called a(n) _______.
24. Materials with radioactive elements that can be extracted during oil and gas extraction from the earth and can get into product streams and become transported in the pipeline are called _______.

Trade Terms

Aboveground span
Abnormal operating condition (AOC)
American National Standards Institute (ANSI)
Breakout tank
Disbondment
Encroachment
Flame ionization detection unit
Highly volatile liquid (HVL)
Incipient fire
Intrinsically safe/explosion-proof
Lower explosive limit (LEL)
Maximum allowable operating pressure (MAOP)
Maximum operating pressure (MOP)
Naturally occurring radioactive materials (NORMs)
One-Call notification system
Pipeline and Hazardous Materials Safety Administration (PHMSA)
Pipeline facility
Pipeline operator
Redundant system
Right-of-way (ROW)
Self-contained breathing apparatus (SCBA)
Supervisory control and data acquisition (SCADA)
Surge pressure
Uninterruptible power system (UPS)

Trade Terms Introduced in This Module

Aboveground span: Length of pipeline that is entirely aboveground and supported by pipe supports.

Abnormal operating condition (AOC): A condition within a pipeline system, identified by the operator, that may indicate (1) a malfunction of a component or (2) a deviation from normal operations that may indicate a condition exceeding design limits or result in a hazard to persons, property, or the environment.

American National Standards Institute (ANSI): An organization that adopts standards for a variety of products, services, and processes in the United States to ensure consistency.

Breakout tank: a tank used to (a) relieve surges in a hazardous liquid pipeline system or (b) receive and store hazardous liquid transported by a pipeline for reinjection and continued transportation by pipeline.

Disbondment: Separation of the protective coating from the pipeline.

Encroachment: Activity carried out by, and/or equipment belonging to outside parties on the pipeline right-of-way (ROW). A particular encroachment might be authorized or not, so each situation requires specific handling.

Flame ionization detection unit: A device that detects some hazardous gases in the atmosphere.

Highly volatile liquid (HVL): A hazardous liquid which will form a vapor cloud when released to the atmosphere and which has a vapor pressure exceeding 276 kPa (40 psia) at 37.8°C (100°F).

Incipient fire: A fire in its beginning stage.

Intrinsically safe/explosion-proof: Describes electrical equipment that is designed specifically for use in explosive or flammable locations so that a spark cannot ignite the atmosphere.

Lower explosive limit (LEL): The lowest concentration of a gas or a vapor in the atmosphere capable of producing fire in the presence of an ignition source. Different products have different LELs.

Maximum allowable operating pressure (MAOP): The maximum pressure at which a pipeline or segment of a pipeline may be operated under this part (gas).

Maximum operating pressure (MOP): The maximum pressure at which a pipeline or segment of a pipeline may be normally operated under this part (liquids).

Naturally occurring radioactive materials (NORMs): Materials with radioactive elements that can be extracted during oil and gas extraction from the earth and can get into product streams and become transported in the pipeline.

One-Call notification system: A centralized communication system established by utilities, government agencies, and operators of underground facilities. This system provides one telephone number for contractors and the general public to call to notify of their intent to use equipment on or near a pipeline. It is also in place for any individual to call if pipeline damage is discovered. The One-Call system is generally set up on a state-by-state basis.

Pipeline and Hazardous Materials Safety Administration (PHMSA): An agency of the US Department of Transportation responsible for developing and enforcing regulations for pipeline operations in the United States.

Pipeline facility: New and existing pipe, rights-of-way and any equipment, facility, or building used in the transportation of hazardous liquids or carbon dioxide, or new and existing pipelines, rights-of-way, and any equipment, facility, or building used in the transportation of gas or in the treatment of gas during the course of transportation.

Pipeline operator: The company that operates the pipeline, as opposed to an individual person who works for the company that operates the pipeline.

Redundant system: A system that duplicates the operations of a system so that, in the event of a failure, a backup system is available.

Right-of-way (ROW): The area along the pipeline which has been legally permitted to be used by the pipeline operator to construct, operate, and maintain the pipeline. The ROW extends the entire length of the pipeline.

Self-contained breathing apparatus (SCBA): A device worn to provide breathable air for an individual in an area where the atmosphere is hazardous.

Supervisory control and data acquisition (SCADA): a computer-based system or systems used by a controller in a control room that collects and displays information about a pipeline facility and may have the ability to send commands back to the pipeline facility.

Surge pressure: Pressure produced by a change in velocity of the moving stream that results from shutting down a pump station or pumping unit, closure of a valve, or any other blockage of the moving stream.

Uninterruptible power system (UPS): A system that provides continuity of power when the normal power supply is interrupted. A UPS provides AC power to the system it serves.

Additional Resources

The following recommended resources can provide additional helpful information related to the requirements for this covered task.

American National Standard Institute (ANSI), 1899 L Street, NW, 11th Floor, Washington, DC 20036. **www.ansi.org**

American Petroleum Institute (API), 1220 L Street, NW, Washington, DC 20005. **www.api.org**

American Society for Mechanical Engineers (ASME), Two Park Avenue, New York, NY 10016-5990. **www.asme.org**

American Society for Testing and Materials (ASTM) Specification D396.

Association of Oil Pipe Lines (AOPL), 1808 Eye Street, NW, Suite 300, Washington, DC 20006. **www.aopl.org**

Code of Federal Regulations 49, Parts 192 and 195. **www.ecfr.gov**

International Society of Automation (ISA), 67 TW Alexander Drive, Research Triangle Park, NC 27709. **www.isa.org**

National Association of Corrosion Engineers (NACE), 1440 South Creek Drive, Houston, TX 77084. **www.nace.org**

National Electrical Code®. **www.nfpa.org**

Occupational Safety and Health Administration (OSHA), 200 Constitution Avenue, NW, Washington, DC 20210. **www.osha.gov**

Pipeline and Hazardous Material Safety Administration (PHMSA), East Building, 2nd Floor, 1200 New Jersey Avenue, SE, Washington, DC 20590. **www.phmsa.dot.gov**

US Bureau of Labor Statistics, Postal Square Building, 2 Massachusetts Avenue, NE, Washington, DC 20212. **www.bls.gov**

US Energy Information Administration, 100 Independence Avenue, SW, Washington DC 20585. **www.eia.gov**

Figure Credits

U.S. Geological Survey/photo by Don Becker, Module Opener

Energy Information Administration, Office of Oil & Gas, Natural Gas Division, Gas Transportation Information System, 2009, Figure 1

Photograph No. 5998083 (Photographer Eric Vance), September 2006; Records of the Environmental Protection Agency, 1944–2006, Record Group 412; National Archives and Records Administration–College Park, Figure 9

Cameron Strother, Southland Safety, Figure 15

Photo courtesy of STV, Figure 34

Courtesy of Liquid Controls, LLC, Figure 35

Photographs and illustrations provided courtesy of Daniel Measurement and Control, Inc. Daniel Senior and Simplex are registered trademarks of Daniel Measurement and Control, Figure 36

Courtesy of Sunoco Logistics, Figure 39

Appendix A

Weather and Natural Disaster Information Sources

Emergency Managers Weather Information. US Department of Commerce, National Oceanic and Atmospheric Administration, National Weather Service, 1325 East-West Highway, Silver Spring, MD 20910.**www.nws.noaa.gov/emwin/index.htm**

National Earthquake Information Center, US Geological Survey. Box 25046, DFC, MS 967 Denver, CO 80225. **www.earthquake.usgs.gov/regional/neic**

National Hurricane Center Tropical Prediction Center. 11691 SW 17th Street, Miami, FL 33165-2149. **www.nhc.noaa.gov**

National Weather Service and National Weather Service Weather Radio, National Oceanic and Atmospheric Administration (NOAA). 1325 East-West Highway, Silver Spring, MD 20910. **www.weather.gov**

Appendix B

Organizations of Interest to the Pipeline Industry

American Gas Association (AGA). 400 N. Capitol Street, NW, Washington, DC 20001. **www.aga.org**

American Petroleum Institute (API). 1220 L Street, NW, Washington, DC 20005. **www.aga.org**

Interstate Natural Gas Association of America (INGAA). 20 F Street, NW, Suite 450, Washington, DC 20001. **www.ingaa.org**

National Institute for Occupational Safety and Health (NIOSH). 395 E Street, SW, Suite 9200, Patriots Plaza Building, Washington, DC 20201. **www.cdc.gov/niosh**

National Response Center (NRC). c/o United States Coast Guard (CG-5335 - Stop 7581), 2100 2nd Street, SW, Washington, DC 20593-0001. **www.nrc.uscg.mil**

Occupational Safety and Health Administration (OSHA). 200 Constitution Avenue, NW, Washington, DC 20210. **www.osha.gov**

Pipeline Hazardous Material Safety Administration (PHMSA). East Building, 2nd Floor, 1200 New Jersey Ave., SE, Washington, DC 20590. **www.phmsa.dot.gov**

Section Review Answer Key

Answer	Section Reference	Objective
Section One		
1. b	1.1.0	1a
2. a	1.2.0	1b
3. d	1.3.2	1c
4. b	1.4.0	1d
5. a	1.5.1	1e
6. c	1.6.0	1f
7. b	1.7.0	1g
Section Two		
1. c	2.1.2	2a
2. b	2.2.0	2b
3. a	2.3.2	2c
4. b	2.4.0	2d
5. a	2.5.1	2e
6. b	2.6.0	2f
7. c	2.7.0	2g

NCCER CURRICULA — USER UPDATE

NCCER makes every effort to keep its textbooks up-to-date and free of technical errors. We appreciate your help in this process. If you find an error, a typographical mistake, or an inaccuracy in NCCER's curricula, please fill out this form (or a photocopy), or complete the online form at **www.nccer.org/olf**. Be sure to include the exact module ID number, page number, a detailed description, and your recommended correction. Your input will be brought to the attention of the Authoring Team. Thank you for your assistance.

Instructors – If you have an idea for improving this textbook, or have found that additional materials were necessary to teach this module effectively, please let us know so that we may present your suggestions to the Authoring Team.

NCCER Product Development and Revision

13614 Progress Blvd., Alachua, FL 32615

Email: curriculum@nccer.org
Online: www.nccer.org/olf

❑ Trainee Guide ❑ Lesson Plans ❑ Exam ❑ PowerPoints Other ______________

Craft / Level: Copyright Date:

Module ID Number / Title:

Section Number(s):

Description:

Recommended Correction:

Your Name:

Address:

Email: Phone:

CT50_0-17

Purge Gas from a Pipeline

OVERVIEW

Segments of gas pipelines are routinely isolated from the main line for inspection or maintenance. When this operation occurs, the natural gas in the line must be safely removed. This module examines the procedure for purging a pipeline of natural gas and ensuring that all hazardous vapors are dispersed and eliminated before inspection or maintenance begins.

Trainees with successful module completions may be eligible for credentialing through the NCCER Registry. To learn more, go to **www.nccer.org** or contact us at 1.888.622.3720. Our website has information on the latest product releases and training, as well as online versions of our *Cornerstone* magazine and Pearson's product catalog.

Your feedback is welcome. You may email your comments to **curriculum@nccer.org**, send general comments and inquiries to **info@nccer.org**, or fill in the User Update form at the back of this module.

This information is general in nature and intended for training purposes only. Actual performance of activities described in this manual requires compliance with all applicable operating, service, maintenance, and safety procedures under the direction of qualified personnel. References in this manual to patented or proprietary devices do not constitute a recommendation of their use.

CT50_0-17
Purge Gas from a Pipeline

Objective

When you have completed this module, you will be able to do the following:

1. Describe how to purge gas from a pipeline (CT50_0-17).
   a. Describe the process of purging gas from a pipeline.
   b. Identify the knowledge and skills required to purge gas from a pipeline.

Performance Task

Under the supervision of the instructor, you should be able to do the following:

1. Purge gas from a pipeline (CT50_0-17).
   - Identify potential abnormal operating conditions that may occur during performance of this CT, and know the appropriate actions to take in response to them.
   - Utilize the appropriate personal protective equipment according to relevant company procedures.
   - Complete a lockout/tagout procedure.
   - Purge the pipeline segment by injecting an inert gas or other substance into the line.
   - Use a vent pipe to direct purged vapors away from any ignition source.
   - Sample the vent stream with a combustible-gas indicator to determine when the pipeline has been purged of all hazardous vapors.
   - Complete appropriate documentation.

Trade Terms

Abnormal operating conditions (AOCs)
Blind flange
Bonding
Combustible-gas indicator (CGI)
Explosive limit (flammable limit)
Hot work
Lockout/tagout procedure
Lower explosive limit (LEL)
Nitrogen pack
Nitrogen purge
Purge
Shut-in
Upper explosive limit (UEL)

Industry Recognized Credentials

If you are training through an NCCER-accredited sponsor, you may be eligible for credentials from NCCER's Registry. The ID number for this module is CT50_0-17. Note that this module may have been used in other NCCER curricula and may apply to other level completions. Contact NCCER's Registry at 888.622.3720 or go to **www.nccer.org** for more information.

Contents

Figures and Tables

1.0.0 PURGE GAS FROM A PIPELINE

Objective

Describe how to purge gas from a pipeline (CT50_0-17).

a. Describe the process of purging gas from a pipeline.
b. Identify the knowledge and skills required to purge gas from a pipeline.

Performance Task

1. Purge gas from a pipeline (CT50_0-17).
   - Identify potential abnormal operating conditions that may occur during performance of this CT, and know the appropriate actions to take in response to them.
   - Utilize the appropriate personal protective equipment according to relevant company procedures.
   - Complete a lockout/tagout procedure.
   - Purge the pipeline segment by injecting an inert gas or other substance into the line.
   - Use a vent pipe to direct purged vapors away from any ignition source.
   - Sample the vent stream with a combustible-gas indicator to determine when the pipeline has been purged of all hazardous vapors.
   - Complete appropriate documentation.

Trade Terms

Abnormal operating conditions (AOCs): Conditions within a pipeline system, identified by the operator, that may indicate (1) a malfunction of a component or (2) a deviation from normal operations that may indicate a condition exceeding design limits or result in a hazard to persons, property, or the environment.

Blind flange: A solid plate used to close the component on which it is attached.

Bonding: A process of electrically connecting across a point of separation of an electrical conductive material (in this case the pipe) to prevent sparking or arcing.

Combustible-gas indicator (CGI): An instrument used to detect various flammable gases and vapors.

Explosive limit (flammable limit): The percentage ratio range of a mixture of air and a flammable gas that will support combustion of the gas. Each gas has its characteristic range, but the typical natural gas explosive limit is 5 to 15 percent gas to 85 to 95 percent air. Explosive limits are also called flammable limits.

Hot work: Work involving burning, welding, or similar operations that create sparks, flame, or heat, and are capable of initiating fires or explosions.

Lockout/tagout procedure: A procedure used to identify sources of energy and hazardous materials that could adversely affect maintenance activities, to then isolate all such sources from the work area, and to ensure that the isolation remains effective until the work is completed.

Lower explosive limit (LEL): The minimum concentration of vapor in the air below which propagation of a flame does not occur in the presence of an ignition source.

Nitrogen pack: To leave a disconnected segment of pipe filled with nitrogen under pressure.

Nitrogen purge: Process of displacing hazardous vapors from a section of pipeline by injecting an inert gas (nitrogen) into the line at a pressure sufficient to force the hazardous vapors out.

Purge: To displace air or hazardous vapor from a pipeline.

Shut-in: The process of isolating a section of pipeline for inspection or maintenance.

Upper explosive limit (UEL): The maximum concentration of vapor in the air above which propagation of a flame does not occur in the presence of an ignition source.

There are times when a pipeline section (*Figure 1*) must be **shut-in** so that it may be purged for inspection or maintenance. Gas is routed around the shut-in section so that there is no interruption in service. When isolating a pipeline section, field personnel must **purge** the pipe segment by injecting an inert gas, such as nitrogen, into the segment. After the pipeline is emptied of gas, explosive vapors may still be present. The line must be monitored until all traces of explosive vapors are removed.

> **CAUTION**
>
> An individual's actions during the performance of a covered task could cause a safety device to activate. The activation represents a deviation from normal operations, resulting in an abnormal operating condition (AOC).

Figure 1 Segment of a pipeline.

1.1.0 Purging Gas from a Pipeline

Flammable gas explodes or catches fire when it is mixed with a certain percentage of air. Combustion can range in violence from deflagration, through detonation, to explosion. Typically, a mixture of gas and air must be 5 to 15 percent gas and 85 to 95 percent air before it reaches its **explosive limit (flammable limit)**. A mixture of gas and air that does not have enough gas to explode is below the **lower explosive limit (LEL)**. That is, it is too lean to explode. A mixture with more than the maximum amount of gas is above the **upper explosive limit (UEL)**, so it is too rich to explode.

Purging establishes a moderately fast flow in the line being purged. This rapid flow helps to ensure that where the air and gas meet they mix as little as possible, thus reducing the potential for explosion. Personnel must maintain a constant purge rate until all the air or gas has been expelled.

To purge a transmission pipeline segment, personnel open valves and use an inert gas, such as nitrogen, to push out the flammable gas. This is called a **nitrogen purge**. Before placing the line back in service, the line must be purged of air by injecting natural gas. A slug of nitrogen gas may first be injected to provide a barrier between the air being removed and the natural gas being injected.

The employee in charge of the purging procedure defines the appropriate process to be followed. All employees engaged in the process must wear the appropriate PPE. Before and during purging operations, these personnel must perform the following tasks:

- Follow the company operations and maintenance (O&M) procedures and be aware of possible hazards along the pipeline right-of-way (ROW) (*Figure 2*).
- Monitor for hazardous vapors throughout the purging process.
- Know and follow the company's **lockout/tagout procedure**.
- Know how to obtain the appropriate work permits, including **hot work** and confined space permits as required.

1.1.1 Procedure

This section describes the basic steps involved in purging gas from a pipeline:

> **WARNING!**
>
> Only trained, qualified individuals may perform pipeline monitoring procedures. Wear all appropriate personal protective equipment (PPE) and stay alert to the possibility of unexpected leaks. Safely eliminate all ignition sources and immediately notify the appropriate personnel in the event of a leak.

Step 1 Identify potential **abnormal operating conditions (AOCs)** that may occur, and know the appropriate actions to take in response to them.

Step 2 Utilize the appropriate personal protective equipment according to relevant company procedures.

Step 3 Complete a lockout/tagout procedure. Ensure that valves, compression, or other devices with a direct impact on the segment being disconnected cannot be accidentally opened or operated.

Step 4 Purge the pipeline segment by injecting an inert gas or other substance into the line. Remove all hazardous vapors from the pipeline to prevent a hazardous atmosphere from remaining in the pipe segment.

Step 5 Use a vent pipe to direct purged vapors away from any ignition source. Ensure that the hazardous vapors vent into the atmosphere for safe dispersion.

Step 6 Sample the vent stream with a combustible-gas indicator to determine when the pipeline has been purged of all hazardous vapors. Ensure that the purging process has removed all traces of explosive vapors.

Step 7 Complete appropriate documentation.

Figure 2 Potential hazards on a pipeline right-of-way.

Local distribution companies may have additional procedures to purge the distribution lines that serve homes and businesses. Their procedures also include a process for quickly and safely restoring service.

Isolating the section to be purged requires following the company's lockout/tagout procedure. This procedure must be followed when hazardous materials could adversely affect maintenance. It includes isolating all such materials from the work area. The isolation must continue until the work is completed. Personnel must close off all valves and other devices that are directly connected to the section being shut-in. Steps must be taken to ensure that these devices are not accidentally opened while the purging operation is occurring.

Sparking or arcing should be prevented during the purging process. **Bonding** provides a safe path for electric current across any gap in the pipeline while the purging is occurring.

Personnel must measure the explosive level of hazardous vapors with a **combustible-gas indicator (CGI)**. Flammable-gas leak detectors should be used as well if available on the segment being purged (*Figure 3*). No one should rely on their sense of smell to detect hazardous vapors for the following reasons:

- The concentration levels may not be high enough to detect dangerous levels by the sense of smell alone.
- Not all product is odorized and therefore cannot be detected by smell.
- A medical condition may affect a person's sense of smell.
- Certain hazardous vapors may deaden the sense of smell.

Figure 3 Flammable-gas leak detector.

1.2.0 Prerequisites, Competencies, and Abnormal Operating Conditions

Personnel must possess a variety of skills in order to successfully complete a pipeline purge. The operation itself requires a number of steps, which the personnel conducting the purge must be competent to perform. Personnel must also be aware of the particular abnormal operating conditions that may occur during the procedure and know the proper responses to them.

1.2.1 Prerequisites and Competencies

WARNING!

Improperly conducted hot work is a major cause of fires and explosions, resulting in property damage, serious injuries, and even death.

Operations personnel must possess basic knowledge and skills in a variety of areas prior to performing a pipeline purge. They must know how to obtain a hot work permit if such work is required on the line.

They must also know and be able to follow the company's lockout/tagout procedures in order to ensure that valves, compressors, or other devices with a direct impact on the segment being disconnected cannot be accidentally opened or operated.

Personnel conducting a purge must also be aware of any hazards that may be found along the right-of-way (ROW), and know and be able to follow all applicable company operations and maintenance (O&M) procedures.

Personnel conducting a pipeline purge must be competent in performing a system shut-in procedure. This involves installing a loop of pipeline so that the gas can be diverted around the shut-in section. The shut-in section is checked to ensure that all gas and air have been purged before any repairs or inspections are performed. Specific steps for shutting-in a pipeline segment, and the order in which the steps are performed, vary from one company to another.

Additional steps are required for a full system shutdown (if one should ever be required) and each company has its own procedures for this process. A full system shutdown may require that the end of the pipe be sealed with a **blind flange** and the line left in a **nitrogen pack** condition.

CAUTION

Conditions on a pipeline can change without warning. It is critical to monitor conditions when a segment is shut-in. Field personnel rely on the control center to monitor pipeline pressure and temperature to ensure they remain within acceptable limits. They also rely on safety devices at the work site to alert them to problems.

It is important to monitor the area for leaks during the entire shut-in process. Field personnel must wear the appropriate personal protective equipment (PPE) while performing this task. Operations personnel must be able to purge the pipeline segment by injecting an inert gas or other substance into the segment in order to remove all hazardous vapors. They must be able to use a vent pipe to direct purged vapors away from any ignition source, and to ensure that the hazardous vapors vent to the atmosphere for safe dispersion. The vent stream must be sampled with a combustible-gas indicator to determine when the pipeline has been purged of all hazardous vapors. The purging operation will follow company O&M procedures and requires the completion of all company-mandated records.

1.2.2 Abnormal Operating Conditions

Personnel conducting a pipeline purge must be able to identify abnormal operating conditions and know the proper response to each should they occur while the purge is in progress. The following are some potential AOCs, and the appropriate action personnel should take in response to them:

- *Offensive odors occur during the purging operation, which adversely affects the surrounding neighborhoods* – Personnel must monitor the operation to ensure that the vapors are not in the explosive range. Purging operations must be suspended if weather conditions do not allow for safe dispersion of the vapors.
- *The ignition of released hazardous liquids or gases causes a fire or explosion* – If the situation cannot be safely controlled by hand, personnel should immediately evacuate the area, limit access, and make the proper notifications, including calling 911 if appropriate.

Additional Resources

It is assumed that you will learn to perform the purging procedure under the direct supervision of your crew leader, foreman, or instructor. For independent study, refer to the recommended Additional Resources at the back of this module.

1.0.0 Section Review

1. Lockout/tagout procedures must be followed when hazardous materials _____.
   a. are inadvertently released
   b. may affect neighboring communities
   c. are present
   d. adversely affect maintenance

2. Prerequisites for performing a gas pipeline purge include knowing how to obtain a hot work permit, knowing how to follow lockout/tagout procedures, knowing how to complete company-mandated records, and knowing _____.
   a. how to monitor pipeline alarms
   b. how to inspect ROWs for hazards
   c. compressor shutdown procedures
   d. proper community notification procedures

SUMMARY

Pipelines are purged to remove hazardous gases prior to inspection or maintenance. Personnel performing purging procedures must know how to obtain hot work permits, follow lockout/tagout procedures, inspect ROWs for hazards, follow company O&M procedures, and complete company-mandated records.

During the purging operation, personnel must be able to complete a lockout/tagout procedure, purge the line with an inert gas, use a vent pipe to direct purged vapors away from ignition sources, sample the vent stream with a combustible-gas indicator, and identify and respond to potential abnormal operating conditions. Natural gas is injected to purge the line of air after work is complete.

The explosive limit is reached when the gas to air ratio reaches the proper percentage. Purging a natural gas line with an inert gas at a moderately fast rate prevents the excess mixing of the gas and air. Prior to purging, the lockout/tagout procedures must be in place and the pipeline must be shut-in, vented, and disconnected from the main line.

The segment being purged must be grounded or bonded to prevent a static spark. Personnel must rely on combustible-gas indicators, rather than their sense of smell, to ensure that all hazardous vapors have been eliminated.

Abnormal operating conditions may include offensive odors affecting surrounding neighborhoods, and fire or explosion caused by the release of hazardous liquids or gas. Vapors should be monitored to ensure that they are not in the explosive range and operations must be suspended if weather conditions prevent their safe dispersal. If a fire or explosion occurs, the area should be evacuated, access prevented, and proper notifications made, which may include calling 911.

Review Questions

1. Purging establishes a moderately fast flow in order to ensure _____.
   a. adequate mixing of the inert gas and the flammable gas
   b. rapid emptying of the line
   c. that the MAOP is not exceeded
   d. that air and gas mix as little as possible

2. To ensure that valves, compressors, or other devices are not operated or opened, it is necessary to _____.
   a. notify the control center prior to inspection or maintenance
   b. follow the company's lockout/tagout procedures
   c. post written warnings near all affected equipment
   d. monitor each device closely while work is being performed

3. What is the typical gas/air ratio when the explosive limit is reached?
   a. 5 to 10 percent gas, 90 to 95 percent air
   b. 5 to 15 percent gas, 85 to 95 percent air
   c. 10 to 20 percent gas, 80 to 90 percent air
   d. 20 to 25 percent gas, 75 to 80 percent air

4. The purpose of a vent pipe is to _____.
   a. ensure that all hazardous vapors have been removed from the pipe
   b. remove excess gas after purging is complete
   c. direct purged vapors away from ignition sources
   d. allow the purging gas to enter the pipe

5. Potential abnormal operating conditions that may occur during purging include fire or explosion and _____.
   a. offensive odors
   b. compressor failure
   c. excess venting
   d. system isolation

Trade Terms Quiz

Fill in the blank with the correct term that you learned from your study of this module.

1. The policy used to isolate sources of energy and hazardous materials from the work area is called _________.
2. Displacing hazardous vapors from a section of pipe by injecting nitrogen is called a(n) _________.
3. The process of isolating a section of pipeline for inspection or repair is called _________.
4. The gas/air mixture range that will support combustion is called _________.
5. A solid plate used to close the component on which it is attached is called a(n) _________.
6. Work involving burning, welding, or similar operations is called _________.
7. To leave a disconnected segment of pipe filled with nitrogen is called _________.
8. When air or hazardous vapors are displaced from a pipeline, this is called a(n) _________.
9. The process of connecting a conductive material across a point of separation to prevent sparking is called _________.
10. The minimum concentration of a vapor in the air below which the propagation of a flame does not occur is called the _________.
11. An instrument used to detect flammable gases and vapors is called a(n) _________.
12. The maximum concentration of a vapor in the air above which the propagation of a flame does not occur is called the _________.
13. Conditions within a pipeline system, identified by the operator, that may indicate (1) a malfunction of a component or (2) a deviation from normal operations that may indicate a condition exceeding design limits or result in a hazard to persons, property, or the environment are known as _________.

Trade Terms

Abnormal operating conditions (AOCs)
Blind flange
Bonding
Combustible-gas indicator (CGI)
Explosive limit (flammable limit)
Hot work
Lower explosive limit (LEL)
Lockout/tagout procedure
Nitrogen pack
Nitrogen purge
Purge
Shut-in
Upper explosive limit (UEL)

Trade Terms Introduced in This Module

Abnormal operating conditions (AOCs): Conditions within a pipeline system, identified by the operator, that may indicate (1) a malfunction of a component or (2) a deviation from normal operations that may indicate a condition exceeding design limits or result in a hazard to persons, property, or the environment.

Blind flange: A solid plate used to close the component on which it is attached.

Bonding: A process of electrically connecting across a point of separation of an electrical conductive material (in this case the pipe) to prevent sparking or arcing.

Combustible-gas indicator (CGI): An instrument used to detect various flammable gases and vapors.

Explosive limit (flammable limit): The percentage ratio range of a mixture of air and a flammable gas that will support combustion of the gas. Each gas has its characteristic range, but the typical natural gas explosive limit is 5 to 15 percent gas to 85 to 95 percent air. Explosive limits are also called flammable limits.

Hot work: Work involving burning, welding, or similar operations that create sparks, flame, or heat, and are capable of initiating fires or explosions.

Lockout/tagout procedure: A procedure used to identify sources of energy and hazardous materials that could adversely affect maintenance activities, to then isolate all such sources from the work area, and to ensure that the isolation remains effective until the work is completed.

Lower explosive limit (LEL): The minimum concentration of vapor in the air below which propagation of a flame does not occur in the presence of an ignition source.

Nitrogen pack: To leave a disconnected segment of pipe filled with nitrogen under pressure.

Nitrogen purge: The process of displacing hazardous vapors from a section of pipeline by injecting an inert gas (nitrogen) into the line at a pressure sufficient to force the hazardous vapors out.

Purge: To displace air or hazardous vapor from a pipeline.

Shut-in: The process of isolating a section of pipeline for inspection or maintenance.

Upper explosive limit (UEL): The maximum concentration of vapor in the air above which propagation of a flame does not occur in the presence of an ignition source.

Additional Resources

The following recommended resources can provide additional helpful information related to the requirements for this covered task:

API Recommended Practice 1161, Third Edition, © 2014 American Petroleum Institute. API Publishing Services, 1220 L Street, NW, Washington, DC 20005. Distributed by Thompson Reuters (Scientific) LLC, **www.techstreet.com**.

NCCER Module CT36_1-17, *Safe Disconnect of Pipeline Facilities*

Purging Principles and Practice. (PDF). American Gas Association. **https://s0.hfdstatic.com/p/losscontrol/files/private/1287777537857.pdf**. Last accessed Sept. 2016.

NCCER Module CT15_1-17, *Visual Inspection of the Surface*

NCCER Module CT65_2-17, *Shutdown of a Pipeline*

NCCER Module 26209-17, *Grounding and Bonding*

Electrical Bonding of Gas Piping Systems. (PDF). American Gas Association. **www.aga.org/sites/default/files/fact_sheet_nfgc_gas_piping_bonding_-_january_7-2015_1.pdf**. Last accessed Sept. 2016.

NCCER Module 64102-02, *Pipeline E&I Safety*

NCCER Module 71101-14, *Field Abnormal Operating Conditions*

NCCER Module 71210-03, *Work Permit Policies*

Figure Credits

arhendrix/Shutterstock.com, Module Opener
© iStockphoto.com/Michal Krakowiak, Figure 1
leungchopan/Shutterstock.com, Figure 2
General Monitors, Figure 3

Section Review Answer Key

Answer	Section Reference	Objective
Section One		
1. d	1.1.1	1a
2. b	1.2.1	1b

NCCER CURRICULA — USER UPDATE

NCCER makes every effort to keep its textbooks up-to-date and free of technical errors. We appreciate your help in this process. If you find an error, a typographical mistake, or an inaccuracy in NCCER's curricula, please fill out this form (or a photocopy), or complete the online form at **www.nccer.org/olf**. Be sure to include the exact module ID number, page number, a detailed description, and your recommended correction. Your input will be brought to the attention of the Authoring Team. Thank you for your assistance.

Instructors – If you have an idea for improving this textbook, or have found that additional materials were necessary to teach this module effectively, please let us know so that we may present your suggestions to the Authoring Team.

NCCER Product Development and Revision

13614 Progress Blvd., Alachua, FL 32615

Email: curriculum@nccer.org

Online: www.nccer.org/olf

❑ Trainee Guide ❑ Lesson Plans ❑ Exam ❑ PowerPoints Other ____________

Craft / Level: Copyright Date:

Module ID Number / Title:

Section Number(s):

Description:

Recommended Correction:

Your Name:

Address:

Email: Phone:

CT51_0-17

Purge Air from a Gas Pipeline

Overview

Segments of gas pipelines are routinely isolated from the main line for inspection or maintenance. When this operation occurs, the natural gas in the line must be safely removed. When the pipeline is ready to return to service, the air inside the segment must be purged with natural gas. This module examines the procedure for purging a pipeline of air and ensuring that all hazardous vapors are dispersed and eliminated before inspection or maintenance begins.

Trainees with successful module completions may be eligible for credentialing through the NCCER Registry. To learn more, go to **www.nccer.org** or contact us at 1.888.622.3720. Our website has information on the latest product releases and training, as well as online versions of our *Cornerstone* magazine and Pearson's product catalog.

Your feedback is welcome. You may email your comments to **curriculum@nccer.org**, send general comments and inquiries to **info@nccer.org**, or fill in the User Update form at the back of this module.

This information is general in nature and intended for training purposes only. Actual performance of activities described in this manual requires compliance with all applicable operating, service, maintenance, and safety procedures under the direction of qualified personnel. References in this manual to patented or proprietary devices do not constitute a recommendation of their use.

CT51_0-17
Purge Air from a Gas Pipeline

Objective

When you have completed this module, you will be able to do the following:

1. Describe how to purge air from a pipeline (CT51_0-17).
   a. Describe the process for purging air from a pipeline.
   b. Identify the prerequisites, competencies, and abnormal operating conditions associated with purging air from a pipeline.

Performance Task

Under the supervision of the instructor, you should be able to do the following:

1. Purge air from a pipeline (CT51_0-17).
   - Identify potential abnormal operating conditions that may occur, and know the appropriate actions to take in response to them.
   - Utilize the appropriate personal protective equipment according to relevant company procedures.
   - Purge the pipeline segment by allowing the air/gas mixture in the pipeline to vent safely into the atmosphere.
   - Use a vent pipe to direct purged vapors away from any ignition source.
   - Sample the atmosphere in the vent stream with a combustible gas indicator or portable gravitometer to determine when the vapor reaches 100% gas (the pipeline has been purged of all air).
   - Complete appropriate documentation and notifications as required by operator's procedures.

Trade Terms

Abnormal Operating Conditions (AOCs)
Blind flange
Bonding
Combustible-gas indicator (CGI)
Explosive limit (flammable limit)
Hot work
Lockout/tagout procedure
Lower explosive limit (LEL)
Nitrogen pack
Nitrogen purge
Purge
Shut-in
Upper explosive limit (UEL)

Industry Recognized Credentials

If you are training through an NCCER-accredited sponsor, you may be eligible for credentials from NCCER's Registry. The ID number for this module is CT51_0-17. Note that this module may have been used in other NCCER curricula and may apply to other level completions. Contact NCCER's Registry at 888.622.3720 or go to **www.nccer.org** for more information.

Contents

Figures

SECTION ONE

1.0.0 PURGE AIR FROM A PIPELINE

Objective

Describe how to purge air from a pipeline (CT51_0-17).

a. Describe the process for purging air from a pipeline.
b. Identify the prerequisites, competencies, and abnormal operating conditions associated with purging air from a pipeline.

Performance Task

Purge air from a pipeline. (CT51_0-17).

- Identify potential abnormal operating conditions that may occur, and know the appropriate actions to take in response to them.
- Utilize the appropriate personal protective equipment according to relevant company procedures.
- Purge the pipeline segment by allowing the air/gas mixture in the pipeline to vent safely into the atmosphere.
- Use a vent pipe to direct purged vapors away from any ignition source.
- Sample the atmosphere in the vent stream with a combustible gas indicator or portable gravitometer to determine when the vapor reaches 100% gas (the pipeline has been purged of all air).
- Complete appropriate documentation and notifications as required by operator's procedures.

Trade Terms

Abnormal operating conditions (AOCs): Conditions within a pipeline system, identified by the operator, that may indicate (1) a malfunction of a component or (2) a deviation from normal operations that may indicate a condition exceeding design limits or result in a hazard to persons, property, or the environment.

Blind flange: A solid plate used to close the component on which it is attached.

Bonding: A process of electrically connecting across a point of separation of an electrical conductive material (in this case, the pipe) to prevent sparking or arcing.

Combustible-gas indicator (CGI): An instrument used to detect various flammable gases and vapors.

Explosive limit (flammable limit): The percentage ratio range of a mixture of air and a flammable gas that will support combustion of the gas. Each gas has its characteristic range, but the typical natural gas explosive limit is 5 to 15 percent gas to 85 to 95 percent air. Explosive limits are also called flammable limits.

Hot work: Work involving burning, welding, or similar operations that create sparks, flame, or heat, and are capable of initiating fires or explosions.

Lockout/tagout procedure: A procedure used to identify sources of energy and hazardous materials that could adversely affect maintenance activities, to then isolate all such sources from the work area, and to ensure that the isolation remains effective until the work is completed.

Lower explosive limit (LEL): The minimum concentration of vapor in the air below which propagation of a flame does not occur in the presence of an ignition source.

Nitrogen pack: To leave a disconnected segment of pipe filled with nitrogen under pressure.

Nitrogen purge: The process of displacing hazardous vapors from a section of pipeline by injecting an inert gas (nitrogen) into the line at a pressure sufficient to force the hazardous vapors out.

Purge: To displace air or hazardous vapor from a pipeline.

Shut-in: The process of isolating a section of pipeline for inspection or maintenance.

Upper explosive limit (UEL): The maximum concentration of vapor in the air above which propagation of a flame does not occur in the presence of an ignition source.

There are times when a pipeline section (*Figure 1*) must be **shut-in** so that it may be purged for inspection or maintenance. Gas is routed around the shut-in section so that there is no interruption in service. When isolating a pipeline section, field personnel must **purge** the pipe segment of natural gas by injecting an inert gas, such as nitrogen, into the segment. When the line is ready to return to service, the air in the line must be purged using natural gas. The purging process removes potentially explosive gas/air mixtures from the segment. After the pipeline is emptied of gas or air, explosive vapors may still be present. The line must be monitored until all traces of explosive vapors are removed.

> **CAUTION**
>
> An individual's actions during the performance of a covered task could cause a safety device to activate. The activation represents a deviation from normal operations, resulting in an abnormal operating condition (AOC).

1.1.0 Purging Air from a Pipeline

Flammable gas explodes or catches fire when it is mixed with a certain percentage of air. Typically, a mixture of gas and air must be 5 to 15 percent gas and 85 to 95 percent air before it reaches its **explosive limit (flammable limit)**. A mixture of gas and air that does not have enough gas to explode is below the **lower explosive limit (LEL)**. That is, it is too lean to explode. A mixture with more than the maximum amount of gas is above the **upper explosive limit (UEL)**, so it is too rich to explode.

Purging establishes a moderately fast flow in the line being purged. This rapid flow helps to ensure that where the gas and air meet, they mix as little as possible, thus reducing the potential for explosion. Personnel must maintain a constant purge rate until the hazardous gas/air mixture has been expelled.

To purge a pipeline segment that has been shut-in, personnel open valves and use an inert gas, such as nitrogen, to push out the flammable gas. This is called a **nitrogen purge**. Before placing the line back in service, the line must be purged of air by injecting natural gas. A slug of nitrogen gas may first be injected to provide a barrier between the air being removed and the natural gas being injected.

Figure 1 Segment of a pipeline.

The procedures for purging air from a pipeline are the same as those for initially purging the gas, with the exception that the vent stream must be sampled with a **combustible-gas indicator (CGI)** to determine when the vapor being vented reaches 100-percent natural gas. Personnel must follow all of the same safety procedures.

The employee in charge of the purging procedure defines the appropriate process to be followed. All employees engaged in the process must wear the appropriate PPE. Before and during purging operations, these personnel must perform the following tasks:

- Follow the company operations and maintenance (O&M) procedures and be aware of possible hazards along the pipeline right-of-way (ROW) (*Figure 2*).
- Monitor for hazardous vapors throughout the purging process.
- Know and follow the company's **lockout/tagout procedure** and know how to obtain the appropriate work permits.

1.1.1 Procedure

This section describes the basic steps involved in purging air from a pipeline when the segment is ready to be returned to service:

> **WARNING!**
>
> Only trained, qualified individuals may perform pipeline purging procedures. Wear all appropriate personal protective equipment (PPE) and stay alert to the possibility of unexpected leaks. Safely eliminate all ignition sources and immediately notify the appropriate personnel in the event of a leak.

Step 1 Identify potential **abnormal operating conditions (AOCs)** that may occur, and know the appropriate actions to take in response to them.

Step 2 Utilize the appropriate personal protective equipment according to relevant company procedures.

Step 3 Purge the pipeline segment by allowing the air/gas mixture in the pipeline to vent safely into the atmosphere. Remove all air from the pipeline to prevent a hazardous atmosphere from remaining in the pipe segment.

Figure 2 Potential hazards on a pipeline right-of-way.

Step 4 Use a vent pipe to direct purged vapors away from any ignition source, ensuring that the hazardous vapors vent into the atmosphere for safe dispersion.

Step 5 Sample the atmosphere in the vent stream with a combustible gas indicator or portable gravitometer to determine when the vapor reaches 100% gas (the pipeline has been purged of all air). This indicates that the air has been vented from the pipeline and only a combustible atmosphere remains. The vapors are not explosive because the mixture is too rich to support combustion.

Step 6 Complete appropriate documentation and notifications as required by operator's procedures.

Local distribution companies may have additional procedures to purge the distribution lines that serve homes and businesses. Their procedures also include a process for quickly and safely restoring service.

Isolating the section to be purged requires following the company's lockout/tagout procedure. This procedure must be followed when hazardous materials could adversely affect maintenance. It includes isolating all such materials from the work area. The isolation must continue until the work is completed. Personnel must close off all valves and other devices that are directly connected to the section being shut-in. Steps must be taken to ensure that these devices are not accidentally opened while the purging operation is occurring.

Sparking or arcing should be prevented during the purging process. **Bonding** provides a safe path for electric current across any gap in the pipeline while the purging is occurring.

Personnel must measure the explosive level of hazardous vapors with a combustible-gas indicator (CGI). Flammable-gas leak detectors should be used as well if available on the segment being purged (*Figure 3*). No one should rely on their sense of smell to detect hazardous vapors for the following reasons:

- The concentration levels may not be high enough to detect dangerous levels by the sense of smell alone.
- Not all product is odorized and therefore cannot be detected by smell.
- A medical condition may affect a person's sense of smell.
- Certain hazardous vapors may deaden the sense of smell.

1.2.0 Prerequisites, Competencies, and Abnormal Operating Conditions

Personnel must possess a variety of skills in order to successfully purge a pipeline of air. The operation itself requires a number of steps which the personnel conducting the purge must be competent to perform. Personnel must also be aware of the particular abnormal operating conditions that

Figure 3 Flammable-gas leak detector.

may occur during the procedure and know the proper responses to them.

1.2.1 Prerequisites and Competencies

Operations personnel must possess basic knowledge and skills in a variety of areas prior to purging a pipeline of air. They must know how to obtain a **hot work** permit if such work is required on the line.

> **WARNING!**
>
> Improperly conducted hot work is a major cause of fires and explosions, resulting in property damage, serious injuries, and even death.

They must also know and be able to follow the company's lockout/tagout procedures in order to ensure that valves, compressors, or other devices with a direct impact on the segment being disconnected cannot be accidentally opened or operated.

Personnel conducting a purge must also be aware of any hazards that may be found along the right-of-way (ROW), and know and be able to follow all applicable company operations and maintenance (O&M) procedures.

Personnel conducting a pipeline purge must be competent in performing a system shut-in procedure. This involves installing a loop of pipeline so that the gas can be diverted around the shut-in section. The shut-in section is checked to ensure that all gas has been purged before any repairs or inspections are performed, and that all air has been purged before the segment is returned to service. Specific steps for shutting-in a pipeline segment, and the order in which the steps are performed, vary from one company to another.

Additional steps are required for a full system shutdown (if one should ever be required) and each company has its own procedures for this process. A full system shutdown may require that the end of the pipe be sealed with a **blind flange** and the line left in a **nitrogen pack** condition.

It is important to monitor the area for leaks during the entire shut-in process. Field personnel must wear the appropriate personal protective equipment (PPE) while performing this task.

> **CAUTION**
>
> Conditions on a pipeline can change without warning. It is critical to monitor conditions when a segment is shut-in. Field personnel rely on the control center to monitor pipeline pressure and temperature to ensure they remain within acceptable limits. They also rely on safety devices at the work site to alert them to problems.

Operations personnel must be able to purge the pipeline segment of air by injecting natural gas into the segment in order to remove all hazardous vapors created by the gas/air mixture. They must be able to use a vent pipe to direct purged vapors away from any ignition source, and to ensure that the hazardous vapors vent to the atmosphere for safe dispersion. They must be able to sample the vent stream with a combustible-gas indicator (CGI) to determine when the pipeline segment has been purged of all hazardous vapors.

Personnel must ensure that after purging the segment of air, the indicator shows that the segment contains 100-percent natural gas. This indicates that all traces of an explosive atmosphere have been removed. Personnel purging air from a pipeline segment must be able to follow all company O&M procedures and be able to complete and record all company-mandated records.

1.2.2 Abnormal Operating Conditions

Personnel purging a pipeline of air must be able to identify abnormal operating conditions (AOCs) and know the proper response to each should they occur while the purge is in progress. The following are some potential AOCs, and the appropriate action personnel should take in response to them:

- *Offensive odors occur during the purging operation that adversely affect the surrounding neighborhoods* – Personnel must monitor the operation to ensure that the vapors are not in the explosive range. Purging operations must be suspended if weather conditions do not allow for safe dispersion of the vapors.
- *The ignition of released hazardous liquids or gases causes a fire or explosion* – If the situation cannot be safely controlled by hand, personnel should immediately evacuate the area, limit access, and make the proper notifications, including calling 911 if appropriate.

Additional Resources

It is assumed that you will learn to perform the purging procedure under the direct supervision of your crew leader, foreman, or instructor. For independent study, refer to the recommended Additional Resources at the back of this module.

1.0.0 Section Review

1. After purging air from the pipeline, the percentage of gas in the line must be _____.

   a. at least 90 percent
   b. greater than 90 percent
   c. greater than 95 percent
   d. 100 percent

2. Prerequisites for performing a gas pipeline purge include knowing how to obtain a hot work permit, knowing how to follow lockout/tagout procedures, knowing how to complete company-mandated records, and knowing _____.

   a. how to monitor pipeline alarms
   b. how to inspect ROWs for hazards
   c. compressor shutdown procedures
   d. proper community notification procedures

SUMMARY

Pipelines are purged to remove hazardous gases prior to inspection or maintenance, and again before the line is returned to service. Personnel performing purging procedures must know how to obtain hot work permits, follow lockout/tagout procedures, inspect ROWs for hazards, follow company O&M procedures, and complete company-mandated records.

The explosive limit is reached when the ratio of gas to air ratio reaches 5 to 15 percent gas and 85 to 95 percent air. Purging a natural gas line with an inert gas at a moderately fast rate prevents the excess mixing of the gas and air.

Prior to purging, the lockout/tagout procedures must be in place and the pipeline must be shut-in, vented, and disconnected from the main line. The segment being purged must be grounded or bonded to prevent a static spark. Personnel must rely on combustible-gas indicators (CGIs), rather than their sense of smell, to ensure that all hazardous vapors have been eliminated.

During the purging operation, personnel must be able to complete a lockout/tagout procedure, purge the line with an inert gas or natural gas as needed, use a vent pipe to direct purged vapors away from ignition sources, sample the vent stream with a CGI, and identify and respond to potential abnormal operating conditions. Natural gas is injected to purge the line of air after work is complete. The vent stream must be monitored with a CGI to verify that the purged segment contains 100-percent natural gas.

Abnormal operating conditions may include offensive odors affecting surrounding neighborhoods, and fire or explosion caused by the release of hazardous liquids or gas. Vapors should be monitored to ensure that they are not in the explosive range and operations should be suspended if weather conditions prevent their safe dispersal. If a fire or explosion occurs, the area should be evacuated, access prevented, and proper notifications made, which may include contacting emergency responders.

Review Questions

1. Purging establishes a moderately fast flow in order to ensure _____.
   a. adequate mixing of the inert gas flammable gases
   b. rapid emptying of the line
   c. that the MAOP is not exceeded
   d. that air and gas mix as little as possible

2. What is the typical gas/air ratio when the explosive limit is reached?
   a. 5 to 10 percent gas, 90 to 95 percent air
   b. 5 to 15 percent gas, 85 to 95 percent air
   c. 10 to 20 percent gas, 80 to 90 percent air
   d. 20 to 25 percent gas, 75 to 80 percent air

3. To ensure that valves, compressors, or other devices are not operated or opened, it is necessary to _____.
   a. notify the control center before inspection or maintenance
   b. follow the company's lockout/tagout procedures
   c. post written warnings near all affected equipment
   d. monitor each device closely while work is being performed

4. The purpose of a vent pipe is to _____.
   a. ensure that hazardous vapors have been removed from the pipe
   b. remove excess gas after purging is complete
   c. direct purged vapors away from ignition sources
   d. allow the purging gas to enter the pipe

5. Potential abnormal operating conditions that may occur during purging include fire or explosion and _____.
   a. offensive odors
   b. compressor failure
   c. excess venting
   d. system isolation

Trade Terms Quiz

Fill in the blank with the correct term that you learned from your study of this module.

1. The policy used to isolate sources of energy and hazardous materials from the work area is called _________.
2. Displacing hazardous vapors from a section of pipe by injecting nitrogen is called a(n) _________.
3. The process of isolating a section of pipeline for inspection or repair is called _________.
4. The gas/air mixture range that will support combustion is called _________.
5. A solid plate used to close the component on which it is attached is called a(n) _________.
6. Work involving burning, welding, or similar operations is called _________.
7. To leave a disconnected segment of pipe filled with nitrogen is called _________.
8. When air or hazardous vapors are displaced from a pipeline, this is called a(n) _________.
9. The process of connecting a conductive material across a point of separation to prevent sparking is called _________.
10. The minimum concentration of a vapor in the air below which the propagation of a flame does not occur is called the _________.
11. An instrument used to detect flammable gases and vapors is called a(n) _________.
12. The maximum concentration of a vapor in the air above which the propagation of a flame does not occur is called the _________.
13. Conditions within a pipeline system, identified by the operator, that may indicate (1) a malfunction of a component or (2) a deviation from normal operations that may indicate a condition exceeding design limits or result in a hazard to persons, property, or the environment are known as _________.

Trade Terms

Abnormal Operating Conditions (AOCs)
Blind flange
Bonding
Combustible-gas indicator (CGI)
Explosive limit (flammable limit)
Hot work
Lockout/tagout
Lower explosive limit (LEL)
Nitrogen pack
Nitrogen purge
Purge
Shut-in
Upper explosive limit (UEL)

Trade Terms Introduced in This Module

Abnormal operating conditions (AOCs): Conditions within a pipeline system, identified by the operator, that may indicate (1) a malfunction of a component or (2) a deviation from normal operations that may indicate a condition exceeding design limits or result in a hazard to persons, property, or the environment.

Blind flange: A solid plate used to close the component on which it is attached.

Bonding: A process of electrically connecting across a point of separation of an electrical conductive material (in this case, the pipe) to prevent sparking or arcing.

Combustible-gas indicator (CGI): An instrument used to detect various flammable gases and vapors.

Explosive limit (flammable limit): The percentage ratio range of a mixture of air and a flammable gas that will support combustion of the gas. Each gas has its characteristic range, but the typical natural gas explosive limit is 5 to 15 percent gas to 85 to 95 percent air. Explosive limits are also called flammable limits.

Hot work: Work involving burning, welding, or similar operations that create sparks, flame, or heat, and are capable of initiating fires or explosions.

Lockout/tagout procedure: A procedure used to identify sources of energy and hazardous materials that could adversely affect maintenance activities, to then isolate all such sources from the work area, and to ensure that the isolation remains effective until the work is completed.

Lower explosive limit (LEL): The minimum concentration of vapor in the air below which propagation of a flame does not occur in the presence of an ignition source.

Nitrogen pack: To leave a disconnected segment of pipe filled with nitrogen under pressure.

Nitrogen purge: The process of displacing hazardous vapors from a section of pipeline by injecting an inert gas (nitrogen) Into the line at a pressure sufficient to force the hazardous vapors out.

Purge: To displace air or hazardous vapor from a pipeline.

Shut-in: The process of isolating a section of pipeline for inspection or maintenance.

Upper explosive limit (UEL): The maximum concentration of vapor in the air above which propagation of a flame does not occur in the presence of an ignition source.

Additional Resources

The following recommended resources can provide additional helpful information related to the requirements for this covered task:

API Recommended Practice 1161, Third Edition, © 2014 American Petroleum Institute. API Publishing Services, 1220 L Street, NW, Washington, DC 20005. Distributed by Thompson Reuters (Scientific) LLC, **www.techstreet.com**.

NCCER Module 71101-14, *Field Abnormal Operating Conditions*.

NCCER Module CT36_1-17, *Safe Disconnect of Pipeline Facilities*

Purging Principles and Practice. (PDF). American Gas Association. **https://s0.hfdstatic.com/p/losscontrol/files/private/1287777537857.pdf**. Last accessed Sept. 2016.

NCCER Module CT15_1-17, *Visual Inspection of the Surface*

NCCER Module CT65_2-17, *Shutdown of a Pipeline*

NCCER Module 26209-17, *Grounding and Bonding*

Electrical Bonding of Gas Piping Systems. (PDF). American Gas Association. **www.aga.org/sites/default/files/fact_sheet_nfgc_gas_piping_bonding_-_january_7-2015_1.pdf**. Last accessed Sept. 2016.

NCCER Module 64102-02, *Pipeline E&I Safety*

NCCER Module 71101-14, *Field Abnormal Operating Conditions*

NCCER Module 71210-03, *Work Permit Policies*

Figure Credits

arhendrix/Shutterstock.com, Module Opener
©iStockphoto.com/Michal Krakowiak, Figure 1
leungchopan/Shutterstock.com, Figure 2
General Monitors, Figure 3

Section Review Answer Key

Answer Section One	Section Reference	Objective
1. d	1.1.0	1a
2. b	1.2.1	1b

NCCER CURRICULA — USER UPDATE

NCCER makes every effort to keep its textbooks up-to-date and free of technical errors. We appreciate your help in this process. If you find an error, a typographical mistake, or an inaccuracy in NCCER's curricula, please fill out this form (or a photocopy), or complete the online form at **www.nccer.org/olf**. Be sure to include the exact module ID number, page number, a detailed description, and your recommended correction. Your input will be brought to the attention of the Authoring Team. Thank you for your assistance.

Instructors – If you have an idea for improving this textbook, or have found that additional materials were necessary to teach this module effectively, please let us know so that we may present your suggestions to the Authoring Team.

NCCER Product Development and Revision

13614 Progress Blvd., Alachua, FL 32615

Email: curriculum@nccer.org
Online: www.nccer.org/olf

❑ Trainee Guide ❑ Lesson Plans ❑ Exam ❑ PowerPoints Other ______________________

Craft / Level: Copyright Date:

Module ID Number / Title:

Section Number(s):

Description:

Recommended Correction:

Your Name:

Address:

Email: Phone:

CT54_0-17

Test Remotely Controlled Shutdown Devices

Overview

Ideally, all equipment and operations should run smoothly on the pipeline and at the stations, but this is not always the case. Equipment may fail, leaks may develop, or other issues may arise that impede normal operations. For that reason, alarms, safety, and shutdown devices are installed and must be monitored and tested on an ongoing basis. This module describes the procedures and identifies the knowledge and skills required for testing remotely controlled shutdown devices.

Trainees with successful module completions may be eligible for credentialing through the NCCER Registry. To learn more, go to **www.nccer.org** or contact us at 1.888.622.3720. Our website has information on the latest product releases and training, as well as online versions of our *Cornerstone* magazine and Pearson's product catalog.

Your feedback is welcome. You may email your comments to **curriculum@nccer.org**, send general comments and inquiries to **info@nccer.org**, or fill in the User Update form at the back of this module.

This information is general in nature and intended for training purposes only. Actual performance of activities described in this manual requires compliance with all applicable operating, service, maintenance, and safety procedures under the direction of qualified personnel. References in this manual to patented or proprietary devices do not constitute a recommendation of their use.

CT54_0-17
Test Remotely Controlled Shutdown Devices

Objective

When you have completed this module, you will be able to do the following:

1. Describe how to test remotely controlled shutdown devices (CT54_0-17).
   a. Describe the process for testing remotely controlled shutdown devices.
   b. Identify the knowledge and skills associated with testing remotely controlled shutdown devices.

Performance Task

Under the supervision of the instructor, you should be able to do the following:

1. Test remotely controlled shutdown devices (CT54_0-17).
   - Identify potential abnormal operating conditions that may occur during performance of this CT, and know the appropriate actions to take in response to them.
   - Utilize the appropriate personal protective equipment according to relevant company procedures.
   - Review the procedure to perform the inspection.
   - Make appropriate notifications prior to beginning the inspection.
   - Notify the control center.
   - Activate the RCSD.
   - Inspect the RCSD.
   - Report and make any necessary repairs.
   - Reset the RCSD.
   - Make appropriate notifications.
   - Complete appropriate documentation as required by operator's procedures.

Trade Terms

Abnormal operating conditions (AOCs)
Blowdown
Driver
Piping and instrumentation diagram (P&ID)
Remotely controlled shutdown device (RCSD)

Industry Recognized Credentials

If you are training through an NCCER-accredited sponsor, you may be eligible for credentials from NCCER's Registry. The ID number for this module is CT54_0-17. Note that this module may have been used in other NCCER curricula and may apply to other level completions. Contact NCCER's Registry at 888.622.3720 or go to **www.nccer.org** for more information.

Contents

Figures

SECTION ONE

1.0.0 TESTING REMOTELY CONTROLLED SHUTDOWN DEVICES

Objective

Describe how to test remotely controlled shutdown devices (CT54_0-17).

a. Describe the process for testing remotely controlled shutdown devices.
b. Identify the knowledge and skills associated with testing remotely controlled shutdown devices.

Performance Task

Test remotely controlled shutdown devices (CT54_0-17).

- Identify potential abnormal operating conditions that may occur during performance of this CT, and know the appropriate actions to take in response to them.
- Utilize the appropriate personal protective equipment according to relevant company procedures.
- Review the procedure to perform the inspection.
- Make appropriate notifications prior to beginning the inspection.
- Notify the control center.
- Activate the RCSD.
- Inspect the RCSD.
- Report and make any necessary repairs.
- Reset the RCSD.
- Make appropriate notifications.
- Complete appropriate documentation as required by operator's procedures.

Trade Terms

Abnormal operating conditions (AOCs): Conditions within a pipeline system, identified by the operator, that may indicate (1) a malfunction of a component or (2) a deviation from normal operations that may indicate a condition exceeding design limits or result in a hazard to persons, property, or the environment.

Blowdown: The venting of accumulated gas to depressurize pipes and equipment.

Driver: A reciprocating engine, turbine engine, or electric motor used to power a compressor.

Piping and instrumentation diagram (P&ID): A diagram layout of the pipe and equipment that shows the flow through the line using standard piping and instrumentation symbols.

Remotely controlled shutdown device (RCSD): An electronic device that can be used to shut down a compressor driver from a control center or other location.

A remotely controlled shutdown device (RCSD) is an electronic device used to shut down a compressor should any abnormal operating conditions (AOCs) occur. When activated, the RCSD stops the compressor driver, and isolates the compressor and associated station piping. Activation also initiates a blowdown, which releases the gas inside the compressor and piping into the atmosphere through a vent stack, to a flare stack to be burned, or into lower-pressure piping. RSCDs must be routinely tested to ensure that they operate as intended.

1.1.0 Remotely Controlled Shutdown Devices

Remotely controlled shutdown devices at compressor stations (shown in *Figure 1*) must be inspected on a regular basis as set forth in 49 *CFR* 192.731. Each RCSD must be inspected and tested at intervals not exceeding 15 months, but at least

Figure 1 Station with two mainline compressors.

once each calendar year, in order to determine that it functions properly. Any RCSD found to be defective must be promptly repaired or replaced.

> **CAUTION**
> An individual's actions during the performance of a covered task could cause a safety device to activate. The activation represents a deviation from normal operations, resulting in an AOC.

1.1.1 Types of Remotely Controlled Shutdown Devices

Remotely controlled shutdown devices have been designed to activate for a variety of specific AOCs. The types of RCSDs that may be found at a typical compressor station may include shutdown devices for the following conditions:

- High engine temperature
- High engine discharge temperature
- High liquid level
- Engine over-speed
- Excessive engine vibration
- Inadequate engine lubrication
- High and low suction pressure

1.1.2 Procedure

Before testing a remotely controlled shutdown device, personnel must review the testing procedure. This review is critical to ensure the integrity of the system, personal safety, the safety of others, and proper inspection. Appropriate PPE is required during the test. The proper sequence for conducting the actual test follows:

Step 1 Identify potential abnormal operating conditions that may occur, and know the appropriate actions to take in response to them.

Step 2 Utilize the appropriate personal protective equipment according to relevant company procedures.

Step 3 Review the procedure to perform the inspection.

Step 4 Make appropriate notifications prior to beginning the inspection. Alert all affected personnel and the control center that the station is being prepared for a test of the RCSD.

Step 5 Notify the control center. The control center should witness and document alarm status.

Step 6 Activate and test the proper operation of the RCSD.

Step 7 Inspect the RCSD. Ensure all safety, shutdown, alarm, and isolation equipment associated with the RCSD operates properly as described in the procedure and on the P&ID(s).

Step 8 Report and make any necessary repairs. Ensure the integrity and proper operation of the RCSD.

Step 9 Reset the RCSD, placing it back in operational mode.

Step 10 Make appropriate notifications. Alert nearby personnel and the control center that the RCSD inspection is complete and the RCSD has been reset and is ready to return to service.

Step 11 Complete appropriate documentation as required by operator's procedures.

> **WARNING!**
> When pipeline equipment must be repaired, always remember to make the proper notifications per company policy and to lockout/tagout the equipment.

During these tests, there may be a gas leak. Personnel should isolate the source of the leak if it is safe to do so, and follow established company safety procedures to protect people, property, and the environment. Once conditions are safe, the leak may be repaired. During repairs, the air must be continuously monitored for hazardous gas.

1.2.0 Prerequisites, Competencies, and Abnormal Operating Conditions

Personnel must possess a variety of skills to test remotely controlled shutdown devices in a safe and efficient manner. The operation itself requires a series of steps with which the personnel testing RCSDs must be competent to perform. Personnel must also be aware of the particular abnormal operating conditions that may occur during the procedure and must know the proper responses to them.

1.2.1 Prerequisites and Competencies

Personnel must possess certain basic knowledge prior to testing remotely controlled shutdown devices. This includes understanding all company safety procedures relating to the procedure, and the company policies for proper use of PPE.

Operations personnel must understand how the RCSD being inspected operates. They must have a thorough knowledge of the correct procedures for testing and inspecting RCSDs, and know the proper equipment to use. They must also have an understanding of the **piping and instrumentation diagram (P&ID)** associated with the equipment being used. They need to be aware of the abnormal operating conditions that may be encountered during the procedure, and the proper responses should any of these conditions occur.

Personnel who are testing and inspecting remotely controlled shutdown devices must be competent to perform all the related tasks associated with an RCSD test. The person or persons who will be conducting the test must first take time to thoroughly review the procedure in order to become familiar with how to perform the test and conduct the inspection. This step is critical for the integrity of the system, personal safety, the safety of others, and for conducting a proper inspection.

Personnel must be able to make all appropriate notifications prior to beginning the inspection. These notifications are needed to alert all affected personnel and the control center that the station is being prepared for a test of the RCSD.

Prior to the test, operations personnel must choose PPE that will ensure personal safety. They must be capable of ensuring that the RCSD being inspected is receiving the proper electrical supply. When the test is ready to commence, personnel must notify the control center that the inspection is underway. This step is critical for the integrity of the system and the safety of others.

To conduct the test, personnel must be able to activate the RCSD in order to test the proper operation of the device, and to ensure that the compressor (*Figure 2*) shuts down properly after activation of the RCSD being inspected. They must be able to conduct a proper inspection of the RCSD to ensure that all safety, shutdown, alarm, and isolation equipment associated with the device operates properly as described in the procedure and on the P&ID(s). The control center should witness and document the status of all alarms during the inspection.

Operations personnel must be able to report on the operation of the inspected RCSD and make any necessary repairs in order to ensure the integrity and proper operation of the device. When the test is complete, they must be able to reset the RCSD and place the device back in operational mode. The appropriate notifications must be made to alert nearby personnel and the control center that the inspection is complete and that the device has been reset and is ready to return to service.

Figure 2 Mainline compressor.

Personnel must also be able to complete and file all necessary documentation as required by company policy and federal regulations.

1.2.2 Abnormal Operating Conditions

Personnel testing remotely controlled shutdown devices must be able to identify abnormal operating conditions (AOCs) that may occur during the procedure. The following are some potential AOCs, and the appropriate action personnel should take in response to them:

- *An unexpected gas release is encountered during the RCSD inspection* – Personnel must isolate the leak if it is safe to do so, and then evacuate the area in order to allow the station to ventilate. The appropriate company personnel must be notified.
- *A fire or explosion occurs during the procedure* – Personnel must secure the area if it is safe to do so. The appropriate company notifications must be made, including calling 911 if necessary.

> **WARNING!**
>
> Proper PPE must be worn and proper safety precautions must be taken if an unexpected release of hazardous liquids or vapors occurs.

Additional Resources

It is assumed that you will learn to perform the purging procedure under the direct supervision of your crew leader, foreman, or instructor. For independent study, refer to the recommended Additional Resources at the back of this module.

1.0.0 Section Review

1. Remotely controlled shutdown devices must be inspected and tested at intervals not to exceed _____.
   a. once a month
   b. every three months
   c. every 15 months
   d. twice per year

2. After an RCSD is repaired and reset, personnel must next _____.
   a. restart the compressor driver
   b. notify the control center that the device is ready to return to service
   c. complete and file all company-mandated documentation
   d. perform a leak check on all systems

SUMMARY

Remotely controlled shutdown devices (RCSDs) are designed to shut down a compressor in the event of an AOC. RCSDs must be inspected at least once each calendar year with no more than 15 months between inspections. Defective RCSDs must be promptly repaired or replaced. RCSDs may activate in the event of conditions such as high engine or discharge temperature, excessive vibration, high liquid levels, inadequate lubrication, engine overspeed, or high or low suction pressure.

The steps for conducting an RCSD test include making notifications that the test is to begin, then activating the device, inspecting the device, making necessary repairs, and finally, making notifications that the device has been reset and is ready to return to service. It is important to monitor for gas leaks while performing the test, inspection, and repairs.

Personnel conducting an RCSD test must be familiar with company safety procedures and know the appropriate PPE to use. They must know how the RCSD operates, know the procedures for conducting the test, and be familiar with the associated P&IDs. They must understand the potential AOCs and how to properly react to them.

It is critical for the personnel conducting the test to first review the procedures. Prior to the test, they must choose proper PPE, ensure the device has adequate electricity, and make proper notifications. To conduct the test, personnel must activate the device, ensure that compressors shut down properly, inspect the device, and make necessary repairs. After the test is complete, they must make notifications that the device has been reset and is ready to return to service, and complete the required documentation.

If a gas leak occurs, personnel must isolate the leak if it is safe to do so, evacuate the area to allow it to ventilate, and make proper notifications. If a fire or explosion occurs, they must secure the area and make appropriate notifications, which may including calling 911.

Review Questions

1. Remotely controlled shutdown devices are _____.
   a. only used if the MAOP is exceeded
   b. used to close the primary gate valve
   c. only used in extreme emergencies
   d. used to shut down a compressor if an AOC occurs

2. According to federal regulations, an RCSD must be tested _____.
   a. after being reset
   b. no more than once each calendar year
   c. bi-monthly
   d. at least once each calendar year

3. After any necessary repairs are made to an RCSD, personnel must _____.
   a. initiate a blowdown
   b. activate the device
   c. notify the control center that the device is ready to return to service
   d. reset the device

4. Prior to testing remotely controlled shutdown devices, personnel must _____.
   a. have an understanding of the P&IDs
   b. shut down the compressor
   c. reset the RCSD
   d. activate the RCSD

5. If a fire or explosion occurs during the RCSD test, personnel must _____,
   a. contain the situation until emergency responders arrive
   b. secure the area if it is safe to do so
   c. immediately evacuate the area
   d. take appropriate steps to mitigate the situation

Trade Terms Quiz

Fill in the blank with the correct term that you learned from your study of this module.

1. A diagram layout of the pipe and equipment that shows the flow through the line is called a(n) ________.
2. Conditions within a pipeline system, identified by the operator, that may indicate (1) a malfunction of a component or (2) a deviation from normal operations that may indicate a condition exceeding design limits or result in a hazard to persons, property, or the environment are known as ________.
3. An electronic device that can be used to shut down a compressor driver is called a(n) ________.
4. The venting of accumulated gas to depressurize equipment is known as ________.
5. An engine or motor used to power a compressor is called a(n) ________.

Trade Terms

Abnormal operating conditions (AOCs)
Blowdown
Driver
Piping and instrumentation diagram (P&ID)
Remotely controlled shutdown device (RCSD)

Trade Terms Introduced in This Module

Abnormal operating conditions (AOCs): Conditions within a pipeline system, identified by the operator, that may indicate (1) a malfunction of a component or (2) a deviation from normal operations that may indicate a condition exceeding design limits or result in a hazard to persons, property, or the environment.

Blowdown: The venting of accumulated gas to depressurize pipes and equipment.

Driver: A reciprocating engine, turbine engine, or electric motor used to power a compressor.

Piping and instrumentation diagram (P&ID): A diagram layout of the pipe and equipment that shows the flow through the line using standard piping and instrumentation symbols.

Remotely controlled shutdown device (RCSD): An electronic device that can be used to shut down a compressor driver from a control center or other location.

Additional Resources

The following recommended resources can provide additional helpful information related to the requirements for this covered task:

API Recommended Practice 1161, Third Edition, © 2014 American Petroleum Institute. API Publishing Services, 1220 L Street, NW, Washington, DC 20005. Distributed by Thompson Reuters (Scientific) LLC, **www.techstreet.com**.

NCCER Module AOCCC-17, *Abnormal Operating Conditions Control Center*

NCCER Module AOCFG-17, *Abnormal Operating Conditions Field and Gas*

NCCER Module 62105-02, *Pipeline Documentation*

NCCER Module 63104-02, *Piping and Mechanical Blueprint Reading*

NCCER Module 63208-02, *Introduction to Gas Compressors*

NCCER Module 63211-02, *Maintain and Repair Drivers*

NCCER Module 63305-02, *Maintain, Troubleshoot, and Repair Gas Compressors*

NCCER Module 64102-02, *Pipeline E&I Safety*

NCCER Module 67102-15, *Basic Pipeline Pneumatics and Equipment*

NCCER Module 67103-15, *Pipeline Communications*

Figure Credits

arhendrix/Shutterstock.com, Module Opener

©iStockphoto.com/westphalia, Figure 1

GTS Productions/Shutterstock.com, Figure 2

Section Review Answer Key

Answer	Section Reference	Objective
Section One		
1. c	1.1.0	1a
2. b	1.2.1	1b

NCCER CURRICULA — USER UPDATE

NCCER makes every effort to keep its textbooks up-to-date and free of technical errors. We appreciate your help in this process. If you find an error, a typographical mistake, or an inaccuracy in NCCER's curricula, please fill out this form (or a photocopy), or complete the online form at **www.nccer.org/olf**. Be sure to include the exact module ID number, page number, a detailed description, and your recommended correction. Your input will be brought to the attention of the Authoring Team. Thank you for your assistance.

Instructors – If you have an idea for improving this textbook, or have found that additional materials were necessary to teach this module effectively, please let us know so that we may present your suggestions to the Authoring Team.

NCCER Product Development and Revision

13614 Progress Blvd., Alachua, FL 32615

Email: curriculum@nccer.org
Online: www.nccer.org/olf

❑ Trainee Guide ❑ Lesson Plans ❑ Exam ❑ PowerPoints Other ____________

Craft / Level: Copyright Date:

Module ID Number / Title:

Section Number(s):

Description:

Recommended Correction:

Your Name:

Address:

Email: Phone:

CT56_0-17

Perform Incremental Pressure Increases to Uprate the MAOP

Overview

The maximum allowable operating pressure (MAOP) of a pipeline may be increased through a process called uprating. This module examines the procedure for uprating the MAOP of a pipeline through incremental pressure increases.

Trainees with successful module completions may be eligible for credentialing through the NCCER Registry. To learn more, go to **www.nccer.org** or contact us at 1.888.622.3720. Our website has information on the latest product releases and training, as well as online versions of our *Cornerstone* magazine and Pearson's product catalog.

Your feedback is welcome. You may email your comments to **curriculum@nccer.org**, send general comments and inquiries to **info@nccer.org**, or fill in the User Update form at the back of this module.

This information is general in nature and intended for training purposes only. Actual performance of activities described in this manual requires compliance with all applicable operating, service, maintenance, and safety procedures under the direction of qualified personnel. References in this manual to patented or proprietary devices do not constitute a recommendation of their use.

CT56_0-17
Perform Incremental Pressure Increases to Uprate the MAOP

Objective

When you have completed this module, you will be able to do the following:

1. Describe how to perform incremental pressure increases to uprate the MAOP (CT56_0-17).
   a. Describe how to perform incremental pressure increases to uprate the MAOP.
   b. Identify the knowledge and skills needed to uprate the MAOP.

Performance Task

Under the supervision of the instructor, you should be able to do the following:

1. Perform incremental pressure increases to uprate the MAOP (CT56_0-17).
   - Identify potential abnormal operating conditions (AOCs) that may occur during performance of this CT, and know the appropriate actions to take in response to them.
   - Utilize the appropriate personal protective equipment according to relevant company procedures.
   - Increase line pressure to the current system MAOP.
   - While monitoring, gradually increase line pressure to first predetermined setting and stop.
   - Record incremental pipeline increases on pressure test form.
   - Perform a leakage survey. If leaks are discovered, repair them before continuing.
   - Continue to incrementally increase line pressure and check for leaks.
   - Increase line pressure to final established pressure.
   - Complete appropriate documentation as required by operator's procedures.

Trade Terms

Abnormal operating conditions (AOCs)
Leak detection
Maximum allowable operating pressure (MAOP)
Pipeline design
Procedure
Segment
Uprating

Industry Recognized Credentials

If you are training through an NCCER-accredited sponsor, you may be eligible for credentials from NCCER's Registry. The ID number for this module is CT56_0-17. Note that this module may have been used in other NCCER curricula and may apply to other level completions. Contact NCCER's Registry at 888.622.3720 or go to **www.nccer.org** for more information.

Contents

Figures and Tables

Section One

1.0.0 Perform Incremental Pressure Increases to Uprate the MAOP

Objective

Describe how to perform incremental pressure increases to uprate the MAOP (CT56_0-17).

a. Describe how to perform incremental pressure increases to uprate the MAOP.
b. Identify the knowledge and skills needed to uprate the MAOP.

Performance Task

Perform incremental pressure increases to uprate the MAOP (CT56_0-17).

- Identify potential abnormal operating conditions (AOCs) that may occur during performance of this CT, and know the appropriate actions to take in response to them.
- Utilize the appropriate personal protective equipment according to relevant company procedures.
- Increase line pressure to the current system MAOP.
- While monitoring, gradually increase line pressure to first predetermined setting and stop.
- Record incremental pipeline increases on pressure test form.
- Perform a leakage survey. If leaks are discovered, repair them before continuing.
- Continue to incrementally increase line pressure and check for leaks.
- Increase line pressure to final established pressure.
- Complete appropriate documentation as required by operator's procedures.

Trade Terms

Abnormal operating conditions (AOCs): Conditions within a pipeline system, identified by the operator, that may indicate (1) a malfunction of a component or (2) a deviation from normal operations that may indicate a condition exceeding design limits or result in a hazard to persons, property, or the environment.

Leak detection: A means of determining pressure drops in a pipeline.

Maximum allowable operating pressure (MAOP): The highest level of internal gauge pressure allowed for a particular pipeline or pipeline segment, as set by the US Department of Transportation.

Pipeline design: Material and engineering requirements necessary to create a safe operating pipeline. Familiarity with 49 CFR 192.101 through 49 CFR 192.149 is required.

Procedure: A step-by-step written set of instructions to logically accomplish a task.

Segment: A specified length of pipeline constructed using a specified grade of pipe.

Uprating: Increasing the maximum allowable operating pressure (MAOP) on an existing pipeline segment.

The maximum allowable operating pressure (MAOP) of a pipeline or pipeline segment may be increased, or uprated, in order to allow more product to be transported through the line over a given time. Uprating may be performed when stronger components are added to the line, or if regulations change allowing for a higher MAOP. The increased MAOP is possible only if the integrity of the line can safely sustain the increased pressure.

Uprating is the procedure by which the line pressure is incrementally increased until the new MAOP is established. Federal regulation 49 *CFR* 192, Subpart K provides the guidelines that must be followed to safely uprate a pipeline or pipeline segment.

1.1.0 Uprating the MAOP

Uprating is the term used to describe the process for increasing the maximum allowable operating pressure (MAOP) on an existing pipeline segment. Various factors, such as pipeline design and operating experience, are considered when setting the MAOP. Engineers who have been trained in pipeline design and safe operating limits are responsible for uprating the MAOP. However, it is useful for all pipeline personnel to have a basic understanding of the uprating process.

1.1.1 The Uprating Process

Before the uprating procedure begins, all repairs, replacements, or alterations in the segment being uprated must be accomplished; components specified in 49 *CFR* 192 Subpart K must be reinforced or anchored; and the segment must be isolated from any adjacent segment that will continue to operate at a lower pressure. Personnel must follow company safety procedures to ensure the safety of those who are performing the uprating procedure.

CAUTION

An individual's actions during the performance of a covered task could cause a safety device to activate. The activation represents a deviation from normal operations, resulting in an abnormal operating condition (AOC).

Step-by-step pressure increases are necessary to safely uprate the MAOP of a pipeline. Before performing the incremental pressure increases to the planned new MAOP, it is essential to increase the pressure in the pipeline segment to the current system MAOP. Doing so provides a starting point for the step-by-step pressure increases. The pipeline pressure must then be increased gradually to the first planned setting (*Figure 1*).

The pipeline segment undergoing uprating must be monitored constantly as the pressure increases. When the pipeline pressure reaches the first planned setting, the process must be halted and the pressure held constant before it is increased to the next setting. During this time, the segment is checked for leaks as specified in 49 *CFR* 192 Subpart K (*Figure 2*).

Figure 1 Electronic pressure transmitter.

If potentially hazardous leaks are discovered, they must be repaired before a further increase in pressure is made. If it is determined that a leak is not potentially hazardous, it does not need to be repaired immediately; however, it must be monitored during subsequent pressure increases to ensure that it does not become hazardous. Gradual, step-by-step pressure increases help to prevent pipeline failure and accidents. Each pressure increase must be recorded on a pressure test form.

1.1.2 Procedure

The procedure for a typical uprating plan involves the following steps:

Step 1 Identify potential abnormal operating conditions (AOCs) that may occur, and know the appropriate actions to take in response to them.

Step 2 Utilize the appropriate personal protective equipment according to relevant company procedures.

Step 3 Before the uprating procedure begins, any repairs or alterations must be completed, specified components must be anchored or reinforced, and the segment being uprated must be isolated.

Step 4 In order to provide a starting point for further incremental pressure increases, steadily increase the line pressure to the current system MAOP.

Figure 2 Pipeline employee inspecting line segment.

Step 5 Gradually increase line pressure while monitoring for a pressure loss which could indicate a potential leak. Gradual pressure increases help prevent premature pipeline failure and accidents.

Step 6 Stop when the first predetermined pressure setting is reached. Record this and subsequent pressure increases on a pressure test form, in compliance with Subpart K of the regulations.

Step 7 During the stop, perform a leak detection survey along the entire line to investigate any leaks which may have occurred during the pressure increase (*Figure 3*). If a leak is discovered that appears to be hazardous, repair it in order to prevent further pipeline damage.

> **WARNING!**
>
> When a valve, manifold, or other pipeline equipment must be repaired, always remember to make the proper notifications per company policy and to lockout/tagout the equipment.

Step 8 Continue to incrementally increase the line pressure to each predetermined stop. After each stop, record the pressure, perform a leak detection survey, and complete any necessary repairs.

Step 9 Gradually increase line pressure to the final established test pressure. This will become the new MAOP. Perform a final leak detection survey and make any necessary repairs.

Step 10 Document the final established MAOP to finalize the uprating record.

Step-by-step increases in pipeline pressure should be continued, with a stop at each setting as planned, until the final planned test pressure is reached. When the final pressure is reached, and the final leak survey and necessary repairs performed, the pipeline is uprated.

1.1.3 Required Documentation

The final established MAOP must be documented, and the following information must be included in accordance with federal regulations:

- The name of the operating company
- The name of the contractor conducting the uprating

Figure 3 Pipeline and valves at a remote station.

- The names of the personnel conducting the uprating test
- The date and time of the test
- The location of the pipeline or the pipeline segment being uprated
- The test medium used
- The test pressures
- The duration of the test
- The uprating plan
- Any leakage and repairs
- Any AOCs and the corrective actions taken

1.2.0 Prerequisites, Competencies, and Abnormal Operating Conditions

Personnel must possess a variety of skills in order to successfully uprate a pipeline. The operation itself requires a number of steps which the personnel conducting the uprating must be competent to perform. Personnel must also be aware of the particular abnormal operating conditions (AOCs) that may occur during the procedure and know the proper responses to them.

1.2.1 Prerequisites and Competencies

Personnel involved in the task of uprating the MAOP of a pipeline or pipeline segment must have knowledge and training on the elements of pipeline design. In addition, they must be aware of the safe operating limits of the line undergoing uprating. For example, a pipeline is only as strong as its weakest component. Each valve, flange, or other component on the line is assigned an American National Standards Institute (ANSI) pressure rating based on their construction (*Table 1*). Personnel performing an uprating must be able to

Table 1. Maximum Operating Pressure

ANSI	MOP (psi)
150	285
300	740
600	1,480
900	2,220
1,500	3,705
2,500	6,170

interpret and apply the ANSI rating in order to determine the MAOP the line can sustain.

Personnel conducting a pipeline MAOP uprating must understand all company safety policies and procedures relating to the uprating. They must also have a basic knowledge of the system being uprated and possess a thorough understanding of the uprating plan.

1.2.2 Abnormal Operating Conditions

Throughout the procedure, personnel must be aware of the AOCs that could develop and know the proper responses should they occur. The following are some potential AOCs, and the appropriate action personnel should take in response to them:

- *Pipeline components leak during the uprating procedure* – The test must be stopped and the line segment repaired. The uprating may continue without making repairs if it is determined that the leak is not potentially hazardous; however, the leak must be monitored continuously throughout the uprating procedure.

WARNING!

Under no circumstances should any person attempt to reach a valve when a vapor cloud is observed in the area.

- *The pipeline MAOP uprating procedure causes the pipe or a component to fail, resulting in a rupture in the line and an unintended release* – The test must be stopped, appropriate notifications must be made, and the necessary repairs performed.

WARNING!

Proper PPE must be worn and proper safety precautions must be taken when dealing with an unexpected release of hazardous liquids or vapors.

Additional Resources

It is assumed that you will learn to perform the purging procedure under the direct supervision of your crew leader, foreman, or instructor. For independent study, refer to the recommended Additional Resources at the back of this module.

1.0.0 Section Review

1. The pipeline uprating procedure involves _____.
   a. increasing the pressure to the new rating, then incrementally lowering it to the original MAOP
   b. quickly and steadily establishing the new MAOP rating
   c. a slow, steady increase in pressure until the new MAOP rating is established
   d. a step-by-step increase in pressure until the new MAOP rating is established

2. Pipeline uprating is usually handled by personnel with knowledge of _____.
   a. pipeline design
   b. pipeline purging
   c. remote control shutdown devices
   d. the SCADA system

SUMMARY

Uprating is the procedure for incrementally increasing a pipeline or segment pressure in order to establish a new maximum allowable operating pressure (MAOP). Personnel performing an uprating must understand pipeline design and the safe operating limits of the line. They must also have knowledge of the company safety policies as well as the details of the uprating plan.

Federal regulation 49 *CFR* 192 Subpart K provides the requirements for uprating a line. The uprating procedure involves incrementally increasing the pressure to the current MAOP, and then continuing the incremental increases until the new MAOP is reached. At each step the pressure is recorded, the line is checked for leaks, and repairs are made if necessary.

Required documentation for the uprating include the pressure at each incremental increase, the medium used, the date, the location of the line, the duration of the procedure, the new MAOP, and the company and personnel conducting the uprating.

During the uprating procedure, there is a potential for pipeline components to leak or for the pipe to rupture. If a leak that is deemed hazardous or a rupture occurs, the test must be stopped and necessary repairs performed. The test may continue if a leak is not deemed hazardous, but the leak must be monitored.

Review Questions

1. Uprating may only be performed by engineers trained in pipeline design and _____.
   a. pipeline leak repair
   b. handling hazardous materials
   c. safe operating limits
   d. lockout/tagout procedures

2. During uprating, initially increasing the pressure to the original MAOP _____.
   a. determines the number of pressure increments needed
   b. ensures that the pipe can withstand the pressure increases
   c. provides a starting point for the incremental pressure increases
   d. may be omitted for smaller line segments

3. When reaching the first planned pressure setting, _____.
   a. maintain that pressure for at least one hour while performing a leak detection survey
   b. the pressure must then be reduced temporally before resuming the pressure increases
   c. continue the pressure increases unless a leak is detected
   d. the process must be halted and the pressure held constant before the next increase

4. Any leakage detected _____.
   a. must be repaired immediately
   b. may be repaired later if not hazardous
   c. may be repaired later if the amount is not significant
   d. must be repaired and the uprating test cancelled

5. The pipeline MAOP uprating procedure may expose _____.
   a. a stuck or inoperable valve
   b. a weakness in the pipe
   c. an inaccurate pressure rating
   d. improper documentation

Trade Terms Quiz

Fill in the blank with the correct term that you learned from your study of this module.

1. The material and engineering requirements necessary to create a pipeline is called _________.
2. Increasing the maximum allowable operating pressure (MAOP) on a line is called _________.
3. A specified length of pipeline is called a(n) _________.
4. A step-by-step written set of instructions is called a(n) _________.
5. A means of determining pressure drops in a pipeline is called _________.
6. The highest level of internal gauge pressure allowed in a pipeline is called _________.
7. Conditions within a pipeline system, identified by the operator, that may indicate (1) a malfunction of a component or (2) a deviation from normal operations that may indicate a condition exceeding design limits or result in a hazard to persons, property, or the environment are known as _________.

Trade Terms

Abnormal operating conditions (AOCs)
Leak detection
Maximum allowable operating pressure (MAOP)
Pipeline design
Procedure
Segment
Uprating

Trade Terms Introduced in This Module

Abnormal operating conditions (AOCs): Conditions within a pipeline system, identified by the operator, that may indicate (1) a malfunction of a component or (2) a deviation from normal operations that may indicate a condition exceeding design limits or result in a hazard to persons, property, or the environment.

Leak detection: A means of determining pressure drops in a pipeline.

Maximum allowable operating pressure (MAOP): The highest level of internal gauge pressure allowed for a particular pipeline or pipeline segment, as set by the US Department of Transportation.

Pipeline design: The material and engineering requirements necessary to create a safe operating pipeline. Familiarity with 49 CFR 192.101 through 49 CFR 192.149 is required.

Procedure: A step-by-step written set of instructions to logically accomplish a task.

Segment: A specified length of pipeline constructed using a specified grade of pipe.

Uprating: Increasing the maximum allowable operating pressure (MAOP) on an existing pipeline segment.

Additional Resources

The following recommended resources can provide additional helpful information related to the requirements for this covered task:

API Recommended Practice 1161, Third Edition, © 2014 American Petroleum Institute. API Publishing Services, 1220 L Street, NW, Washington, DC 20005. Distributed by Thompson Reuters (Scientific) LLC, **www.techstreet.com**.

49 *CFR* 192, Subpart K, **www.ecfr.gov**

NCCER Module AOCCC-17, *Abnormal Operating Conditions Control Center*

NCCER Module AOCFG-17, *Abnormal Operating Conditions Field & Gas*

NCCER Module 62105-02, *Pipeline Documentation*

NCCER Module 62206-14, *Leakage Survey*

NCCER Module 62304-02, *Pipeline Repair*

NCCER Module 63104-02, *Piping and Mechanical Blueprint Reading*

NCCER Module 63204-02, *Inspect and Repair Valves*

NCCER Module 64102-02, *Pipeline E&I Safety*

NCCER Module 67102-15, *Basic Pipeline Pneumatics and Equipment*

NCCER Module 67103-15, *Pipeline Communications*

NCCER Module 71101-14, *Field Abnormal Operating Conditions*

Figure Credits

arhendrix/Shutterstock.com, Module Opener

Tim Mills, Figure 1

Sarapinas Valeriy/Shutterstock.com, Figure 2

Jim Parkin/Shutterstock.com, Figure 3

Section Review Answer Key

Answer	Section Reference	Objective
Section One		
1. d	1.1.1	1a
2. a	1.2.1	1b

NCCER CURRICULA — USER UPDATE

NCCER makes every effort to keep its textbooks up-to-date and free of technical errors. We appreciate your help in this process. If you find an error, a typographical mistake, or an inaccuracy in NCCER's curricula, please fill out this form (or a photocopy), or complete the online form at **www.nccer.org/olf**. Be sure to include the exact module ID number, page number, a detailed description, and your recommended correction. Your input will be brought to the attention of the Authoring Team. Thank you for your assistance.

Instructors – If you have an idea for improving this textbook, or have found that additional materials were necessary to teach this module effectively, please let us know so that we may present your suggestions to the Authoring Team.

NCCER Product Development and Revision
13614 Progress Blvd., Alachua, FL 32615
Email: curriculum@nccer.org
Online: www.nccer.org/olf

❏ Trainee Guide ❏ Lesson Plans ❏ Exam ❏ PowerPoints Other ______________

Craft / Level: Copyright Date:

Module ID Number / Title:

Section Number(s):

Description:

Recommended Correction:

Your Name:

Address:

Email: Phone:

CT57_0-17

Operate Odorant Equipment

Overview

Pure natural gas is not detectable by smell, so an odorant is added by using odorization equipment to make the gas detectable in the event of a leak. The odorant intensity level is routinely monitored to ensure that it remains within limits. This module describes the procedures for operating odorant equipment.

Trainees with successful module completions may be eligible for credentialing through the NCCER Registry. To learn more, go to **www.nccer.org** or contact us at 1.888.622.3720. Our website has information on the latest product releases and training, as well as online versions of our *Cornerstone* magazine and Pearson's product catalog.

Your feedback is welcome. You may email your comments to **curriculum@nccer.org**, send general comments and inquiries to **info@nccer.org**, or fill in the User Update form at the back of this module.

This information is general in nature and intended for training purposes only. Actual performance of activities described in this manual requires compliance with all applicable operating, service, maintenance, and safety procedures under the direction of qualified personnel. References in this manual to patented or proprietary devices do not constitute a recommendation of their use.

CT57_0-17
Operate Odorant Equipment

Objective

When you have completed this module, you will be able to do the following:

1. Describe how to operate odorant equipment (CT57_0-17).
   a. Describe the process and procedure for odorizing natural gas.
   b. Identify the knowledge and skills associated with operating odorant equipment.

Performance Task

Under the supervision of the instructor, you should be able to do the following:

1. Operate odorant equipment (CT57_0-17).
   - Identify potential abnormal operating conditions that may occur during performance of this CT, and know the appropriate actions to take in response to them.
   - Utilize the appropriate personal protective equipment according to relevant company procedures.
   - Maintain odorizer according to manufacturer recommendations, including the pump, valves, bonnet, regulators, control panel, battery (if present), and filters.
   - Compare the operation of the odorizer with its design limits.
   - Check odorant tank levels.
   - Check supply gas pressure.
   - Determine and set injection rate per company procedure.
   - Determine if the injection of odorant is on a consistent basis.
   - Complete appropriate documentation as required by operator's procedures.

Trade Terms

Abnormal operating conditions (AOCs)
Bypass-type odorizer
Electronic odorizer
Lower explosive limit (LEL)
Odorant
Odorizer
Odorometer
Pitot tube
Wick-type odorizer

Industry Recognized Credentials

If you are training through an NCCER-accredited sponsor, you may be eligible for credentials from NCCER's Registry. The ID number for this module is CT57_0-17. Note that this module may have been used in other NCCER curricula and may apply to other level completions. Contact NCCER's Registry at 888.622.3720 or go to **www.nccer.org** for more information.

Contents

Figures

SECTION ONE

1.0.0 OPERATING ODORANT EQUIPMENT

Objective

Describe how to operate odorant equipment (CT57_0-17).

a. Describe the process and procedure for odorizing natural gas.
b. Identify the knowledge and skills associated with operating odorant equipment.

Performance Task

Operate odorant equipment (CT57_0-17).

- Identify potential abnormal operating conditions (AOCs) that may occur during performance of this CT, and know the appropriate actions to take in response to them.
- Utilize the appropriate personal protective equipment according to relevant company procedures.
- Describe how the odorizer functions.
- Compare the operation of the odorizer with its design limits.
- Measure the intensity level of the odorant.
- Determine if the injection of odorant is on a consistent basis.
- Complete appropriate documentation as required by operator's procedures.

Trade Terms

Abnormal operating conditions (AOCs): Conditions within a pipeline system, identified by the operator, that may indicate (1) a malfunction of a component or (2) a deviation from normal operations that may indicate a condition exceeding design limits or result in a hazard to persons, property, or the environment.

Bypass-type odorizer: An odorizer that diverts the main gas stream through a mechanical device to disperse the liquid odorant into the gas.

Electronic odorizer: An odorizer that uses an electronic processor to provide an odorant flow measurement signal.

Lower explosive limit (LEL): The minimum concentration of vapor in the air below which propagation of a flame does not occur in the presence of an ignition source.

Odorant: A liquid added to natural gas as a safety feature to make it smell like rotten eggs.

Odorizer: A piece of equipment used to inject an odorant into the gas stream.

Odorometer: An instrument that is used to measure the odorant level in a gas.

Pitot tube: A tube through which odorant is dispersed as the gas flow passes the opening.

Wick-type odorizer: A type of odorization equipment that uses the flow of gas across a wick or pitot tube to pick up liquid odorant for dispersion into the gas stream.

The typical natural gas customer would not be able to detect a gas leak were it not for the fact that an odor-containing chemical is routinely added to the otherwise odorless gas. The addition of this chemical results in an odor that can be detected in concentrations as small as 0.1 parts per billion, making even small leaks readily detectable by someone with an average sense of smell.

CAUTION

An individual's actions during the performance of a covered task could cause a safety device to activate. The activation represents a deviation from normal operations, resulting in an abnormal operating condition (AOC).

1.1.0 Odorizing Natural Gas

Natural gas is odorless. For safety reasons, federal regulation 49 *CFR* 192.625 requires that a person with a normal sense of smell must be able to easily detect natural gas when the gas concentration is only 20 percent of its **lower explosive limit (LEL)**. An **odorant**, usually a chemical called *mercaptan*, is added to the natural gas stream in distribution lines to provide the odor that makes the gas detectable. Very small amounts of mercaptan are able to impart a distinctive, readily recognizable odor to a large volume of natural gas.

The LEL of natural gas is reached when the gas makes up about 5 percent of the gas/air mixture. A small amount of mercaptan (or other odorant) is regularly injected into the natural gas stream so that the ratio of odorant to natural gas creates an odor that will be detectable when the gas concentration is one-fifth of the LEL, or only 1 percent of natural gas present in the surrounding air.

Generally, local distribution companies are responsible for adding the odorant, which gives the gas a distinctive rotten-egg smell to aid in leak detection by the customer. The amount of odorant to be injected can vary from place to place, even within the same system, depending on operating conditions.

Personnel responsible for injecting odorant into a natural-gas stream should read and follow the safety data sheet (SDS) for the odorant being used. In addition, employees must follow established company procedures for injecting odorant, and must wear the appropriate PPE.

1.1.1 Operating Odorant Equipment

There are three types of **odorizer** injection equipment that may be used to inject the odorant (*Figure 1*). A **bypass-type odorizer** diverts the main gas stream through a mechanical device to disperse the liquid odorant into the gas. A **wick-type odorizer** uses the flow of gas across a wick or **pitot tube** to pick up liquid odorant for dispersion in the gas stream. An **electronic odorizer** uses an onboard microprocessor to provide a flow measurement signal and generate data that is stored in an audit trail.

CAUTION

To prevent false readings, odorant units, hoses, regulators and other accessories should be checked and smelled before making a test. Smelly hoses should be replaced, and any other accessories that smell should be replaced or cleaned to remove the smell.

1.1.2 Procedure

The procedure for operating odorant equipment involves the following steps:

Step 1 Identify potential abnormal operating conditions (AOCs) that may occur, and know the appropriate actions to take in response to them.

Step 2 Utilize the appropriate personal protective equipment according to relevant company procedures.

Step 3 Describe how the odorizer functions.

Step 4 Compare the operation of the odorizer with its design limits.

Step 5 Measure the intensity level of the odorant.

Step 6 Complete appropriate documentation as required by operator's procedures.

Figure 1 Odorant injection station.

1.2.0 Prerequisites, Competencies, and Abnormal Operating Conditions

Personnel assigned to operate odorant equipment must have a working understanding of company safety procedures, company policies for the proper use of PPE, and company policies and procedures for responding to emergencies. They must also have a basic understanding of the gas

History of Formal Odorization

Adding odorant in order to detect odorless gases goes back to 1880 when a German scientist named Von Quaglio added ethyl mercaptan to water gas, a fuel made up of carbon dioxide and hydrogen, which was used for residential heating and lighting. Mercaptan itself is harmless and is even part of the human metabolism. However, it has a pungent odor, which some have described as the stench of rotting cabbages, due to the sulfur found in the compound.

Germany began formally odorizing natural gas in 1918, and other countries soon followed. Odorization guidelines first appeared in the American Society of Mechanical Engineers (ASME) guidelines for natural gas pipelines in 1951. By the 1970s, odorization of natural gas was part of the federal code of regulations.

pipeline system operation and flow characteristics. While specific procedures may vary from company to company, a basic set of knowledge and skills is required to perform the task of operating odorant equipment.

1.2.1 Prerequisites and Competencies

The personnel operating odorant equipment must be able to do the following:

- Describe how the particular odorizer stationed at the local facility functions.
- Ensure that the odorant being used is the proper odorant for the intended service.
- Determine the safety aspects of the particular odorant as detailed in the safety data sheet (SDS).
- Understand the operational characteristics of the odorizer, and be able to compare the day-to-day operation of the unit to its design limits in order to verify that it is functioning properly.
- Determine the amount of odorant to use and the rate of injection.
- Ensure that the odorant is being injected in a smooth and consistent manner.
- Calculate the ratio between odorant usage and the volume of gas being transported in order to ensure that the odorant level remains within $^1/_5$ (20 percent) of the LEL of the gas being transported.
- Determine if the routine injection of odorant by the odorizer is consistent over time, and that the resultant odor intensity levels do not vary widely.
- Routinely monitor the odorant level in the gas being transported (CT58_0-17).
- Know how to prepare the appropriate records documenting the injection of odorant.

1.2.2 Abnormal Operating Conditions

Personnel monitoring odorant levels must also be able to identify potential **abnormal operating conditions (AOCs)** and be able to take the appropriate actions. If an unexpected gas release occurs, first evaluate the situation and make the appropriate emergency notifications per company policy. Be prepared to work with the emergency response team if appropriate.

Additional Resources

It is assumed that you will learn to perform the purging procedure under the direct supervision of your crew leader, foreman, or instructor. For independent study, refer to the recommended Additional Resources at the back of this module.

1.0.0 Section Review

1. A person with a normal sense of smell must be able to detect the presence of natural gas when the level is 20 percent of its _____.
   a. maximum detectable level
   b. lower explosive limit (LEL)
   c. upper explosive limit
   d. minimum injection level

2. Personnel must be able to determine the safety aspects of the odorant used as specified in _____.
   a. the odorant safety handbook
   b. the odorizer specifications
   c. company policy
   d. the safety data sheet

SUMMARY

An odorant is added to natural gas in order to make it detectable when the amount of gas in the air is 5 percent of the LEL of natural gas. Local distribution companies are usually responsible for odorant injection. Three types of odorizers are used to inject the odorant into the gas stream. Bypass-type odorizers divert the gas stream through the instrument to disperse the odorant. Wick-type odorizers inject odorant into the stream through a wick or pitot tube. Electronic odorizers use a microprocessor to deliver an injection signal.

Personnel must know company safety, PPE, and emergency procedures. They must understand the pipeline system operations and flow characteristics.

Personnel must be able to describe how their odorizer functions, understand the operational characteristics of the equipment, and compare the day-to-day operation to the design limits in order to ensure it functions properly. They must know the proper odorant to use, the safety aspects of it, the amount to use, and the injection rate. Odorant should be injected in a smooth and consistent manner so that the ratio of usage to gas volume ensures a level within 20 percent of the LEL. The injection of odorant should be consistent over time and the odor intensity level should not vary widely.

If an unexpected gas leak occurs, personnel must evaluate the situation and make proper notifications.

Review Questions

1. Mercaptan is added to natural gas in order to _____.
   a. enhance the odor of the gas
   b. raise the odor intensity level of the gas
   c. moderate the intensity level of the gas
   d. provide an odor to the gas

2. One advantage to using an electronic odorizer is that it _____.
   a. is more efficient than the other types of odorizers
   b. provides an audit trail of the odorant injection
   c. uses less odorant than the other types of odorizers
   d. is easier to operate and maintain

3. Personnel must ensure that the odorant being used _____.
   a. has the proper odor intensity level
   b. has no safety issues as specified on the SDS
   c. is the proper odorant to use for the intended service
   d. can be injected at the proper rate

Trade Terms Quiz

Fill in the blank with the correct term that you learned from your study of this module.

1. A liquid added to give a smell to natural gas is called a(n) _________.
2. An odorizer that uses an electronic processor to provide an odorant flow measurement signal is called a(n) _________.
3. Conditions within a pipeline system, identified by the operator, that may indicate (1) a malfunction of a component or (2) a deviation from normal operations that may indicate a condition exceeding design limits or result in a hazard to persons, property, or the environment are known as _________.
4. A tube through which odorant is dispersed into the gas flow is called a(n) _________.
5. The minimum concentration of vapor in the air below which propagation of a flame does not occur in the presence of an ignition source is called the _________.
6. A piece of equipment used to inject an odorant into the gas stream is called a(n) _________.
7. A type of odorizer that uses a flow of gas across a pitot tube to disperse the odorant is called a(n) _________.
8. A type of odorization equipment that diverts the main gas through a mechanical device to disperse the odorant is called a(n) _________.

Trade Terms

Abnormal operating conditions (AOCs)
Bypass-type odorizer
Electronic odorizer
Lower explosive limit (LEL)
Odorant
Odorizer
Pitot tube
Wick-type odorizer

Trade Terms Introduced in This Module

Abnormal operating conditions (AOCs): Conditions within a pipeline system, identified by the operator, that may indicate (1) a malfunction of a component or (2) a deviation from normal operations that may indicate a condition exceeding design limits or result in a hazard to persons, property, or the environment.

Bypass-type odorizer: An odorizer that diverts the main gas stream through a mechanical device to disperse the liquid odorant into the gas.

Electronic odorizer: An odorizer that uses an electronic processor to provide an odorant flow measurement signal.

Lower explosive limit (LEL): The minimum concentration of vapor in the air below which propagation of a flame does not occur in the presence of an ignition source.

Odorant: A liquid added to natural gas as a safety feature to make it smell like rotten eggs.

Odorizer: A piece of equipment used to inject an odorant into the gas stream.

Pitot tube: A tube through which odorant is dispersed as the gas flow passes the opening.

Wick-type odorizer: A type of odorization equipment that uses the flow of gas across a wick or pitot tube to pick up liquid odorant for dispersion into the gas stream.

Additional Resources

The following recommended resources can provide additional helpful information related to the requirements for this covered task:

API Recommended Practice 1161, Third Edition, © 2014 American Petroleum Institute. API Publishing Services, 1220 L Street, NW, Washington, DC 20005. Distributed by Thompson Reuters (Scientific) LLC, **www.techstreet.com**.

NCCER Module AOCCC-17, *Abnormal Operating Conditions Control Center*

NCCER Module AOCFG-17, *Abnormal Operating Conditions Field & Gas*

NCCER Module CT58_0-17, *Monitor Odorant Levels*

NCCER Module 62105-14, *Introduction to Pipeline Documents*

NCCER Module 63104-02, *Piping and Mechanical Blueprint Reading*

NCCER Module 67102-15, *Basic Pipeline Pneumatics and Equipment*

NCCER Module 67103-15, *Pipeline Communications*

Figure Credits

arhendrix/Shutterstock.com, Module Opener
Photo by GPL Odorizers, Figure 1

Section Review Answer Key

Answer	Section Reference	Objective
Section One		
1. b	1.1.0	1a
2. d	1.2.0	1b

NCCER CURRICULA — USER UPDATE

NCCER makes every effort to keep its textbooks up-to-date and free of technical errors. We appreciate your help in this process. If you find an error, a typographical mistake, or an inaccuracy in NCCER's curricula, please fill out this form (or a photocopy), or complete the online form at **www.nccer.org/olf**. Be sure to include the exact module ID number, page number, a detailed description, and your recommended correction. Your input will be brought to the attention of the Authoring Team. Thank you for your assistance.

Instructors – If you have an idea for improving this textbook, or have found that additional materials were necessary to teach this module effectively, please let us know so that we may present your suggestions to the Authoring Team.

NCCER Product Development and Revision

13614 Progress Blvd., Alachua, FL 32615

Email: curriculum@nccer.org
Online: www.nccer.org/olf

❑ Trainee Guide ❑ Lesson Plans ❑ Exam ❑ PowerPoints Other ____________

Craft / Level: Copyright Date:

Module ID Number / Title:

Section Number(s):

Description:

Recommended Correction:

Your Name:

Address:

Email: Phone:

CT58_0-17

Monitor Odorant Level

Overview

Pure natural gas is not detectable by smell so an odorant is added making the gas detectable in the event of a leak. The odorant intensity level is routinely monitored to ensure that it remains within limits. This module describes the procedures for monitoring odorant levels.

Trainees with successful module completions may be eligible for credentialing through the NCCER Registry. To learn more, go to **www.nccer.org** or contact us at 1.888.622.3720. Our website has information on the latest product releases and training, as well as online versions of our *Cornerstone* magazine and Pearson's product catalog.

Your feedback is welcome. You may email your comments to **curriculum@nccer.org**, send general comments and inquiries to **info@nccer.org**, or fill in the User Update form at the back of this module.

This information is general in nature and intended for training purposes only. Actual performance of activities described in this manual requires compliance with all applicable operating, service, maintenance, and safety procedures under the direction of qualified personnel. References in this manual to patented or proprietary devices do not constitute a recommendation of their use.

CT58_0-17
Monitor Odorant Level

Objective

When you have completed this module, you will be able to do the following:

1. Describe how to monitor odorant levels (CT58_0-17).
   a. Describe the process and procedure for odorizing natural gas.
   b. Identify the knowledge and skills associated with monitoring odorant levels.

Performance Task

Under the supervision of the instructor, you should be able to do the following:

1. Monitor odorant levels (CT58_0-17).
   - Identify potential abnormal operating conditions that may occur during performance of this CT, and know the appropriate actions to take in response to them.
   - Utilize the appropriate personal protective equipment according to relevant company procedures.
   - Turn on, calibrate, and operate odorant level measuring instrument to measure odorant intensity level per company procedures and manufacturer's recommendations.
   - Determine the amount of odorant used and calculate the ratio between usage and the volume of gas transported.
   - Complete appropriate documentation as required by operator's procedures.

Trade Terms

Abnormal operating conditions (AOCs)
Bypass-type odorizer
Electronic odorizer
Lower explosive limit (LEL)
Odorant
Odorizer
Odorometer
Pitot tube
Wick-type odorizer

Industry Recognized Credentials

If you are training through an NCCER-accredited sponsor, you may be eligible for credentials from NCCER's Registry. The ID number for this module is CT58_0-17. Note that this module may have been used in other NCCER curricula and may apply to other level completions. Contact NCCER's Registry at 888.622.3720 or go to **www.nccer.org** for more information.

Contents

Figures

SECTION ONE

1.0.0 MONITORING ODORANT LEVELS

Objective

Describe how to monitor odorant levels (CT58_0-17).

a. Describe the process and procedure for odorizing natural gas.
b. Identify the knowledge and skills associated with monitoring odorant levels.

Performance Task

Monitor odorant levels (CT58_0-17).

- Identify potential abnormal operating conditions that may occur during performance of this CT, and know the appropriate actions to take in response to them.
- Utilize the appropriate personal protective equipment according to relevant company procedures.
- Turn on, calibrate, and operate odorant level measuring instrument to measure odorant intensity level per company procedures and manufacturer's recommendations.
- Determine the amount of odorant used and calculate the ratio between usage and the volume of gas transported.
- Complete appropriate documentation as required by operator's procedures.

Trade Terms

Abnormal operating conditions (AOCs): Conditions within a pipeline system, identified by the operator, that may indicate (1) a malfunction of a component or (2) a deviation from normal operations that may indicate a condition exceeding design limits or result in a hazard to persons, property, or the environment.

Bypass-type odorizer: An odorizer that diverts the main gas stream through a mechanical device to disperse the liquid odorant into the gas.

Electronic odorizer: An odorizer that uses an electronic processor to provide an odorant flow measurement signal.

Lower explosive limit (LEL): The minimum concentration of vapor in the air below which propagation of a flame does not occur in the presence of an ignition source.

Odorant: A liquid added to natural gas as a safety feature to make it smell like rotten eggs.

Odorizer: A piece of equipment used to inject an odorant into the gas stream.

Odorometer: An instrument that is used to measure the odorant level in a gas.

Pitot tube: A tube through which odorant is dispersed as the gas flow passes the opening.

Wick-type odorizer: A type of odorization equipment that uses the flow of gas across a wick or pitot tube to pick up liquid odorant for dispersion into the gas stream.

The typical natural gas customer would not be able to detect a gas leak were it not for the fact that an odor-containing chemical is routinely added to the otherwise odorless gas. The addition of this chemical produces an odor that can be detected in concentrations as small as 0.1 parts per billion, making even small leaks readily detectable by someone with an average sense of smell.

CAUTION

An individual's actions during the performance of a covered task could cause a safety device to activate. The activation represents a deviation from normal operations, resulting in an abnormal operating condition (AOC).

1.1.0 Odorizing Natural Gas

Natural gas is odorless. For safety reasons, federal regulation 49 *CFR* 192.625 requires that a person with a normal sense of smell must be able to easily detect natural gas when the gas concentration is only 20 percent of its **lower explosive limit (LEL)**. An **odorant**, usually a chemical called *mercaptan,* is added to the natural gas stream in distribution lines to provide the odor that makes the gas detectable. Very small amounts of mercaptan are able to impart a distinctive, readily recognizable odor to a large volume of natural gas. Generally, local distribution companies are responsible for adding the odorant, which gives the gas a distinctive rotten-egg smell to aid in leak detection by the customer.

The LEL of natural gas is reached when the gas makes up about 5 percent of the gas/air mixture. A small amount of mercaptan (or other odorant) is regularly injected into the natural gas stream so that the ratio of odorant to natural gas creates an odor that will be detectable when the gas concentration is one-fifth of the LEL, or only 1 percent of natural gas present in the surrounding air. Personnel responsible for injecting or monitoring odorant must wear the appropriate PPE.

A **bypass-type odorizer** diverts the main gas stream through a mechanical device to disperse the liquid odorant into the gas (an**odorizer** is a piece of equipment used to inject an odorant into the gas stream). A **wick-type odorizer** uses the flow of gas across a wick or **pitot tube** to pick up liquid odorant for dispersion in the gas stream. An **electronic odorizer** uses an onboard microprocessor to provide a flow measurement signal and generate data that is stored in an audit trail. The amount of odorant to be injected can vary from place to place, even within the same system, depending on operating conditions. An odorant injection station is shown in *Figure 1*.

Adding odorant in order to detect odorless gasses goes back to 1880 when a German scientist named Von Quaglio added ethyl mercaptan to water gas, a fuel made up of carbon dioxide and hydrogen, which was used for residential heating and lighting. Mercaptan itself is harmless and is even part of human metabolism; however, it has a pungent odor which some have described as the stench of rotting cabbages due to the sulfur found in the compound.

Germany began formally odorizing natural gas in 1918, and other countries soon followed. Odorization guidelines first appeared in the American Society of Mechanical Engineers (ASME) guidelines for natural gas pipelines in 1951. By the 1970s, odorization of natural gas was part of the federal code of regulations.

Figure 1 Odorant injection station.

1.1.1 Monitoring Odorant Levels

Once odorant has been injected into the gas, personnel must monitor it to ensure that proper ratios are maintained. Various factors can dilute the odorant to below mandated levels. For example, the steel in a new pipeline can absorb and react with the odorant in the gas in a process called *odor fade*. New lines have to be pre-odorized or "pickled" in order to maintain acceptable odorant levels at the distribution end of the pipeline.

The type of equipment used for monitoring odorant depends on company policy. Generally, personnel use an **odorometer**, which allows personnel to determine whether the ratio of the odorant usage to the volume of gas is within acceptable limits. Odorometers must be properly calibrated to produce valid measurements, and personnel must follow the manufacturer's recommendations for proper use of the meters.

CAUTION

To prevent false readings, odorant units, hoses, regulators and other accessories should be checked and smelled before making a test. Smelly hoses should be replaced, and any other accessories that smell should be replaced or cleaned to remove the smell.

1.1.2 Procedure

The procedure for monitoring odorant levels involves the following steps:

Step 1 Identify potential abnormal operating conditions that may occur during performance of this CT, and know the appropriate actions to take in response to them.

Step 2 Utilize the appropriate personal protective equipment according to relevant company procedures.

Step 3 Turn on, calibrate, and operate odorant level measuring instrument to measure odorant intensity level per company procedures and manufacturer's recommendations. Set up the instrument to take a valid reading of the odorant intensity level. Ensure sufficient tests are performed at appropriate locations. Ensure the instrument is being used in an appropriate manner to produce valid results.

Step 4 Determine the amount of odorant used and calculate the ratio between usage and the volume of gas transported. Determine if the amount of odorant injected per volume of natural gas is within a specific ratio.

Step 5 Complete appropriate documentation as required by operator's procedures.

1.2.0 Prerequisites, Competencies, and Abnormal Operating Conditions

Personnel assigned to monitor odorant levels must have a working understanding of company safety procedures, company policies for the proper use of personal protective equipment, and company policies and procedures for responding to emergencies. They must also have a basic understanding of the gas pipeline system operation and flow characteristics.

1.2.1 Prerequisites and Competencies

While specific procedures may vary from company to company, personnel performing odorant monitoring must typically be able to do the following:

- Select the most desirable point or points at which to measure the odorant intensity level. Typically, this is the point of lowest flow.
- Understand how to turn on, calibrate, and operate the odorant level measuring instrument to measure odorant intensity levels per company procedures and manufacturer's recommendations. The measuring instruments are typically field instruments; however, laboratory instruments to measure odorant intensity levels may also be used.
- Know how to set up the instrument so that it takes a valid reading of the odorant intensity level. Ensure sufficient tests are performed at appropriate locations, and that the instrument is being used in an appropriate manner to produce valid results.
- Be able to determine the amount of odorant used and the rate at which it is being injected, and to calculate the ratio between odorant usage and the volume of gas being transported.
- Understand how to determine if the amount of odorant injected per volume of natural gas is within a specific ratio as defined by regulations and company policy.
- Know how to document the results of the tests using the appropriate records.

1.2.2 Abnormal Operating Conditions

Personnel monitoring odorant levels must also be able to identify potential abnormal operating conditions (AOCs) and be able to take the appropriate actions. If an unexpected gas release occurs, first evaluate the situation and make the appropriate emergency notifications per company policy. Be prepared to work with the emergency response team if appropriate.

Additional Resources

It is assumed that you will learn to perform the purging procedure under the direct supervision of your crew leader, foreman, or instructor. For independent study, refer to the recommended Additional Resources at the back of this module.

1.0.0 Section Review

1. A person with a normal sense of smell must be able to detect the presence of natural gas when the level is 20 percent of its _____.
   a. maximum detectable level
   b. lower explosive limit (LEL)
   c. upper explosive limit
   d. minimum injection level

2. Once an odorant is injected into a gas, personnel must monitor it in order to ensure that _____.
   a. the gas/air ratio remains below 5 percent
   b. it does not leak into the atmosphere
   c. the proper ratios are maintained
   d. the odorant doesn't exceed 5 percent of the odorant/gas mixture

SUMMARY

An odorant is added to natural gas in order to make it detectable when the amount of gas in the air is 5 percent of the LEL of natural gas. Local distribution companies are usually responsible for odorant injection. Three types of odorizers are used to inject the odorant into the gas stream. Bypass-type odorizers divert the gas stream through the instrument to disperse the odorant. Wick-type odorizers inject odorant into the stream through a wick or pitot tube. Electronic odorizers use a microprocessor to deliver an injection signal.

An odorometer is used to determine if the ratio of the odorant usage to the volume of gas is within acceptable limits. Odorometers must be calibrated to produce valid results, and personnel must follow manufacturer's recommendations for their use.

Personnel must know company safety, PPE, and emergency procedures. They must understand the pipeline system operations and flow characteristics. The most desirable location to monitor the odorant intensity level is at the point of lowest flow.

Personnel must know how to turn on, set up, calibrate, and operate the odorant monitoring equipment. They must be able to determine the amount of odorant used and the rate at which it is injected, and to calculate the ratio between odorant usage and the volume of gas being transported. They must determine if the amount of odorant injected per volume of natural gas is within a specific ratio.

If an unexpected gas leak occurs, personnel must know how to evaluate the situation and make proper notifications.

Review Questions

1. Mercaptan is added to natural gas in order to _____.
   a. enhance the odor of the gas
   b. raise the odor intensity level of the gas
   c. moderate the intensity level of the gas
   d. provide an odor to the gas

2. The most desirable location for measuring the odorant intensity level is _____.
   a. where it is being injected
   b. at the point of lowest flow
   c. where the flow is most easily accessible
   d. at the point of highest flow

3. Personnel assigned to monitor odorant levels must have a basic understanding of the gas pipeline system operations and _____.
   a. be familiar with flow characteristics
   b. be able to monitor usage rates
   c. know how to use odorizer injection equipment
   d. be able to detect the presence of gas by using their sense of smell

Trade Terms Quiz

Fill in the blank with the correct term that you learned from your study of this module.

1. A liquid added to give a smell to natural gas is called a(n) ________.
2. An odorizer that uses an electronic processor to provide an odorant flow measurement signal is called a(n) ________.
3. Conditions within a pipeline system, identified by the operator, that may indicate (1) a malfunction of a component or (2) a deviation from normal operations that may indicate a condition exceeding design limits or result in a hazard to persons, property, or the environment are known as ________.
4. A tube through which odorant is dispersed into the gas flow is called a(n) ________.
5. The minimum concentration of vapor in the air below which propagation of a flame does not occur in the presence of an ignition source is called the ________.
6. A piece of equipment used to inject an odorant into the gas stream is called a(n) ________.
7. A type of odorizer that uses a flow of gas across a pitot tube to disperse the odorant is called a(n) ________.
8. An instrument used to measure the odorant level in a gas is called a(n) ________.
9. A type of odorization equipment that diverts the main gas through a mechanical device to disperse the odorant is called a(n) ________.

Trade Terms

Abnormal operating conditions (AOCs)
Bypass-type odorizer
Electronic odorizer
Lower explosive limit (LEL)
Odorant
Odorizer
Odorometer
Pitot tube
Wick-type odorizer

Trade Terms Introduced in This Module

Abnormal operating conditions (AOCs): Conditions within a pipeline system, identified by the operator, that may indicate (1) a malfunction of a component or (2) a deviation from normal operations that may indicate a condition exceeding design limits or result in a hazard to persons, property, or the environment.

Bypass-type odorizer: An odorizer that diverts the main gas stream through a mechanical device to disperse the liquid odorant into the gas.

Electronic odorizer: An odorizer that uses an electronic processor to provide an odorant flow measurement signal.

Lower explosive limit (LEL): The minimum concentration of vapor in the air below which propagation of a flame does not occur in the presence of an ignition source.

Odorant: A liquid added to natural gas as a safety feature to make it smell like rotten eggs.

Odorizer: A piece of equipment used to inject an odorant into the gas stream.

Odorometer: An instrument that is used to measure the odorant level in a gas.

Pitot tube: A tube through which odorant is dispersed as the gas flow passes the opening.

Wick-type odorizer: A type of odorization equipment that uses the flow of gas across a wick or pitot tube to pick up liquid odorant for dispersion into the gas stream.

Additional Resources

The following recommended resources can provide additional helpful information related to the requirements for this covered task:

API Recommended Practice 1161, Third Edition, © 2014 American Petroleum Institute. API Publishing Services, 1220 L Street, NW, Washington, DC 20005. Distributed by Thompson Reuters (Scientific) LLC, **www.techstreet.com**.

NCCER Module AOCCC-17, *Abnormal Operating Conditions Control Center*

NCCER Module AOCFG-17, *Abnormal Operating Conditions Field & Gas*

NCCER Module CT57_0-17, *Operate Odorant Equipment*

NCCER Module 62105-14, *Introduction toPipeline Documentations*

NCCER Module 63104-02, *Piping and Mechanical Blueprint Reading*

NCCER Module 67102-15, *Basic Pipeline Pneumatics and Equipment*

NCCER Module 67103-15, *Pipeline Communications*

Figure Credits

arhendrix/Shutterstock.com, Module Opener

Photo by GPL Odorizers, Figure 1

Section Review Answer Key

Answer	Section Reference	Objective
Section One		
1. b	1.1.0	1a
2. c	1.1.1	1b

NCCER CURRICULA — USER UPDATE

NCCER makes every effort to keep its textbooks up-to-date and free of technical errors. We appreciate your help in this process. If you find an error, a typographical mistake, or an inaccuracy in NCCER's curricula, please fill out this form (or a photocopy), or complete the online form at **www.nccer.org/olf**. Be sure to include the exact module ID number, page number, a detailed description, and your recommended correction. Your input will be brought to the attention of the Authoring Team. Thank you for your assistance.

Instructors – If you have an idea for improving this textbook, or have found that additional materials were necessary to teach this module effectively, please let us know so that we may present your suggestions to the Authoring Team.

NCCER Product Development and Revision

13614 Progress Blvd., Alachua, FL 32615

Email: curriculum@nccer.org
Online: www.nccer.org/olf

❑ Trainee Guide ❑ Lesson Plans ❑ Exam ❑ PowerPoints Other ____________________

Craft / Level: Copyright Date:

Module ID Number / Title:

Section Number(s):

Description:

Recommended Correction:

Your Name:

Address:

Email: Phone:

CT63_1-17

Start-Up of a Liquid Pipeline (Field)

Overview

A newly installed pipeline, or one that has been shut down for an emergency or for maintenance, must go through a prescribed start-up procedure. This module describes the procedure for starting up a liquid pipeline.

Trainees with successful module completions may be eligible for credentialing through the NCCER Registry. To learn more, go to **www.nccer.org** or contact us at 1.888.622.3720. Our website has information on the latest product releases and training, as well as online versions of our *Cornerstone* magazine and Pearson's product catalog.

Your feedback is welcome. You may email your comments to **curriculum@nccer.org**, send general comments and inquiries to **info@nccer.org**, or fill in the User Update form at the back of this module.

This information is general in nature and intended for training purposes only. Actual performance of activities described in this manual requires compliance with all applicable operating, service, maintenance, and safety procedures under the direction of qualified personnel. References in this manual to patented or proprietary devices do not constitute a recommendation of their use.

CT63_1-17
START-UP OF A LIQUID PIPELINE (FIELD)

Objective

When you have completed this module, you will be able to do the following:

1. Describe how to start up a liquid pipeline (CT63_1-17).
   a. Describe how to start up a liquid pipeline.
   b. Identify the prerequisites, competencies, and abnormal operating conditions associated with starting up a liquid pipeline.

Performance Task

Under the supervision of the instructor, you should be able to do the following:

1. Start up a liquid pipeline (CT63_1-17).
   - Identify potential abnormal operating conditions that may occur during performance of this CT, and know the appropriate actions to take in response to them.
   - Utilize the appropriate personal protective equipment according to relevant company procedures.
   - Verify that flows, pressures, and product characteristics are within normal range.
   - Ensure that upstream and downstream stations are prepared to bring the pipeline up before beginning the start-up procedure.
   - Confirm that the flow path is correct.
   - Maintain orderly start-up of the line by activating units beginning at the origination point and proceeding downstream to the delivery point. (Activate units at downstream stations upon arrival of a pressure rise from the upstream station.)
   - Complete appropriate documentation as required by operator's procedures.

Trade Terms

Abnormal operating conditions (AOCs)
Drag reducing agent (DRA)
Human machine interface (HMI)
Maximum operating pressure (MOP)
Procedure
Steady state
Supervisory Control and Data Acquisition (SCADA)

Industry Recognized Credentials

If you are training through an NCCER-accredited sponsor, you may be eligible for credentials from NCCER's Registry. The ID number for this module is CT63_1-17. Note that this module may have been used in other NCCER curricula and may apply to other level completions. Contact NCCER's Registry at 1.888.622.3720 or go to **www.nccer.org** for more information.

Contents

Figures

Section One

1.0.0 Starting Up a Liquid Pipeline

Objective

Describe how to start up a liquid pipeline (CT63_1-17).

a. Describe how to start up a liquid pipeline.
b. Identify the knowledge and skills associated with starting up a liquid pipeline.

Performance Task

Start up a liquid pipeline (CT63_1-17).

- Identify potential abnormal operating conditions that may occur during performance of this CT, and know the appropriate actions to take in response to them.
- Utilize the appropriate personal protective equipment according to relevant company procedures.
- Verify that flows, pressures, and product characteristics are within normal range.
- Ensure that upstream and downstream stations are prepared to bring the pipeline up before beginning the start-up procedure.
- Confirm that the flow path is correct.
- Maintain orderly start-up of the line by activating units beginning at the origination point and proceeding downstream to the delivery point. (Activate units at downstream stations upon arrival of a pressure rise from the upstream station.)
- Complete appropriate documentation as required by operator's procedures.

Trade Terms

Abnormal operating conditions (AOCs): Conditions within a pipeline system, identified by the operator, that may indicate (1) a malfunction of a component or (2) a deviation from normal operations that may indicate a condition exceeding design limits or result in a hazard to persons, property, or the environment.

Drag reducing agent (DRA): A chemical injected into a pipeline that reduces the internal friction by decreasing turbulence, allowing the product to flow more freely.

Human machine interface (HMI): A software application that presents information to the operator about the state of a process and accepts and implements an operator's control instructions. Typically, information is displayed in a graphic form.

Maximum operating pressure (MOP): The highest pressure at which a pipeline (or segment of a pipeline) may be normally operated.

Procedure: A set of step-by-step instructions for logically accomplishing a task.

Steady state: A condition on an active or flowing pipeline where pressure and flows are relatively constant over time, and are comparable to historic operational data for that segment.

Supervisory Control and Data Acquisition (SCADA): A computer-based system used remotely by a controller in a control room that displays information about a pipeline facility, and which may have the ability to send commands back to the facility.

A liquid pipeline start-up procedure begins with the configuration of the intended flow path in accordance with applicable operating procedures. The procedure ends when the control center assumes control of the operation.

1.1.0 Facility Start-Up Procedures

Starting up any segment of a pipeline is required by 49 *CFR* 195.402 to be done in a manner designed to ensure operation within limits prescribed in section 195.406. The law requires that special consideration be given to the hazardous liquids being transported, along with variations in altitude, pressure monitoring, and control devices. The operations controller is responsible for proper start-up.

1.1.1 Performing Normal Start-Up

Operations will have a set of guidelines that must be followed in order to start up the operation of a particular segment or pipeline, usually outlined in a company manual. These guidelines may be adjusted as conditions change. Basic responsibilities before and during start-up include the following:

- Ensuring that the correct product will be pumped by double checking the schedule
- Ensuring pump stations (*Figure 1*) are ready for restart
- Minimizing the time main-line units are in critical speed ranges

Figure 1 Main-line pumps with electric drivers.

- Ensuring that the maximum operating pressure (MOP) is not exceeded
- Logging all pertinent data
- Ensuring that appropriate notifications are made

CAUTION

An individual's actions during the performance of a covered task could cause a safety device to activate. The activation represents a deviation from normal operations, resulting in an abnormal operating condition (AOC).

1.1.2 Procedures for Normal Start-Up

Pipelines must be started in an appropriate sequence to ensure that a successful operation is achieved and that limits are maintained throughout the line; this configuration must be communicated with the control center early in the process, allowing start-up to be confirmed, and verifying its compliance with applicable operating procedures. The operations controller must maintain communications with or monitor all stations until the line is running at steady state. The following steps outline the general procedure for starting up a liquid pipeline:

NOTE

Procedures for starting up a pipeline will vary widely depending on the situation. Always follow company-specific procedures for performing start-up.

Step 1 Identify potential abnormal operating conditions (AOCs) that may occur, and know the appropriate actions to take in response to them.

Step 2 Utilize the appropriate personal protective equipment (PPE) according to relevant company procedures.

Step 3 Ensure that upstream and downstream stations are prepared for start-up. This includes confirming equipment availability.

Step 4 Confirm that all main-line and station valves (*Figure 2*) have been returned to normal position and that the flow path is correct.

Step 5 Start a suction booster pump at the source of supply.

Step 6 Start the main-line unit.

Step 7 Notify the next station downstream that the main-line unit is running.

Step 8 Start the main-line units at stations on the line that will immediately be affected.

Step 9 Adjust suction setpoints at specified stations downstream.

Step 10 When setpoints have been adjusted, notify the next station downstream that the main-line unit at that station is coming on-line.

Step 11 Ramp up the main-line unit to the desired flow rate while watching for the rise in suction pressures downstream. An electronic pressure transmitter is shown in *Figure 3*.

Figure 2 Pipeline valves.

Figure 3 Electronic pressure transmitter.

Step 12 When the increase in suction pressure at each consecutive station is observed, notify each consecutive station downstream that the main-line unit is coming on-line.

Step 13 If applicable, as soon as the throughput has stabilized in the line, notify the downstream stations that another main-line unit is being brought on-line.

Step 14 Notify affected downstream stations that as soon as suction is sufficient, their main-line units should be brought on-line.

Step 15 Monitor until flow has stabilized and the pipeline is in steady state operation.

Step 16 Start any required **drag reducing agent (DRA)** injections at this time. A DRA injection control valve is shown in *Figure 4*.

Step 17 Make certain that the computer logs, incident reports, event notifications, log book entries, and all other documentation have been completed.

1.1.3 Performing Start-Up Following Maintenance

There are times when the pipeline must be shut down for extended periods while maintenance work is completed. Whether the reason is an unexpected valve closure or a release, as soon as it is repaired and tested, the pipeline must be started. Start-up after maintenance is the same as a normal start-up, with the exception that the operations controller must make certain that all procedures are completed and all systems are ready for the start-up.

Figure 4 DRA injection control valve.

> **NOTE**
> If the pipeline has been shut down due to a possible release, it should not be restarted until all details relating to the release have been verified, corrected, and resolved.

1.1.4 Performing Start-Up Following Abnormal Operating Conditions

If the pipeline was shut down for an AOC, the control room operator must perform an investigation of the problem and try to determine exactly what went wrong. A technical team may have to physically survey the suspect area. After the problem is completely cleared and testing has been completed, all systems are set up again. Operations personnel follow the same procedure as for a normal start-up; only the documentation differs. Event reports might have to be prepared and filed if there was a significant cause.

> **NOTE**
> In emergency conditions such as a possible release, the pipeline operations controller has full responsibility and authority to make shutdown decisions and take appropriate actions. These actions are to be governed by the company release policy above all other considerations.

1.2.0 Prerequisites, Competencies, and Abnormal Operating Conditions

Personnel must possess a variety of skills in order to successfully start up a pipeline. The operation itself requires a number of steps, which the personnel conducting the start-up must be competent to perform. Personnel must also be aware of the particular abnormal operating conditions (AOCs) that may occur during the procedure and know the proper responses to them.

1.2.1 Prerequisites and Competencies

A pipeline start-up involves a large number of personnel, from those at the control center, to those at pump and valve stations, to those at the receiving facility. Communications between them can be made by telephone or radio, as well as through the **Supervisory Control and Data Acquisition (SCADA)** system or the **human machine interface (HMI)**.

Field personnel performing a pipeline start-up must be able to recognize and respond to visual or audible alarms generated by a SCADA or HMI system (*Figure 5*) that indicate an operating parameter has been exceeded. If field personnel are unable to respond and mitigate the situation, the control center may have to take action.

Field personnel must be able to communicate the start-up configuration to the control center. This includes providing verification that the local piping is aligned and the intended flow path is configured in accordance with applicable operating procedures, and that the pumps and other equipment are in a ready state. The control center should confirm that the remote flow path has been configured. This communication allows the control room and the local operators to confirm that the start-up procedure is ready to commence.

Figure 5 HMI display on a smartphone.

To initiate the start-up, field personnel start the pumps in accordance with written operating procedures, and then monitor pressure and flow rates. Once a steady state condition has been reached, field personnel must communicate with the control center to verify that the center has assumed control of the operation.

1.2.2 Abnormal Operating Conditions

Prior to carrying out pipeline start-up procedures, personnel must be able to identify potential AOCs and know the proper response to each, should they occur. The following are some potential AOCs, and the appropriate action personnel should take in response to them:

- *Loss of SCADA communications to control room or interrupted electrical service occurs* – Personnel should first ensure that the appropriate backup systems are activated. They must then initiate troubleshooting procedures, and take appropriate action to mitigate the situation.

> **NOTE**
> In emergency conditions such as a possible release, the pipeline operations controller has full responsibility and authority to make shutdown decisions and take appropriate actions. These actions are to be governed by the company release policy above all other considerations.

- *Activation of a safety device, such as a pressure relief valve or a remotely controlled shutdown device (RCSD) occurs* – Personnel should investigate the cause and take appropriate action to mitigate the situation and make proper notification to affected personnel. If this AOC occurs, personnel would receive a visual or audible indication through the HMI or another system.
- *An unexplained flow rate deviation, such as high flow, low flow, or no flow occurs* – Personnel should investigate the cause and take appropriate action to mitigate the situation and make proper notification to affected personnel. If this AOC occurs, personnel would receive a visual or audible indication through the HMI or another system.
- *An unexplained pressure deviation, such as an increase, decrease, or lack of pressure occurs* – Personnel should investigate the cause and take appropriate action to mitigate the situation and make proper notification to affected personnel. If this AOC occurs, personnel would receive a visual or audible indication through the HMI or another system.

- *An unintended status change, such as a change in unit status or valve position occurs* – Personnel should investigate the cause and take appropriate action to mitigate the situation and make proper notification to affected personnel. If this AOC occurs, personnel would receive a visual or audible indication through the HMI or another system.

Additional Resources

It is assumed that you will learn to perform these procedures under the direct supervision of your crew leader, foreman, or instructor. For independent study, refer to the recommended Additional Resources at the back of this module.

1.0.0 Section Review

1. The start-up procedure for a liquid pipeline ends when _____.
   a. the line pressure reaches 80 percent of MOP
   b. the control center assumes control
   c. all main-line units are operating normally
   d. an acceptable flow rate has been established

2. What must field personnel be able to communicate to the control center?
   a. The correct suction setpoints
   b. The number of main-line units that will be activated
   c. The start-up configuration
   d. The maximum operating pressure

SUMMARY

A liquid pipeline start-up is done in accordance with 49 *CFR* 195.402, which requires that special consideration be given to the transported hazardous liquids. Guidelines for start-up include ensuring the correct product is pumped, pumping stations are ready, minimizing the time main-line units are in critical speed range, ensuring the MOP is not exceeded, ensuring that all data is logged, and making all notifications.

Communication between start-up personnel is conducted by radio or telephone, as well as through the SCADA and HMI systems. Personnel should be able to recognize and respond to alarms generated by the SCADA or HMI. Field personnel align the valves and pipes, verify the flow path, and inform control that the pipeline is configured for start-up. Personnel start the pumps and monitor the pressure and flow rates. The control center assumes control when a steady state is reached.

If SCADA communication or electrical power is lost, personnel must activate backup systems and take action to mitigate the situation. If an AOC occurs, personnel must investigate the cause and take action to mitigate the situation. All required notifications must be made.

Review Questions

1. 49 *CFR* 195.402 requires that a start-up be done in a manner designed to ensure that _____.
   a. all main-line units are functioning at capacity
   b. a steady state flow is achieved once main-line units are activated
   c. the MOP is reached in a safe and efficient manner
   d. operations are within limits as specified in 49*CFR*195.406

2. A basic responsibility before and during start-up includes minimizing the time _____.
   a. main-line units are in critical speed ranges
   b. main-line units are brought on-line
   c. the system is at maximum operating pressure
   d. the suction pressure is in the critical range

3. One way of ensuring that the correct product will be pumped can be accomplished by _____.
   a. running an API line velocity test
   b. verifying with the One Call system
   c. double checking the schedule
   d. contacting the shipper

4. Prior to start-up, the control center must confirm that _____.
   a. the suction setpoints are correct
   b. the remote flow path has been configured
   c. a steady state condition exists
   d. all alarm systems have been checked and reset

5. If the loss of SCADA communication occurs, which of the following must be done first?
   a. Notify the control center
   b. Take appropriate action to mitigate the situation
   c. Initiate troubleshooting procedures
   d. Ensure the appropriate backup systems are activated

Trade Terms Quiz

Fill in the blank with the correct term that you learned from your study of this module.

1. A chemical injected into the pipeline to reduce internal friction is called a(n) _________.
2. The highest pressure at which a pipeline can be operated is called the _________.
3. A system that displays information about a pipeline facility is called _________.
4. A software application that presents information about the state of a process is called _________.
5. A set of step-by-step instructions for accomplishing a task is called a(n) _________.
6. Conditions within a pipeline that may indicate a malfunction of a component or a condition exceeding design limits are called _________.
7. A condition on an active pipeline where pressure and flow are relatively constant over time is known as _________.

Trade Terms

Abnormal operating conditions (AOCs)
Drag reducing agent (DRA)
Human machine interface (HMI)
Maximum operating pressure (MOP)
Procedure
Steady state
Supervisory Control and Data Acquisition (SCADA)

Trade Terms Introduced in This Module

Abnormal operating conditions (AOCs): Conditions within a pipeline system, identified by the operator, that may indicate (1) a malfunction of a component or (2) a deviation from normal operations that may indicate a condition exceeding design limits or result in a hazard to persons, property, or the environment.

Drag reducing agent (DRA): A chemical injected into a pipeline that reduces the internal friction by decreasing turbulence, allowing the product to flow more freely.

Human machine interface (HMI): A software application that presents information to the operator about the state of a process and accepts and implements an operator's control instructions. Typically, information is displayed in a graphic form.

Maximum operating pressure (MOP): The highest pressure at which a pipeline (or segment of a pipeline) may be normally operated.

Procedure: A set of step-by-step instructions for logically accomplishing a task.

Steady state: A condition on an active or flowing pipeline where pressure and flows are relatively constant over time, and are comparable to historic operational data for that segment.

Supervisory Control and Data Acquisition (SCADA): A computer-based system used remotely by a controller in a control room that displays information about a pipeline facility, and which may have the ability to send commands back to the facility.

Additional Resources

The following recommended resources can provide additional helpful information related to the requirements for this covered task:

API Recommended Practice 1161, Third Edition, © 2014 American Petroleum Institute. API Publishing Services, 1220 L Street, NW, Washington, DC 20005. Distributed by Thompson Reuters (Scientific) LLC, **www.techstreet.com**.

NCCER Module AOCCC-17, *Abnormal Operating Conditions Control Center*

NCCER Module AOCFG-17, *Abnormal Operating Conditions Field and Gas*

NCCER Module 60102-15, *Basic Pipeline Hydraulics and Equipment*

NCCER Module 60103-15, *Pipeline Communications*

NCCER Module 60106-15, *Monitoring Pipeline Operations*

NCCER Module 62105-02, *Pipeline Documentation*

NCCER Module 64106-02, *Pipeline Operations*

NCCER Module 64307-02, *SCADA*

NCCER Module CT63_2-17, *Shutdown of a Liquid Pipeline (Field)*

NCCER Module CT63_3-17, *Monitor Pressures, Flows, Communications, and Line Integrity and Maintain Them Within Allowable Limits on a Liquid Pipeline System (Field)*

NCCER Module CT63_4-17, *Locally Operate Valves on a Liquid Pipeline System*

49 *CFR* 192, **www.ecfr.gov**.

49 *CFR* 195, **www.ecfr.gov**.

Figure Credits

© iStockphoto.com/Pro-syanov, Module Opener

Benson County Farmers Press, Figure 1

© iStockphoto.com/westphalia, Figure 2

Tim Mills, Figures 3 and 4

ABB, Figure 5

Section Review Answer Key

Answer	Section Reference	Objective
Section One		
1. b	1.0.0	1a
2. c	1.2.1	1b

NCCER CURRICULA — USER UPDATE

NCCER makes every effort to keep its textbooks up-to-date and free of technical errors. We appreciate your help in this process. If you find an error, a typographical mistake, or an inaccuracy in NCCER's curricula, please fill out this form (or a photocopy), or complete the online form at **www.nccer.org/olf**. Be sure to include the exact module ID number, page number, a detailed description, and your recommended correction. Your input will be brought to the attention of the Authoring Team. Thank you for your assistance.

Instructors – If you have an idea for improving this textbook, or have found that additional materials were necessary to teach this module effectively, please let us know so that we may present your suggestions to the Authoring Team.

NCCER Product Development and Revision

13614 Progress Blvd., Alachua, FL 32615

Email: curriculum@nccer.org
Online: www.nccer.org/olf

❑ Trainee Guide ❑ Lesson Plans ❑ Exam ❑ PowerPoints Other ______________

Craft / Level: Copyright Date:

Module ID Number / Title:

Section Number(s):

Description:

Recommended Correction:

Your Name:

Address:

Email: Phone:

CT63_2-17

Shutdown of a Liquid Pipeline (Field)

Overview

Pipelines may need to be shut down for a number of reasons, including for scheduled maintenance or in response to an emergency. There are prescribed shutdown procedures for these situations. This module describes the procedures for shutting down a liquid pipeline.

Trainees with successful module completions may be eligible for credentialing through the NCCER Registry. To learn more, go to **www.nccer.org** or contact us at 1.888.622.3720. Our website has information on the latest product releases and training, as well as online versions of our *Cornerstone* magazine and Pearson's product catalog.

Your feedback is welcome. You may email your comments to **curriculum@nccer.org**, send general comments and inquiries to **info@nccer.org**, or fill in the User Update form at the back of this module.

This information is general in nature and intended for training purposes only. Actual performance of activities described in this manual requires compliance with all applicable operating, service, maintenance, and safety procedures under the direction of qualified personnel. References in this manual to patented or proprietary devices do not constitute a recommendation of their use.

CT63_2-17

Shutdown of a Liquid Pipeline (Field)

Objective

When you have completed this module, you will be able to do the following:

1. Describe how to shut down a liquid pipeline (CT63_2-17).
   a. Describe how to shut down a liquid pipeline.
   b. Identify prerequisites, competencies, and abnormal operating conditions (AOCs) associated with shutting down a liquid pipeline.

Performance Task

Under the supervision of the instructor, you should be able to do the following:

1. Shut down a liquid pipeline (CT63_2-17).
   - Identify potential abnormal operating conditions (AOCs) that may occur during performance of this CT, and know the appropriate actions to take in response to them.
   - Utilize the appropriate personal protective equipment according to relevant company procedures.
   - Ensure familiarity with desired pressure to keep on a static line in accordance with company policies.
   - Shut down pump units or compressor units beginning at the origination point and proceeding downstream to the destination. Shut down in conjunction with valve closures at intermediate and terminal delivery points in accordance with company policies.
   - Complete appropriate documentation as required by operator's procedures.

Trade Terms

Abnormal operating conditions (AOCs)

Alarm
Drag reducing agent (DRA)
Human machine interface (HMI)
Maximum operating pressure (MOP)
Packed
Pressure surge
Procedure
Static state
Steady state
Supervisory Control and Data Acquisition (SCADA)

Industry Recognized Credentials

If you are training through an NCCER-accredited sponsor, you may be eligible for credentials from NCCER's Registry. The ID number for this module is CT63_2-17. Note that this module may have been used in other NCCER curricula and may apply to other level completions. Contact NCCER's Registry at 1.888.622.3720 or go to **www.nccer.org** for more information.

Contents

Figures

SECTION ONE

1.0.0 SHUTTING DOWN A LIQUID PIPELINE

Objective

Describe how to shut down a liquid pipeline (CT63_2-17).

a. Describe how to shut down a liquid pipeline.
b. Identify the prerequisites, competencies, and abnormal operating conditions (AOCs) associated with shutting down a liquid pipeline.

Performance Task

Shut down a liquid pipeline (CT63_2-17).

- Identify potential abnormal operating conditions (AOCs) that may occur during performance of this CT, and know the appropriate actions to take in response to them.
- Utilize the appropriate personal protective equipment according to relevant company procedures.
- Ensure familiarity with desired pressure to keep on a static line in accordance with company policies.
- Shut down pump units or compressor units beginning at the origination point and proceeding downstream to the destination. Shut down in conjunction with valve closures at intermediate and terminal delivery points in accordance with company policies.
- Complete appropriate documentation as required by operator's procedures.

Trade Terms

Abnormal operating conditions (AOCs): Conditions within a pipeline system, identified by the operator, that may indicate (1) a malfunction of a component or (2) a deviation from normal operations that may indicate a condition exceeding design limits or result in a hazard to persons, property, or the environment.

Alarm: A visual or audible alert generated by the SCADA or HMI that indicates that a parameter has been exceeded.

Drag reducing agent (DRA): A chemical injected into a pipeline that reduces the internal friction by decreasing turbulence, allowing the product to flow more freely.

Human machine interface (HMI): A software application that presents information to the operator about the state of a process and accepts and implements an operator's control instructions. Typically, information is displayed in a graphic form.

Maximum operating pressure (MOP): The highest pressure at which a pipeline (or segment of a pipeline) may be normally operated.

Packed: A condition in the line where product vaporization and product mixing is reduced or eliminated. Line pack is reached when minimum pressures are held throughout the line section.

Pressure surge: A wave resulting when a fluid in motion is forced to stop or change direction suddenly. This commonly occurs when a valve is suddenly closed at the end of a pipeline system and a pressure wave propagates in the pipe.

Procedure: A step-by-step set of instructions for logically accomplishing a task.

Static state: The state of an inactive or shutdown pipeline where product is not flowing.

Steady state: A condition on an active or flowing pipeline where pressure and flow rates are relatively constant over time and are comparable to historical operational data for that segment.

Supervisory Control and Data Acquisition (SCADA): A computer-based system used remotely by a controller in a control room that displays information about a pipeline facility, and which may have the ability to send commands back to the facility.

A liquid pipeline shutdown procedure begins by first identifying the portion of the pipeline system to be shut down, confirming the intended shutdown with the control center, and ensuring that the necessary valves are aligned. The pressure and flow rates are monitored during the procedure to ensure that the operation remains within safe limits. The shutdown procedure ends when the portion of the pipeline to be shut down reaches a static state or steady state condition, or when the control center assumes control of the operation.

1.1.0 Shutdown Sequences

All shutdown sequences are company-, pipeline-, and product-specific, and exact procedures are outlined in company manuals. Company shutdown procedures and instructions take into consideration the specific conditions, the products involved, and the characteristics of the pipeline.

1.1.1 Performing a Scheduled Shutdown

Although every pipeline system is different, a planned shutdown involves most or all of the same basic actions and general procedures.

CAUTION

An individual's actions during the performance of a covered task could cause a safety device to activate. The activation represents a deviation from normal operations, resulting in an abnormal operating condition (AOC).

Approximately 15 minutes before the actual shutdown is to take place, begin provisions for shutdown, including all stations and sites. This includes notifying the control center of the intended shutdown configuration, identifying pumping units to be shut down, and all other relevant preparation activities covered in company/operator procedures. After all notifications and verifications have been made, shutdown can begin. A typical shutdown procedure generally includes some or all the following steps:

Step 1 Identify potential **abnormal operating conditions (AOCs)** that may occur during performance of this CT, and know the appropriate actions to take in response to them.

Step 2 Utilize the appropriate personal protective equipment (PPE) according to relevant company procedures.

Step 3 Idle all active pumping units. By idling the main-line pump, the **pressure surge** associated with closing the main-line valve is reduced.

Step 4 Adjust pressure and flow regulators to remain within operating pressure limits.

Step 5 Notify the originating location to begin adjusting product flow into the pipeline as required in order to keep the pipeline within operating limits.

Step 6 Notify the receiving terminal to adjust back pressure if necessary to keep the pipeline in a **packed** condition, to maintain a positive static line pressure, or to keep natural gas liquids (NGLs) and highly volatile liquids (HVLs) in liquid phase during shutdown operations.

At the scheduled shutdown time, follow these steps:

Step 1 Confirm that all locations are prepared to shut down the pipeline.

Step 2 Notify the originating location to stop all pumping and close their main-line block valve. Inform the originating location of the approximate restart time, if available.

Step 3 Notify all injection sites, including portable units, to stop operations.

Step 4 Shut down all pumping units (*Figure 1*) in sequence, from the originating terminal to the receiving terminal.

Step 5 Shut any remotely controlled pipeline gate valves and block valves (*Figure 2*) from the control center.

Step 6 Direct personnel to close any manually operated block valves.

Step 7 Notify the receiving terminal that the pipeline is shut down and that the station upstream of them is blocked.

Figure 1 Main-line pumps with electric drivers.

Figure 2 Pipeline valve with both manually operated and remote controls.

Step 8 Notify all manned locations that their block valve is now closed. Give approximate re-open time, if available.

Step 9 Complete appropriate documentation (including any records, logs, or forms) as required by operator's procedures.

1.1.2 Monitoring a Scheduled Shutdown

It is important to monitor the pipeline after a planned shutdown, because conditions can change without warning. It is possible that a rate-of-change alarm will not operate accurately with the pipeline shut down. Pipeline pressure must be monitored to verify that it is remaining within limits. Pressure could increase due to product expansion from heat. A pressure transmitter is shown in *Figure 3*.

If the line was shut down in a packed condition, maximum operating pressure (MOP) could be exceeded. The pipeline should be monitored to verify that it is remaining in a packed condition so that the leak detection system will operate accurately. If pressure were allowed to lower, a release would not be as obvious and might be undetected for some time. If the line is filled with an NGL or HVL, it must be kept in a packed condition to keep the product in a liquid phase. This is important with many refined products, as well as crude oil in high elevations. If the pressure is gradually changing and the change is not noticed, serious problems might be developing. If not noticed, they will not be discovered until start-up when the situation could cause problems.

Figure 3 Electronic pressure transmitter.

1.1.3 Performing a Shutdown for Maintenance

Pipelines must be shut down at times for maintenance due to major equipment failure. Some causes of required shutdowns include main-line pump failure, main-line valve failure, relief system failure, a release, or other major equipment failure affecting the main flow of the pipeline. Although it may be possible to keep the pipeline flowing at or near normal for a short time, maintenance will be scheduled when a crew is available. When all work permits are in order and the crew is on site, operations personnel shut the line down in a normal fashion. Operations personnel maintain constant contact with the maintenance crew until the work is finished.

1.1.4 Performing an Emergency Shutdown

Events occurring on the pipeline that present an immediate hazard to persons or property require an emergency shutdown. An event might be caused by a wrong action due to human error, mechanical failure, electrical failure, or accident on the pipeline. It might also be caused by threats from weather or civil disobedience. Regardless of the reason, the line must be shut down quickly and safely. Each company has a standard operating procedure for such an action, and every operator must be familiar with this procedure. Once the need is confirmed, decisive action must be taken immediately.

NOTE

In emergency conditions such as a possible release, the pipeline controller has full responsibility and authority to make shutdown decisions and take appropriate actions. These actions are to be governed by the company release policy above all other considerations.

Operations personnel must isolate the pipeline by safely shutting down the originating station and continuing downstream in a safe manner while minimizing product contamination until all pumps are off line. Once the pressure is sufficiently reduced, the operations personnel will then close all block valves to complete the shutdown. The responsible control center, field personnel, and all affected parties should be notified.

Threats from natural disaster or civil disorders come without warning. Such affected facilities must be isolated. Once the facilities are isolated, pipeline shutdown may follow.

> **NOTE**
> If the pipeline is shut down due to a possible release, it should not be restarted until all details relating to the release have been verified, corrected, and resolved.

1.2.0 Prerequisites, Competencies, and Abnormal Operating Conditions

Personnel must possess a variety of skills in order to successfully shut down a pipeline. The operation itself requires a number of steps, which the personnel conducting the shutdown must be competent to perform. Personnel must also be aware of the particular abnormal operating conditions (AOCs) that may occur during the procedure and know the proper responses to them.

1.2.1 Prerequisites and Competencies

A pipeline shutdown involves a large number of personnel, from those at the control center, to those at pump and valve stations, and at the receiving facility. Communications between them can be made by telephone or radio, as well as through the **Supervisory Control and Data Acquisition (SCADA)** system or the **human machine interface (HMI)** (*Figure 4*).

> **NOTE**
> Communications between pipeline personnel are specific to the pipeline. For this reason, local operating procedures will dictate notification procedures.

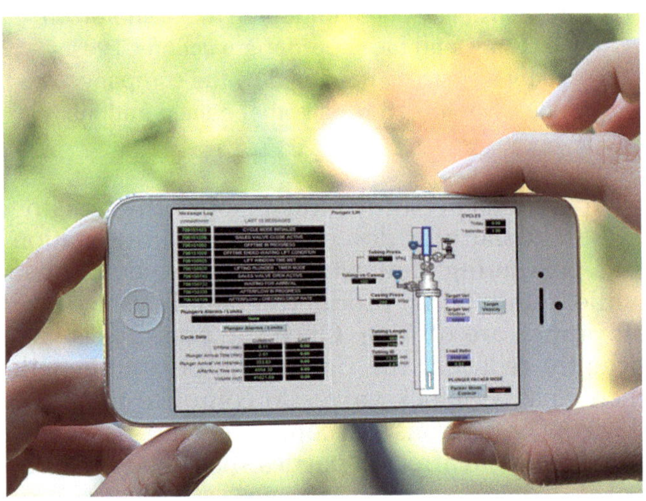

Figure 4 HMI display on a smartphone.

Operations personnel must be able to communicate the shutdown configuration to the control center. This includes providing verification that the local piping is aligned and the intended flow path is configured in accordance with applicable operating procedures. The control center should confirm that the remote flow path has been configured. This communication allows the control room and the local operators to confirm that the shutdown procedure is ready to commence.

Operations personnel must know how to initiate the shutdown by first identifying which, if any, pumping units will be shut down. They must also know how to begin shutting down the line in accordance with the written operating procedures. These procedures may include steps to maintain the appropriate line pressure to minimize contaminating the product. Line pressure and flow rates must be monitored during the shutdown procedure. Personnel must know how to make necessary adjustments to the pressure and flow in order to achieve a static or steady state condition at shutdown. After the line has been shut down, personnel must be able to verify that the valve status is correct, and that a static or steady state condition has been achieved.

Personnel must have a working knowledge of pipeline hydraulics and the maximum operating pressure (MOP) of the pipeline system. The characteristics of the fluid flow in a pipeline may impact shutdown operations. These characteristics include the elevation profile of the line; characteristics of the product in the line, including any **drag reducing agent (DRA)**; and any operational changes that have been made. Operations personnel must understand that a damaging pressure surge may occur if a valve is suddenly closed at the end of a pipeline system. This could occur as valves are manually closed during the shutdown procedure. Personnel should be familiar with company policies on operations of the particular pipeline system, including the proper unit setup for the desired rate.

1.2.2 Abnormal Operating Conditions

Prior to carrying out pipeline shutdown procedures, personnel must be aware of the various AOCs that could be encountered during a pipeline shutdown, and be able to respond appropriately should one occur.

Field personnel performing a pipeline shutdown must be able to recognize, interpret, and respond to visual or audible alarms generated by a SCADA or HMI system that indicate an operating parameter has been exceeded. If they are unable to respond to and mitigate the situation, the control center may have to take action. The following are some potential AOCs, and the appropriate action personnel should take in response to them:

- A fire or explosion occurs – Personnel should immediately evacuate the area and make proper notifications, including notifying 911 as appropriate.
- An exposed pipe or missing/damaged pipeline support occurs – Personnel should make notifications as required by company policy.
- Damage to the pipeline system occurs – Personnel should minimize or eliminate pressures and/or flows as appropriate and make necessary notifications as required by company policy.
- A failure of communication devices, SCADA, or electrical services occurs – Personnel should follow troubleshooting procedures and make contacts as required by company policy.
- Unexpected release of hazardous liquid or carbon dioxide occurs – Personnel should evacuate the area and contact control center to isolate or shut down the line.

Other AOCs that may occur during a liquid pipeline shutdown include the following:

- The activation of a safety device, such as a pressure relief valve or a remotely controlled shutdown device (RCSD).
- An unexplained flow rate deviation, such as high flow, low flow, or no flow.
- An unexplained pressure deviation, such as an increase, decrease, or lack of pressure.
- An unintended status change, such as a change in unit status or valve position.

Should one of these AOCs occur, a visual or audible indication will be received through the HMI or another system. If this happens, personnel must investigate the cause of the AOC, take appropriate actions to mitigate the situation (if possible), and make any necessary notifications. Always be familiar with and utilize company-specific procedures regarding potential AOCs.

Additional Resources

It is assumed that you will learn to perform these procedures under the direct supervision of your crew leader, foreman, or instructor. For independent study, refer to the recommended Additional Resources at the back of this module.

1.0.0 Section Review

1. A liquid pipeline shutdown begins by ensuring that _____.
   a. the suction pressure is sufficient
   b. all main-line units are at idle
   c. the line is in a static or steady state
   d. the necessary valves are aligned

1. If operations personnel are unable to respond to and mitigate an AOC, _____.
   a. the shutdown must be suspended
   b. the control center may have to take action
   c. a technical team must take action
   d. notification must be made to the appropriate government agency

SUMMARY

A liquid pipeline shutdown begins by identifying the pipeline system to be shut down, confirming with the control center, and ensuring valves are aligned. The procedure ends when a static or steady state is reached. Shutdown sequences are company-, pipeline-, and product-specific. Begin provisioning for shutdown at all stations and sites 15 minutes prior to the actual shutdown and make all notifications and verifications. Preparing for shutdown begins by idling pumps, adjusting pressure and flow regulators, notifying originating location to begin adjusting product flow, and notifying the receiving terminal to adjust backpressure to keep the line in a packed condition.

Begin the shutdown by confirming that all locations are prepared, notifying the originating location to stop pumping and close their block valve, and notifying all injection stations to stop operations. Next, shut down all pumping units in sequence from beginning to end, and shut all gate and block valves. Complete the shutdown by notifying the receiving terminal and all stations that the pipeline is shut down and their block valves are closed. Complete all documentation. Monitor the line after shutdown to ensure pressure is within limits and not increasing due to heat expansion.

Communications between shutdown personnel is conducted by radio or telephone, as well as through the SCADA and HMI systems. Personnel should be able to recognize and respond to alarms generated by the SCADA or HMI. Operations personnel align the valves and pipes, verify the flow path, and inform control that the pipeline is configured for shutdown. Care must be taken to prevent a pressure surge due to quickly closing a valve. Personnel shut down the pumps and monitor the pressure and flow rates. The control center assumes control when a static or steady state condition is reached.

If SCADA communication or electrical power is lost, personnel must activate backup systems and take action to mitigate the situation. Other AOCs that could occur include the activation of a safety device, an unexplained flow rate or pressure deviation, or an unintended change in status. If an AOC occurs, personnel must investigate the cause and take action to mitigate the situation. All required notifications must be made.

Review Questions

1. To ensure operations within safe limits during shutdown, it is necessary to monitor the _____.
   a. pressure and pump speed
   b. flow rates and pump speed
   c. pressure and flow rates
   d. pump speed and valve alignment

2. Approximately 15 minutes before the actual shutdown, begin provisioning for the shutdown by _____.
   a. idling all active pumping units
   b. ensuring all notifications and verifications have been made
   c. adjusting all flow and pressure regulators to within operating limits
   d. gradually reducing line pressure and flow rates

3. Back pressure at the receiving station must be adjusted in order to _____.
   a. slowly reduce the flow rates
   b. establish an unpacked condition in the line
   c. prevent the line from exceeding the MOP
   d. keep the line in a packed condition

4. After the shutdown is complete, notify all manned locations _____.
   a. to open their block valve
   b. that their block valve is closed
   c. to close their block valve
   d. that their block valve is open

5. Monitoring the line after a planned shutdown _____.
   a. is important because conditions may change
   b. is required by federal regulations
   c. is required only after an AOC event
   d. may be necessary in certain situations

6. After the line has been shut down, personnel must verify that the valve status is correct, and that _____.
   a. all alarms have been checked and/or reset
   b. a static or steady state has been reached
   c. the flow rates have been reduced
   d. the pressure has been stabilized

7. Should a pipeline parameter be exceeded during shutdown, personnel _____.
   a. will be notified by the control center
   b. must immediately reduce the line pressure
   c. will receive a visual or audible alarm
   d. must activate the backup system

Trade Terms Quiz

Fill in the blank with the correct term that you learned from your study of this module.

1. A chemical injected into the pipeline to reduce internal friction is called a _________.
2. The greatest pressure at which a pipeline can be operated is called the _________.
3. A system that displays information about a pipeline facility is called _________.
4. A software application that presents information about the state of a process is called _________.
5. A step-by-step set of instructions to accomplish a task is called a _________.
6. Conditions within a pipeline system that may indicate a malfunction of a component or a condition exceeding design limits are called _________.
7. A wave inside a pipeline caused by the sudden closing of a valve is called a _________.
8. The condition inside an inactive pipeline where product is not flowing is called _________.
9. An alert signaling that a parameter has been exceeded is called an _________.
10. The condition inside a pipeline where pressure and flow rates are relatively constant is called _________.
11. The condition inside a pipeline where product vaporization and mixing is reduced or eliminated is called _________.

Trade Terms

Abnormal operating conditions (AOCs)
Alarm
Drag reducing agent (DRA)
Human machine interface (HMI)
Maximum operating pressure (MOP)
Packed
Pressure surge
Procedure
Static state
Steady state
Supervisory control and data acquisition (SCADA)

Trade Terms Introduced in This Module

Abnormal operating conditions (AOCs): Conditions within a pipeline system, identified by the operator, that may indicate (1) a malfunction of a component or (2) a deviation from normal operations that may indicate a condition exceeding design limits or result in a hazard to persons, property, or the environment.

Alarm: A SCADA or HMI generated visual or audible alert that indicates that a parameter has been exceeded.

Drag reducing agent (DRA): A chemical injected into a pipeline to reduce the internal friction by reducing turbulence allowing the product flow more freely.

Human machine interface (HMI): A software application that presents information to the operator about the state of a process and accepts and implements an operator's control instructions. Typically, information is displayed in a graphic form.

Maximum operating pressure (MOP): The maximum pressure at which a pipeline or segment of a pipeline may be operated.

Packed: A condition in the line where product vaporization and product mixing is reduced or eliminated. Line pack is reached when minimum pressures are held throughout the line section.

Pressure surge: A wave resulting when a fluid in motion is forced to stop or change direction suddenly. This commonly occurs when a valve is suddenly closed at the end of a pipeline system and a pressure wave propagates in the pipe.

Procedure: A step-by-step written set of instructions to logically accomplish a task.

Static state: The state of an inactive or shutdown pipeline where product is not flowing.

Steady state: The point when pressure and flows are relatively constant over time and comparable to historic operational data for that segment. Steady state refers to a condition on an active or flowing pipeline.

Supervisory Control and Data Acquisition: A computer-based system that is used by a controller in a control room which displays information about a pipeline facility, and may have the ability to send commands back to the facility.

Additional Resources

This module is intended as a thorough resource for task training. The following reference works are suggested for further study.

API Recommended Practice 1161, Third Edition, © 2014 American Petroleum Institute. API Publishing Services, 1220 L Street, NW, Washington, DC 20005. Distributed by Thompson Reuters (Scientific) LLC, **www.techstreet.com**.

NCCER Module AOCCC-17, *Abnormal Operating Conditions Control Center*

NCCER Module AOCFG-17, *Abnormal Operating Conditions Field and Gas*

NCCER Module 66102-02, *Liquid Pipeline General Abnormal Operating Conditions*.

NCCER Module 71101-14, *Field Abnormal Operating Conditions*.

NCCER Module 64106-02, *Pipeline Operations*.

NCCER Module 64307-02, *SCADA*.

NCCER Module 60103-15, *Pipeline Communications*.

NCCER Module 60102-17, *Basic Pipeline Hydraulics and Equipment*.

NCCER Module 60106-15, *Monitoring Pipeline Operations*.

NCCER Module 62105-02, *Pipeline Documentation*.

NCCER Module CT63_1-17, *Startup of a Liquid Pipeline*.

NCCER Module CT63_4-17, *Remotely Operate Valves*.

49 *CFR* 192, **www.ecfr.gov**.

49 *CFR* 195, **www.ecfr.gov**.

Figure Credits

© iStockphoto.com/Pro-syanov, Module Opener

Benson County Farmers Press, Figure 1

© iStockphoto.com/ekipaj, Figure 2

Tim Mills, Figure 3

ABB, Figure 4

Section Review Answer Key

Answer	Section Reference	Objective
Section One		
1. d	1.0.0	1a
2. b	1.2.2	1b

NCCER CURRICULA — USER UPDATE

NCCER makes every effort to keep its textbooks up-to-date and free of technical errors. We appreciate your help in this process. If you find an error, a typographical mistake, or an inaccuracy in NCCER's curricula, please fill out this form (or a photocopy), or complete the online form at **www.nccer.org/olf**. Be sure to include the exact module ID number, page number, a detailed description, and your recommended correction. Your input will be brought to the attention of the Authoring Team. Thank you for your assistance.

Instructors – If you have an idea for improving this textbook, or have found that additional materials were necessary to teach this module effectively, please let us know so that we may present your suggestions to the Authoring Team.

NCCER Product Development and Revision
13614 Progress Blvd., Alachua, FL 32615
Email: curriculum@nccer.org
Online: www.nccer.org/olf

❑ Trainee Guide ❑ Lesson Plans ❑ Exam ❑ PowerPoints Other ____________

Craft / Level: Copyright Date:

Module ID Number / Title:

Section Number(s):

Description:

Recommended Correction:

Your Name:

Address:

Email: Phone:

CT63_3-17

Monitor Pressures, Flows, Communications, and Line Integrity and Maintain Them within Allowable Limits on a Liquid Pipeline System (Field)

Overview

The control center functions as the heart of the pipeline, receiving vital information and sending commands to ensure the system operates smoothly. Remote sensors send information about line pressure, valve alignment, flow rates, and other changes in system status. Safety devices and alarms warn the center of potentially hazardous situations. Most of the control center operator's time is spent monitoring the many facets of the pipeline in operation. Monitoring the entire system is critical for keeping ongoing operations in a steady state. The sooner field personnel become aware of a situation, the more likely a solution can be found before the problem becomes critical. For this reason, personnel working in the field must maintain communications with the control center at all times to effectively monitor the pipeline system.

Trainees with successful module completions may be eligible for credentialing through the NCCER Registry. To learn more, go to **www.nccer.org** or contact us at 1.888.622.3720. Our website has information on the latest product releases and training, as well as online versions of our *Cornerstone* magazine and Pearson's product catalog.

Your feedback is welcome. You may email your comments to **curriculum@nccer.org**, send general comments and inquiries to **info@nccer.org**, or fill in the User Update form at the back of this module.

This information is general in nature and intended for training purposes only. Actual performance of activities described in this manual requires compliance with all applicable operating, service, maintenance, and safety procedures under the direction of qualified personnel. References in this manual to patented or proprietary devices do not constitute a recommendation of their use.

CT63_3-17
Monitor Pressures, Flows, Communications, and Line Integrity and Maintain Them within Allowable Limits on a Liquid Pipeline System (Field)

Objectives

When you have completed this module, you will be able to do the following:

1. Explain how to monitor alarms, and describe pipeline monitoring systems.
   a. Describe how to respond to and monitor pipeline alarms and abnormal operating conditions.
   b. Describe the purpose of computational pipeline monitoring (CPM).
2. Explain how to monitor pipeline parameters, operations, and safety device alarms (CT63_3-17).
   a. Identify the types of safety devices.
   b. Identify types of detection systems.
   c. Describe pipeline purging and purge monitoring.
   d. Describe the general procedure and skills associated with monitoring a pipeline.
3. Explain how remote locations and weather conditions are monitored.
   a. Describe how remote video surveillance is performed.
   b. Describe how aerial surveillance is performed.
   c. Describe how weather conditions are monitored.

Performance Task

Under the supervision of the instructor, you should be able to do the following:

1. Monitor pressures, flows, communications, and line integrity and maintain them within allowable limits on a liquid pipeline system (field).
   - Identify potential abnormal operating conditions (AOCs) that may occur during performance of this CT, and know the appropriate actions to take in response to them.
   - Utilize the appropriate personal protective equipment (PPE) according to relevant company procedures.
   - Verify that the flows, pressures, and product characteristics are within normal range.
   - Continuously monitor, by whatever means available (SCADA, pressure gauge, meter), the pressures and flow rates of the pipeline to look for any changes.
   - If pressures, flows, and line balance are not within normal range, determine the cause of variation from normal and correct the situation if possible.
   - Complete appropriate documentation as required by operator's procedures.

Trade Terms

Abnormal operating conditions (AOCs)
Alarm
Flammable gas detection systems
IR detector
Ionization detector
Monitor
Purging
Slug
Steady state
Supervisory Control and Data Acquisition (SCADA)
Thermal detector
UV detector
Voted system

Industry Recognized Credentials

If you are training through an NCCER-accredited sponsor, you may be eligible for credentials from NCCER's Registry. The ID number for this module is CT63_3-17. Note that this module may have been used in other NCCER curricula and may apply to other level completions. Contact NCCER's Registry at 1.888.622.3720 or go to **www.nccer.org** for more information.

Contents

Figures

Section One

1.0.0 Pipeline Monitoring Systems and Alarms

Objective

Explain how to monitor alarms, and describe pipeline monitoring systems.

a. Describe how to respond to and monitor pipeline alarms and abnormal operating conditions.
b. Explain the purpose of computational pipeline monitoring (CPM).

Trade Terms

Abnormal operating conditions (AOCs):Conditions within a pipeline system, identified by the operator, that may indicate (1) a malfunction of a component or (2) a deviation from normal operations that may indicate a condition exceeding design limits or result in a hazard to persons, property, or the environment.

Alarms: Status events from a pipeline monitoring system (such as SCADA) that result from a problem, abnormality, or emergency that occurs in a pipeline system. All alarms have a mandatory response (or action) associated with them.

Monitor: An instrument that reports the performance of a control device or signals if unusual conditions appear in a system. For example, an S&W monitor provides a mechanical means of preventing contaminated oil from entering the pipeline. In common usage, the term also applies to the activities of operators and other personnel observing and tracking conditions on the pipeline.

Steady state: The point when pressures and flows are relatively constant over time.

Supervisory Control and Data Acquisition (SCADA): A control system used in the pipeline industry that collects and assembles data, and also allows operators to control the pipeline remotely.

The control center must continuously monitor every pipeline system. Monitoring is accomplished through the Supervisory Control and Data Acquisition (SCADA) computer system. Microwave, satellite, commercial, cellular telephone, and radio equipment are also used to carry on voice communications.

Pipeline equipment malfunctions can cause conditions to reach operating limits quickly. Although abnormal conditions do not occur frequently, it is the responsibility of control center personnel to use these systems to monitor pipeline status, detect valve closures, identify station shutdowns, react to high-pressure alarms, and begin taking immediate corrective action. Affected personnel must be notified of the condition. Although no manual can cover all the possible abnormal scenarios that can occur, control center personnel must look at the condition and apply company procedures to the method of response. Workers in the control center must be familiar enough with the pipeline system, the current operating conditions, company procedures, equipment being operated upstream, product streams, and environmental factors to be able to immediately identify problems.

Since it is important that field personnel understand how the monitoring and control systems operate, most of the content in this module is written from the perspective of the control center operator. The active role of field personnel in monitoring a pipeline is to maintain effective communication with the control center at all times. For this reason, field personnel must have a basic understanding of control center activities.

> **NOTE**
> Abnormal operating conditions must be detected early and corrective action taken quickly.

Control center personnel must monitor pipeline parameters, be able to assess the need for corrective action, and initiate the appropriate action quickly. They do this by monitoring the screens and analog readings in the control center closely (*Figure 1*). Digital and analog instruments provide an ongoing display of pressure, flow, and many other conditions in the pipeline from all the stations in the line. Control center personnel must be attentive to any unexplained status change. If any pipeline parameters are not normal, they must determine the cause and take necessary steps to bring that area back to normal. They may need to seek assistance from, and make notifications to responsible personnel per company procedure.

Pipeline procedures comply with *49 CFR 195.402* to provide immediate response and mitigation of pipeline hazards. Control center personnel are provided with standard operating procedures for handling, or are given the principles for handling, most situations of this nature. With the use of knowledge and good judgment, control

Figure 1 Typical control center console.

center personnel may be allowed to deviate from procedure to protect personnel, environment, and property in cases of pipeline emergency.

1.1.0 Monitoring Alarms and Responding to Abnormal Operating Conditions

Pressure and flow are the two primary indicators of normal, as opposed to abnormal, operating conditions. You must be familiar with what normal pressure levels and flow rates are on your pipeline, though alarms will often alert you if these levels are not normal. Serious consequences may result if maximum operating pressure (MOP) is exceeded.

Abnormal readings in pressure and flow rates could indicate a possible release or a warning of approaching equipment shutdown. For example, if pressure begins to build at one point in the line, the SCADA readings show it immediately. Control center personnel should notice even the slightest pressure fluctuation, because **abnormal operating conditions (AOCs)** might be indicated. A rise in pressure at some point could indicate an unintended closure of valves that will quickly lead to abnormal conditions that could jeopardize pipeline operations. If an unintended valve closure is not detected and corrected immediately, an emergency condition is created. Such a condition, if not corrected, may endanger public safety, the environment, and pipeline operations.

Control center personnel must attempt to reopen the valve immediately. The situation might be corrected by sending an OPEN command by computer, or by calling field personnel in a manned location who will have access to a valve control panel, shown in *Figure 2*. This is an attempt to control the condition before pressure builds. If the valve cannot be re-opened quickly,

Figure 2 Valve control panel.

a pipeline shutdown must be initiated. Personnel in the control center must use their best judgment based on knowledge of the facilities and experience.

If it is determined that there is enough time to adjust pipeline operations and change pressure and flow rates before reaching MOP, sites must be notified to reduce or stop flow. The pipeline must then be monitored closely until conditions have been corrected and normal operations can be resumed.

The following are some other potential AOCs that personnel must be able to identify, and the appropriate action personnel should take in response to them:

- *An exposed pipe or missing/damaged pipeline support occurs* – Personnel should make notifications as required by company policy.
- *Activation of a safety device, such as a pressure relief valve or a remotely controlled shutdown device (RCSD) occurs* – Personnel should investigate the cause of the AOC, take appropriate actions to mitigate the situation (if possible), and make any necessary notifications.
- *A failure of communication devices, SCADA, or electrical services occurs* – Personnel should follow troubleshooting procedures and make contacts as required by company policy.

- *A fire or explosion occurs* – Personnel should immediately evacuate the area and make proper notifications, including notifying 911 as appropriate.
- *Damage to the pipeline system occurs* – Personnel should minimize or eliminate pressures and/or flows as appropriate and make necessary notifications as required by company policy.
- *Unexpected release of hazardous liquid or carbon dioxide occurs* – Personnel should evacuate the area and contact control center to isolate or shut down the line.
- *An unexplained flow rate deviation, such as high flow, low flow, or no flow, occurs* – Personnel should investigate the cause of the AOC, take appropriate actions to mitigate the situation (if possible), and make any necessary notifications.
- *An unexplained pressure deviation, such as an increase, decrease, or lack of pressure occurs* – Personnel should investigate the cause of the AOC, take appropriate actions to mitigate the situation (if possible), and make any necessary notifications.
- *An unintended status change, such as a change in unit status or valve position occurs* – Personnel should investigate the cause of the AOC, take appropriate actions to mitigate the situation (if possible), and make any necessary notifications.

1.1.1 Monitoring after Corrections

Control center personnel must monitor the line closely after corrections are made. When abnormal conditions have ended at critical locations, checks must be made for any deviations from normal. Pump stations, origin and receiving locations, injections and deliveries, and meter stations must be watched closely to verify that the pipeline is safe and has returned to a steady state.

1.1.2 Monitoring Communications

Control center personnel monitor the conditions within the communications system as required by *49 CFR 195.402* because it is the control line for the pipeline system. Normally, the SCADA commands the system, but if partial or total loss of communication is experienced, action must be taken to control the system safely. It is up to control center personnel to recognize the signs of communication failure and initiate corrective actions as soon as possible.

Always monitor communications and be alert for any element of communication failure. If you do not have communication capabilities, seek assistance from and make notifications to the proper personnel, per company procedures.

1.1.3 General Observation

Control center personnel must monitor all aspects of systems operations at the same time. Safety devices and alarms are designed to operate and reduce or eliminate hazardous situations. Activation of such devices demands serious attention from control center personnel. They must verify every warning and alarm condition, which may include the following:

- Low pressure
- No pressure
- High pressure
- Hazardous atmosphere
- High sump
- High tank level
- Low tank level
- Equipment shutdown
- Seal failure

1.2.0 Computational Pipeline Monitoring

Computational pipeline monitoring (CPM) is a computerized leak detection program required by *49 CFR 195.444*. With this program, computers carry on a continuous program of monitoring pressure, flow rate, line volume balance, and variations between status points. Unplanned deviations, which may indicate a pipeline leak or discharge, will cause an alarm to sound. Control center personnel must investigate these alarms as quickly as possible.

1.2.1 CPM Theory

Line balance calculations are made to compare the difference between the volume received into the system with the volume being delivered. Calculations are adjusted for product retained in relief tanks, removed for other uses, and any other factor that would affect the final total. This data is also averaged in order to eliminate the effects of bad data, which may occasionally occur. If there is a discrepancy, the computer will detect it and alert personnel to investigate a possible release. The CPM program can also gather the information the SCADA system collects and create a simulated model of the hydraulic state of the pipeline. This gives the control center the capability of controlling the pipeline and seeing the results from this program in real time.

1.2.2 Negative Pressure Wave Program

Negative pressure wave (wave alert) monitors use negative pressure wave detectors in combination with the computer to display the exact locations of line breaks. When a line break occurs, there is a sudden drop in pressure, followed immediately by a return to pressure. These sudden changes result in a wave that travels in both directions through the pipe at the speed of sound, growing smaller as it travels. With sensors at status points located every few miles, the location of the release is calculated and known within seconds (*Figure 3*). It is then displayed in graphic form on a screen in the control center.

> **NOTE**
> As with all alarm systems, false alarms are possible with the negative pressure wave program. For example, field personnel calibrating a pressure transmitter may trigger a false CPM alarm. Regardless of this possibility, all CPM alarms must be investigated as quickly as possible.

1.2.3 Trending and Interpreting Data

Trending is the computer operation that demonstrates the ongoing conditions on the pipeline and illustrates any changes that occur during a specified period. With calculations of pipeline pressure, flow, line balance, and line volume balance, CPM programs can demonstrate the trend of change on the screens. These trends are often shown graphically as hydraulic profiles, line volume calculations, and pressure readings at the various points. The computer also calculates and demonstrates the rate of change taking place so that control center personnel can observe a problem that might be developing.

Figure 3 Electronic pressure transmitter.

Operating parameters are defined and programmed into the computer. As information is collected from the status points being polled, the computer performs various tests on each parameter. First, the electrical signal is tested to make certain that it falls into the permissible electrical signal range. A signal is posted on the computer screen and printout to show any out-of-range condition. The computer compares suction pressure against discharge pressure and against preset limits. The computer compares the present pressure reading with the previous base value. Transmitted set points are compared with the value entered by control center personnel. Flow rate is scanned and compared to the previous values.

Any unexplained deviation in these calculations reveals the possibility of a leak on the pipeline. Alarms are sounded, displayed in the alarm window of the controller's computer, and recorded in the alarm log for each of the individual deviations mentioned. Control center personnel must begin investigation of the alarm's cause as quickly as possible. The alarm will clear when the condition that caused the alarm ceases.

> **NOTE**
> When a communications outage occurs with one of the key pressure and flow rate CPM transmitters, control center personnel should be concerned that the lost status points are still being used to calculate CPM, which will affect the accuracy of information and alarms received.

1.2.4 Release Alarm Documentation

Control center personnel are required to log all release alarms on the release alarm record. All pertinent data must be logged into the logbooks of pipeline control center personnel. After the event has been successfully controlled, appropriate personnel must decide if an investigation and analysis report needs to be filled out or if further action needs to be taken. The appropriate personnel must examine any available records of the incident to see if the maximum operating pressure (MOP) has been exceeded. If so, management and government agencies must be notified, and completion of all required paperwork must be confirmed. Entries must be inserted in computer logs, incident reports, event notifications, and control center logbooks. All other related documentation must be checked for accuracy and filed.

Additional Resources

It is assumed that you will learn to perform these procedures under the direct supervision of your crew leader, foreman, or instructor. For independent study, refer to the recommended Additional Resources at the back of this module.

1.0.0 Section Review

1. In case of a pipeline emergency that threatens people, property, or the environment, pipeline personnel may deviate from established procedure _____.
   a. only where specified in state or federal regulations
   b. with prior authorization from operations
   c. only when the situation becomes critical
   d. with the use of knowledge and good judgment

2. The operation in which the computer compares both past and present conditions to the established set points in order to detect potential problems that may be developing is called _____.
   a. computational pipeline monitoring
   b. negative pressure wave program
   c. trending and interpreting data
   d. SCADA simulated model

SECTION TWO

2.0.0 MONITORING PIPELINE PARAMETERS, OPERATIONS, AND SAFETY DEVICE ALARMS

Objective

Explain how to monitor pipeline parameters, operations, and safety device alarms (CT63_3-17).

a. Identify the types of safety devices.
b. Identify types of detection systems.
c. Describe pipeline purging and purge monitoring.
d. Describe the general procedure and skills associated with monitoring a pipeline.

Performance Task

1. Monitor pressures, flows, communications, and line integrity and maintain them within allowable limits on a liquid pipeline system (CT63_3-17).
   - Identify potential abnormal operating conditions (AOCs) that may occur during performance of this CT, and know the appropriate actions to take in response to them.
   - Utilize the appropriate personal protective equipment (PPE) according to relevant company procedures.
   - Verify that the flows, pressures, and product characteristics are within normal range.
   - Continuously monitor, by whatever means available (SCADA, pressure gauge, meter), the pressures and flow rates of the pipeline to look for any changes.
   - If pressures, flows, and line balance are not within normal range, determine the cause of variation from normal and correct the situation if possible.
   - Complete appropriate documentation as required by operator's procedures.

Trade Terms

Flammable gas detection systems: A gas detection device that senses the amount of hydrocarbon gases in the atmosphere and warns when a concentration of gas has become high enough to cause an explosion in the presence of an ignition source.

Ionization detector: A photoelectric detection device used to signal an alarm when smoke is detected in a location. Commonly known as a smoke detector.

IR detector: A sensor that detects the presence of hazardous gas by sending an infrared beam to a receiver and measuring the amount of IR waves that are absorbed by the gas.

Purging: Replacing the atmosphere in a pipeline by injection of an inert substance to prevent formation of explosive mixtures.

Slug: A small amount of inert gas used in purging operations. It is injected between the original gas and the purging gas. The slug acts as a barrier to reduce or prevent mixing of the two gases.

Thermal detector: A heat-sensing device that warns of the presence of flame by detecting a sudden rise in temperature. Also known as a flame detector or bolometer.

UV detector: A sensor that detects the presence of fire by detecting the ultraviolet rays given off by incipient (beginning) fires or flames.

Voted system: A type of alarm system layout used for flame detection and gas detection. It uses very sensitive detectors and delayed-action relays. Such systems will not alarm until at least two or more detectors are tripped, depending on how the voted area is set up.

The potential for fire, hazardous vapors, or leaks to develop exists throughout the pipeline and pipeline facilities. Given the size of the facilities, personnel may not be present when these situations develop. Instead, a series of detection and safety devices are installed, which monitor the area and either respond independently to suppress the situation, or to signal operations that an abnormal situation has occurred. When pipelines are opened for maintenance, hazardous vapors may remain within the pipe and must be purged from the line. **Purging** operations must be continuously monitored in order to detect potentially unsafe conditions that may arise.

> **WARNING!**
> Never ignore any safety device alarm. If you ever receive an alarm with which you are not familiar, contact the appropriate personnel, per company policy.

2.1.0 Types of Safety Devices

Safety devices are found throughout the pipeline and storage facilities. These devices operate to alert, prevent, or relieve abnormal conditions, per company policy. Such devices may include the following:

- Hazardous atmosphere devices
- Thermal relief devices
- Temperature control devices
- Detection systems
- Monitoring controls

> **WARNING!** Never ignore any safety device alarm. If you ever receive an alarm with which you are not familiar, contact the appropriate personnel as directed by company policy.

2.1.1 Hazardous Atmosphere Devices

The lower explosive limit (LEL) detector, thermal detector, and **UV detector** or **IR detector** (*Figure 4*) are all tools that will continuously sample the air for hazardous atmosphere and monitor the area for high heat and flame at each pump station. When the detectors sense an adverse condition, signals are sent to the computer system that activate warning lights and sound an alarm. **Flammable gas detection systems** are used to provide early detection and warning of the presence of hazardous atmospheres so that appropriate safety steps can be taken. A flammable gas leak detector is shown in *Figure 5*.

Gas detection devices monitor gas levels from 0 to 100 percent. By law, they must be set to alarm when gas levels reach 25 percent of the LEL. Most alarms are set to warn at 20 percent of the LEL.

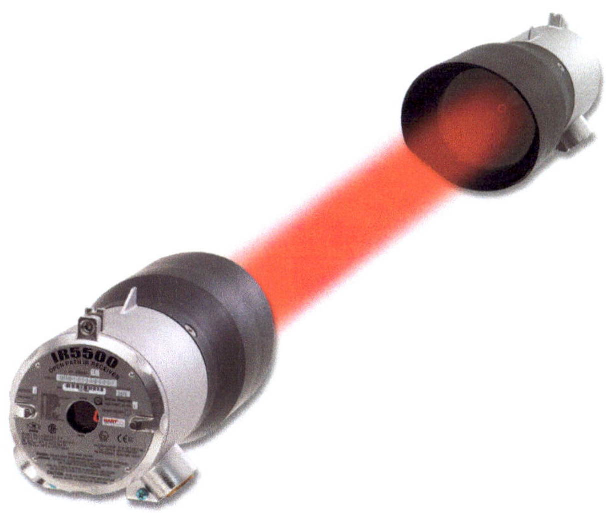

Figure 4 Infrared gas detector (IR detector).

Figure 5 Flammable gas leak detector.

> **NOTE** LEL detectors only measure levels between 0 and 100 percent of the lower explosive limit. If measurement of higher concentrations is expected, a different type of equipment must be used.

2.1.2 Thermal Relief Devices

Liquids expand when heated. If liquid that is trapped within a line expands too much it could rupture the pipe. The purpose of thermal relief devices is to relieve pressure increase due to heat expansion. Such devices may include the following:

- Manifold thermal relief
- Bypass relief
- Pump case temperature relief device

2.1.3 Temperature Control Devices

The mainline and booster pump thermal relief system provides temperature relief or shutdown to the mainline and booster pumps for temperatures that might exceed their design limits. In some applications, a temperature-controlled bypass valve opens when pump case temperature exceeds its set point. In other applications, shutdown will occur.

2.1.4 Monitoring Controls

When an alarm is first activated, a status indicator will flash. Control center personnel must respond to the alarm. Warning and danger alarms remain active (latched) until they are reset or cleared from the control room or field.

There are different levels of warning, which may vary by company policy. Examples of typical warning levels are high priority, medium priority, and low priority.

2.2.0 Detection Systems

Detection systems are used to detect the presence of fire and hazardous gas levels in the atmosphere. These detection systems can also be designed to signal alarms, discharge a chemical fire suppressant to suppress the fire, or shut down key equipment. The purpose of these systems is to provide the first line of defense for personal safety, followed by the protection of field equipment and facilities. Major functions performed by the fire detection and suppression systems are as follows:

- Detecting smoke, flames, and excessive temperature
- Alerting personnel to a hazardous atmosphere or fire condition
- Automatically activating the emergency ventilation system
- Isolating hydrocarbon sources
- Automatically shutting down equipment
- Automatically discharging a chemical suppressant in a hazardous atmosphere/fire zone
- Automatically blowing down large-volume natural gas systems
- Manually activating a chemical suppressant and/or foam in a hazardous atmosphere/fire zone

Detection systems are also used to detect leaks in the pipe, or in the valves, pumps, and gaskets within the pipeline. When a leak is detected, a signal is sent to the control center showing the location. Maintenance personnel can then quickly respond to correct the situation.

2.2.1 Smoke Detectors

An **ionization detector** will detect products of combustion in the incipient (beginning) stage before the release of visible smoke or open flames. They do this by detecting relatively small amounts of smoke in the air. Ionization detectors are especially effective for detecting fire in electrical conduit cable trays in hallways. They are used for smoke detection in office buildings, including crawlspaces, generator rooms, control rooms, and other areas with concentrations of electrical equipment. An ionization alarm and light appears on the control panel and reveals the location of the fire.

> **Did You Know?**
>
> **Halon**
>
> The use of Halon in fire suppression systems was banned by the US Environmental Protection Agency (EPA) because it was found to damage the Earth's ozone layer.

2.2.2 Flame Detectors

Some stations are equipped with ultraviolet (UV) sensors for fire detection (*Figure 6*). The UV flame detection system detects the ultraviolet rays produced by the flames of incipient fires rather than rays produced by naturally occurring light. UV systems have an advantage over smoke-detecting devices like the ionization detector, as they typically provide a nearly instantaneous fire detection response of less than 25 milliseconds. Maximum sensitivity gives opportunity for false alarms, and in sensitive areas, a **voted system** can be set up so that at least two different, overlapping sensors must alarm before any action can be taken.

When the UV system has been activated, the signal is sent to the control panel, where an alarm and lights show the exact area of the fire. Some UV alarms set off an immediate shutdown per company policy. The fire panel alarm must be reset before restarting the unit or station.

Figure 6 UV flame detector.

UV Sensors

Some UV sensors set to maximum sensitivity can detect a gasoline pan fire 4 square feet (0.37 m^2) in area at a distance of 55 feet (15.24 m).

The UV system operates on a line-of-sight principle and is hindered by objects between it and the fire. Such situations must be considered when designing the system. Fires that produce heavy smoke may obliterate the UV detector's ability to detect fire. Other causes of false alarms include welding, electrical arc sources, camera flashes, and radiation such as X-rays or gamma rays.

2.2.3 Thermal Flame Detectors

A **thermal detector** will close controls that are normally open in reaction to rapidly rising temperature. Thermal detectors are used with graduated temperature settings. The spread of settings allows for special applications. The early warning rate compensation feature closes a thermal detector before reaching its fixed set point. Thermal detectors must never be painted or coated.

2.2.4 Liquid Leak Detector

Liquid leak detectors are installed per company policy underneath pipelines. They are composed of a second pipe (possibly PVC) with perforations or slots. Hydrocarbon-sensing devices are installed along the length of the second pipe, with air circulating through it. In the event of a release, the product permeates the soil around the second pipe and enters the perforations. The hydrocarbon-sensing devices immediately detect and relay an alarm to the hazardous gas monitor in the pump room and/or control center panel. Because the sensor locations are charted, control center personnel can determine the exact location of the leak. By noting the relationship of other sensors being tripped near the first one, personnel can gain some insight into the possible extent of the leak.

2.2.5 Seal Leak Detectors

Seal leak detectors are used on large mainline pumps and possibly on other high-speed devices. Seals are made of very durable materials, but due to high pressures, age, and wear, they do fail. If seal failure is not discovered in time, very serious damage can result to equipment. A seal leak detector consists of a collector located under the end of the shaft, with a tube running to a sump. A liquid-flow detector is located in a tube or a catch basin. When any flow is detected, the alarm is sent to the pump room panel and/or control center and should be immediately investigated per company policy.

2.3.0 Purging

There are times when a pipeline section or segment must be opened for inspection or maintenance. Pipelines and storage tanks often contain a combustible gas associated with the product being moved or stored. If the line or tank is opened to the outside air, a mixture of gas and oxygen within flammable limits can form, creating a potential explosion hazard. Purging is the process by which one gas in an enclosed area is deliberately displaced by another in order to significantly reduce this possibility. When a section of pipeline is unpacked, drained, and opened for inspection or maintenance, hazardous or explosive vapors remain and must be purged in order to create a safe working environment. When that section of line is ready to be closed and repacked, the line has to again be purged of air to prevent an explosive mixture of oxygen and flammable gasses from forming once product is reintroduced.

Company polices and plans for purging operations have been developed and must be followed by all involved. Personal safety is paramount during any purging operation and appropriate personal protective equipment (PPE) must be provided and used. While these activities might not be performed in the control center, they must always maintain constant contact with workers in the field who do carry out these activities.

2.3.1 Purging Procedures

Purging can be accomplished in a variety of ways, for a variety of reasons. For example, flammable vapors remaining in a line can be purged by injecting a relatively inert gas such as nitrogen. Air can be purged from a line using a natural gas before product is reintroduced. Some mixing during purging is inevitable; however, by maintaining a moderately fast flow rate, the mixing is kept to a minimum. Dilution, or mixing, of the purge gas with the original content often results in the need to inject up to 1.5 to 2.5 times as much purge gas in order to successfully complete the purge. Mixing can also be reduced by injecting a small quantity of inert gas, called a **slug**, between the original gas or air content and the entering gas or air. The slug and the gas ahead of it are pushed through the pipe by the injected purge gas.

There are many factors to consider in order to complete a successful purge. Pipe diameter, injection inlet diameter, injection flow rate, and the density differential between the original gas and the purging gas are but a few.

The control center operator must constantly monitor the pipeline while purges are being carried out (*Figure 7*). This is why field personnel must always notify the control center when work is beginning and ending. If the pipeline is required to remain in operation during maintenance, operating pressure must be reduced.

2.4.0 Monitoring Pipeline Operations

There are a variety of skills that personnel must possess in order to monitor pipeline operations (such as purging) to ensure normal facility operation. Personnel monitoring pipeline operations must have a working knowledge of pipeline hydraulics or pneumatics and the pipeline elevations, if applicable, as well as the product characteristics for the pipeline. Personnel should also be familiar with the normal operating pressures and flows and the maximum operating pressure (MOP) requirements for the pipeline system.

Personnel should be familiar with company emergency response procedures and have an overall understanding of the pipeline system and process.

The following steps outline the general procedure for monitoring pipeline operations:

Step 1 Identify potential abnormal operating conditions that may occur, and know the appropriate actions to take in response to them.

Figure 7 Solar-powered line monitor and transmitter.

Step 2 Utilize the appropriate personal protective equipment (PPE) according to relevant company procedures.

Step 3 Verify that the flows, pressures, and product characteristics are within normal range.

Step 4 Set appropriate operating limits such as pressure and flow rate while ensuring that MOP is not exceeded.

Step 5 Continuously monitor, by whatever means available (SCADA, pressure gauge, meter), the pressures and flow rates of the pipeline to look for any changes. Monitor storage facilities. Verify continuous communications and that the data is accurate and updated.

Step 6 If pressures, flows, and line balance are not within normal range, determine the cause of variation from normal and correct the situation if possible. Correct the situation using methods such as starting/stopping pumps and units, and using adjustable control devices. Notify and seek assistance from proper personnel if necessary.

Step 7 Complete appropriate documentation as required by operator's procedures.

During performance of this covered task, all appropriate records must be completed and all permits, drawings, work orders, and records of the operation must be recorded. Design drawings must be updated to reflect any changes in the design or construction of the pipeline. Other items that may be recorded include the following:

- Beginning and end times
- Calculations of liquid displacement volumes
- Ambient conditions

The control center operator must return the line to normal operation very cautiously. Product must be introduced at the upper end of the section that has been displaced from the downstream end. On long segments of pipeline, venting may be required at several locations.

Did You Know?

Nitrogen

Inert gases are so-called because they do not, in general, react with other chemicals or gases such as oxygen. Nitrogen is commonly used as a purge gas. Historical documents are often sealed in containers pressurized with nitrogen in order to prevent the degradation that can be caused by exposure to the air.

Additional Resources

It is assumed that you will learn to perform these procedures under the direct supervision of your crew leader, foreman, or instructor. For independent study, refer to the recommended Additional Resources at the back of this module.

2.0.0 Section Review

1. Most hazardous atmosphere detection devices are set to activate when _____.
   a. the presence of an explosive atmosphere is detected
   b. an atmosphere reaches the lower explosive limit
   c. an atmosphere reaches 20 percent of the lower explosive limit
   d. an atmosphere reaches 25 percent of the lower explosive limit

2. A UV fire detection system has an advantage over an ionization detector because _____.
   a. it is not affected by thick smoke
   b. it can instantly detect the presence of a fire
   c. only a small amount of smoke is needed for activation
   d. it can detect fires that are not in the line of sight

3. The purging of a pipeline is necessary _____.
   a. only when the pipe is opened for maintenance
   b. only when the pipe is closed after maintenance
   c. when the pipe is opened and when it is closed
   d. only when hazardous vapors are present

SECTION THREE

3.0.0 MONITORING REMOTE PIPELINE AND FACILITY LOCATIONS

Objective

Explain how remote locations and weather conditions are monitored.

a. Describe how remote video surveillance is performed.
b. Describe how aerial surveillance is performed.
c. Describe how weather conditions are monitored.

Due to the size of many pipelines, pipeline companies must use alternative means to efficiently monitor the line, the associated facilities, and any weather that could affect operation. These alternative means include using remote video, monitoring by aircraft, and performing ground inspections. *Figure 8* shows a remote pipeline installation with a monitoring and transmission station on the right.

3.1.0 Remote Video Surveillance

Facilities provide 24-hour security and supplemental electronic surveillance devices (cameras and monitors), as appropriate. Cameras are positioned on critical equipment and locations so that security and operations personnel can see a problem even if the SCADA is down. Being able to see all parts of the facility is important to the pump station control center operator. Cameras and monitors can show if a valve is open or closed or any number of other conditions, depending on application. Cameras and monitors can reveal serious problems such as equipment failure, physical intrusions, and other security risks. They are also important when the weather is too inclement for outside work.

3.2.0 Aerial Surveillance

Aircraft can be used when monitoring the pipeline. Company staff must take timely and appropriate actions when reports of abnormal conditions are received and must maintain a record of the pilot's observations and the company response as required by company policy. The status or disposition of the item reported must be recorded. The factors in determining whether the observation is a relevant action item must be recorded.

3.3.0 Monitoring Weather Conditions

Operations personnel must monitor weather conditions on a regular basis. Inclement weather such as a severe electrical storm, tornado, hurricane, or blizzard can be a serious event for operations personnel. Attempts to remain outside and continue normal operations could result in serious bodily injury or death. Activities may be altered and shelter sought as warranted. Complete shutdown may be necessary in more severe cases until con-

Figure 8 Remote pipeline installation.

trol of the pipeline can be assured and it is safe for personnel to return to normal duties.

> **WARNING!**
> Always seek shelter during severe weather.

Many operations depend on moderate weather conditions. High winds, rain, snow, ice, and cold temperatures may seriously hinder some operations. In some locations, the affecting condition might be the approach of a severe weather front or a hurricane. Pipeline personnel must monitor weather constantly to be able to protect personnel, equipment, and product.

A severe electrical storm system contains the danger of deadly lightning and wind that could endanger personnel and facilities. In the case of a hurricane or tornado, shutdown of the pipeline should be considered.

Weather reports are available on weather-related web sites for the use of all personnel along the pipeline. Many stations have small weather monitors installed so that conditions can be accurately monitored locally, including items such as the following:

- Outside temperature
- Inside temperature
- Wind direction and speed (with high wind alarm)
- Wind chill factor (with low wind chill alarm)
- Daily rainfall
- Barometric pressure
- Humidity (inside, with alarm)

Additional Resources

It is assumed that you will learn to perform these procedures under the direct supervision of your crew leader, foreman, or instructor. For independent study, refer to the recommended Additional Resources at the back of this module.

3.0.0 Section Review

1. Cameras are important in remote or inaccessible areas because they can show if _____.
   a. hazardous gas is present
   b. a valve is open or closed
   c. a line is venting properly
   d. a negative wave has developed

2. When aerial surveillance detects an abnormal condition, company staff must first _____.
   a. verify the situation using the SCADA system
   b. contact management and government agencies
   c. take timely and appropriate action
   d. reduce line pressure and/or flow rate

3. With regard to weather, a pipeline _____.
   a. can only operate in moderate weather
   b. can operate in any weather condition
   c. cannot operate if lightning is present
   d. can be affected by severe weather

Summary

Pipeline facilities are equipped with a variety of sensors, alarms, controls, and safety devices. Monitoring the status of the facility is accomplished through the Supervisory Control and Data Acquisition (SCADA) computer system. It is the primary job of the control center to monitor the system and to react in an appropriate and timely manner if any abnormalities are detected. Field personnel must remain in constant contact with the control center to accomplish this. If the SCADA system goes down, immediate and appropriate action must be taken to maintain safe and normal operation of the pipeline.

Safety devices and alarms are installed within the facility to report, prevent, or minimize the effects of fire, leaks, over-pressure, and hazardous gases. These include thermal relief devices, ionization detectors, ultraviolet (UV) detectors, infrared (IR) detectors, thermal flame detectors, liquid leak detectors, and seal leak detectors. Any alerts or alarms produced by these devices indicate an abnormal operating condition, and must be responded to in accordance with company procedures.

In a number of situations, pipelines must be purged of either air or hazardous gas. Hazardous gas is accomplished by injecting an inert gas such as nitrogen into the line, and air is purged by injecting natural gas. A small amount of inert gas, called a slug, may be injected to separate the gas or air being purged and the gas doing the purging.

Purging and other pipeline operations must be monitored carefully to ensure conditions are normal throughout the process. Pipeline facilities are also remotely monitored through means such as video and aerial surveillance. In addition, weather conditions in the area of a pipeline must be continuously monitored, and responses to severe weather must be followed according to company/operator procedures.

Review Questions

1. If any pipeline parameters are not normal, field personnel must _____.
   a. immediately shut down the line
   b. determine the cause and take necessary steps
   c. isolate the portion of the line experiencing abnormal conditions
   d. wait for reports from field personnel before acting

2. If an unintended valve closure occurs, field personnel must first _____.
   a. investigate the cause of the closure
   b. initiate a pipeline shutdown
   c. notify the control center to investigate the situation
   d. immediately attempt to reopen the valve

3. A rise in pressure in the line could indicate _____.
   a. a leak in the line
   b. an increase in the flow rate
   c. an unintended valve closure
   d. impending equipment failure

4. The two primary indicators of normal, as opposed to abnormal, operating conditions are pressure and _____.
   a. atmosphere
   b. gravity
   c. flow
   d. temperature

5. The computational pipeline monitoring (CPM) system allows field personnel to _____.
   a. send open and shut signals to selected valves
   b. control the pipeline and see the results in real time
   c. monitor hazardous gas releases
   d. control flame or gas detectors

6. The computer operation that compares suction and discharge pressure against preset limits, and compares present data to past data is called _____.
   a. trending
   b. computational pipeline monitoring
   c. negative pressure wave monitoring
   d. steady state

7. A system that provides early detection and warning of the presence of hazardous atmospheres is a/an _____.
   a. ESD system
   b. UV protector
   c. flammable gas detection system
   d. thermal relief device

8. At each pump station, three types of detectors that continuously sample the air for hazardous atmosphere and monitor the area for high heat and flame are lower explosive limit, thermal, and _____.
   a. ultraviolet
   b. gas
   c. high sump
   d. liquid leak

9. The purpose of thermal relief devices is to _____.
   a. relieve temperature increase
   b. warn of increased pressure
   c. relieve pressure increases
   d. warn of increased temperature

10. A major function performed by the fire detection and suppression system is isolating _____.
    a. a liquid leak
    b. a seal leak
    c. hydrocarbon sources
    d. oxygen

11. Products of combustion in the beginning stage are detected by what types of detectors?
    a. Gas detectors
    b. Ionization detectors
    c. Liquid leak detectors
    d. Seal leak detectors

12. What kind of systems are set up so that at least two different, overlapping sensors must alarm before any action can be taken?
    a. Separate
    b. Combined
    c. Gas
    d. Voted

13. A seal detector is made up of a tube running to a sump and a _____.
    a. pump
    b. bypass
    c. monitor
    d. collector

14. What substance is commonly used to purge hazardous vapors from the line?
    a. Oxygen
    b. Air
    c. Nitrogen
    d. Natural gas

15. If a pipeline is to remain in operation during maintenance, then _____.
    a. flow rate must be reduced
    b. operating pressure must be reduced
    c. safety parameters must be reset
    d. management must be notified

16. What action should be considered if severe weather, such as a hurricane or tornado, is approaching a pipeline?
    a. Reduction in flow rate
    b. Shutdown
    c. Call 911
    d. Purging procedures

Trade Terms Quiz

Fill in the blank with the correct term that you learned from your study of this module.

1. Conditions within a pipeline system that may indicate a malfunction of a component or a condition exceeding design limits that may result in a hazard to persons, property, or the environment are called ________.
2. Prevention of the formation of explosive mixtures by replacing a pipeline's atmosphere with an inert substance is called ________.
3. A sensor that detects the presence of fire by detecting ultraviolet rays is called a(n) ________.
4. An instrument that reports the performance of a control device or signals unusual conditions in a system is called a(n) ________.
5. Gas detection systems that sense the amount of hydrocarbon gas and warn when gas has reached a dangerous concentration are called ________.
6. A type of alarm system layout using sensitive detectors and delayed action relays to detect flame and gas is called a(n) ________.
7. The point when pressures and flows are relatively constant over time is called a(n) ________.
8. A heat-sensing device which warns of flame by detecting a sudden rise in temperature is called a(n) ________.
9. A small amount of gas inserted between gas being purged from a line and the gas doing the purging is called a(n) ________.
10. A device that detects hazardous gas using infrared waves is called a(n) ________.
11. Status events that result from a problem, abnormality, or emergency within a pipeline system are called ________.
12. A photoelectric detection device used to signal an alarm when smoke is detected in a location is called a(n) ________.
13. A control system used in the pipeline industry that collects and assembles data and also allows operators to control the pipeline remotely is called a(n) ________.

Trade Terms

Abnormal operating conditions (AOCs)
Alarms
Flammable gas detection systems
IR detector
Ionization detector
Monitor
Purging
Slug
Steady state
Supervisory Control and Data Acquisition (SCADA)
Thermal detector
UV detector
Voted System

Trade Terms Introduced in This Module

Abnormal operating conditions (AOCs): Conditions within a pipeline system, identified by the operator, that may indicate (1) a malfunction of a component or (2) a deviation from normal operations that may indicate a condition exceeding design limits or result in a hazard to persons, property, or the environment.

Alarms: Status events from a pipeline monitoring system (such as SCADA) that result from a problem, abnormality, or emergency that occurs in a pipeline system. All alarms have a mandatory response (or action) associated with them.

Flammable gas detection systems: A gas detection device that senses the amount of hydrocarbon gases in the atmosphere and warns when a concentration of gas has become high enough to cause an explosion in the presence of an ignition source.

IR detector: A sensor that detects the presence of hazardous gas by sending an infrared beam to a receiver and measuring the amount of IR waves that are absorbed by the gas.

Ionization detector: A photoelectric detection device used to signal an alarm when smoke is detected in a location. Commonly known as a smoke detector.

Monitor: An instrument that reports the performance of a control device or signals if unusual conditions appear in a system. For example, an S&W monitor provides a mechanical means of preventing contaminated oil from entering the pipeline. In common usage, the term also applies to the activities of operators and other personnel observing and tracking conditions on the pipeline.

Purging: Replacing the atmosphere in a pipeline by injection of an inert substance to prevent formation of explosive mixtures.

Slug: A small amount of inert gas used in purging operations which is injected between the original gas and the purging gas. The slug acts as a barrier reducing or preventing mixing of the two gases.

Steady state: The point when pressures and flows are relatively constant over time.

Supervisory Control and Data Acquisition (SCADA): A control system used in the pipeline industry that collects and assembles data, and also allows operators to control the pipeline remotely.

Thermal detector: A heat-sensing device that warns of the presence of flame by detecting a sudden rise in temperature. Also known as a flame detector or bolometer.

UV detector: A sensor that detects the presence of fire by detecting the ultraviolet rays given off by incipient (beginning) fires or flames.

Voted system: A type of alarm system layout used for flame detection and gas detection employing very sensitive detectors and delayed action relays. Such systems will not alarm until at least two or more detectors are tripped, depending on how the voted area is set up.

Additional Resources

The following recommended resources can provide additional helpful information related to the requirements for this covered task:

API Recommended Practice 1161, Third Edition, © 2014 American Petroleum Institute. API Publishing Services, 1220 L Street, NW, Washington, DC 20005. Distributed by Thompson Reuters (Scientific) LLC, **www.techstreet.com**.

NCCER Module AOCCC-17, *Abnormal Operating Conditions Control Center*

NCCER Module AOCFG-17, *Abnormal Operating Conditions Field and Gas*

49 *CFR* 192, **www.ecfr.gov**.

49 *CFR* 195, **www.ecfr.gov**.

The following websites offer resources for products and training:

American Petroleum Institute, **www.api.org**.

Pipeline and Hazardous Materials Safety Administration, **www.ops.dot.gov**.

Transportation Safety Institute (US Department of Transportation), **www.tsi.dot.gov**.

Figure Credits

© iStockphoto.com/Huyangshu, Module Opener

INSAGO/Shutterstock.com, Figure 1

Christian Lagerek/Shutterstock.com, Figure 2

Tim Mills, Figures 4, 8, 9

© iStockphoto.com/ekipaj, Figure 5

Jim Parkin/Shutterstock.com, Figure 6

General Monitors, Figure 7

Section Review Answer Key

Answer	Section Reference	Objective
Section One		
1. d	1.0.0	1
2. c	1.2.3	1b
Section Two		
1. c	2.1.1	2a
2. b	2.2.2	2b
3. c	2.3.0	2c
Section Three		
1. b	3.1.0	3a
2. c	3.2.0	3b
3. d	3.3.0	3c

NCCER CURRICULA — USER UPDATE

NCCER makes every effort to keep its textbooks up-to-date and free of technical errors. We appreciate your help in this process. If you find an error, a typographical mistake, or an inaccuracy in NCCER's curricula, please fill out this form (or a photocopy), or complete the online form at **www.nccer.org/olf**. Be sure to include the exact module ID number, page number, a detailed description, and your recommended correction. Your input will be brought to the attention of the Authoring Team. Thank you for your assistance.

Instructors – If you have an idea for improving this textbook, or have found that additional materials were necessary to teach this module effectively, please let us know so that we may present your suggestions to the Authoring Team.

NCCER Product Development and Revision

13614 Progress Blvd., Alachua, FL 32615

Email: curriculum@nccer.org
Online: www.nccer.org/olf

❑ Trainee Guide ❑ Lesson Plans ❑ Exam ❑ PowerPoints Other ______________

Craft / Level: Copyright Date:

Module ID Number / Title:

Section Number(s):

Description:

Recommended Correction:

Your Name:

Address:

Email: Phone:

CT63_4-17

Locally Operate Valves on a Liquid Pipeline System

Overview

A variety of valve types are used to control and direct the flow of product through the pipeline system. This module examines the procedures for locally operating pipeline valves.

Trainees with successful module completions may be eligible for credentialing through the NCCER Registry. To learn more, go to **www.nccer.org** or contact us at 1.888.622.3720. Our website has information on the latest product releases and training, as well as online versions of our *Cornerstone* magazine and Pearson's product catalog.

Your feedback is welcome. You may email your comments to **curriculum@nccer.org**, send general comments and inquiries to **info@nccer.org**, or fill in the User Update form at the back of this module.

This information is general in nature and intended for training purposes only. Actual performance of activities described in this manual requires compliance with all applicable operating, service, maintenance, and safety procedures under the direction of qualified personnel. References in this manual to patented or proprietary devices do not constitute a recommendation of their use.

CT63_4-17
Locally Operate Valves on a Liquid Pipeline System

Objectives

When you have completed this module, you will be able to do the following:

1. Describe how to locally operate valves on a liquid pipeline system (CT63_4-17).
   a. Describe how to locally operate valves.
   b. Identify the knowledge and skill associated with locally operating valves.

Performance Task

Under the supervision of the instructor, you should be able to do the following:

1. Locally operate valves on a liquid pipeline system (CT63_4-17).
   - Identify potential abnormal operating conditions (AOCs) that may occur during performance of this CT, and know the appropriate actions to take in response to them.
   - Utilize the appropriate personal protective equipment (PPE) according to relevant company procedures.
   - Verify that flows, pressures, and product characteristics are within normal range.
   - Identify the valve(s) to be operated, and identify the intended flow path.
   - Operate the valve in accordance with company policy and procedures.
   - Check that the desired result was achieved. (If it was not, contact appropriate personnel.)
   - Complete appropriate documentation as required by operator's procedures.

Trade Terms

Abnormal operating condition (AOC)
Actuator
Ball valves
Check valve
Gate valves
Lower explosive limit (LEL)
Manifold
Operator
Plug valves
Pressure surge
Valves

Industry Recognized Credentials

If you are training through an NCCER-accredited sponsor, you may be eligible for credentials from NCCER's Registry. The ID number for this module is CT63_4-17. Note that this module may have been used in other NCCER curricula and may apply to other level completions. Contact NCCER's Registry at 1.888.622.3720 or go to **www.nccer.org** for more information.

Contents

Figures

Section One

1.0.0 Locally Operate Valves on a Liquid Pipeline System

Objective

Describe how to locally operate valves on a liquid pipeline system (CT63_4-17).

a. Describe how to locally operate valves.
b. Identify the prerequisites, competencies, and abnormal operating conditions (AOCs) associated with locally operating valves.

Performance Task

Locally operate valves on a liquid pipeline system (CT63_4-17).

- Identify potential abnormal operating conditions (AOCs) that may occur during performance of this CT, and know the appropriate actions to take in response to them.
- Utilize the appropriate personal protective equipment (PPE) according to relevant company procedures.
- Verify that flows, pressures, and product characteristics are within normal range.
- Identify the valve(s) to be operated, and identify the intended flow path.
- Operate the valve in accordance with company policy and procedures.
- Check that the desired result was achieved. (If it was not, contact appropriate personnel.)
- Complete appropriate documentation as required by operator's procedures.

Trade Terms

Abnormal operating condition (AOC): A condition within a pipeline system, identified by the operator, that may indicate (1) a malfunction of a component or (2) a deviation from normal operations that may indicate a condition exceeding design limits or result in a hazard to persons, property, or the environment.

Actuator: A device used to operate a valve using electrical, pneumatic, or hydraulic means. Actuators are often used for the remote control or sequencing of pipeline valves.

Ball valves: Valves that use a spherical closure element which is rotated through 90° to open or close the valve.

Check valve: A valve that is held open by the fluid flow in one direction and closes when the flow is stopped or reversed.

Gate valves: Valves that contain a closing element consisting of a wedge or slab positioned between two seating surfaces, which can be moved into or out of the flow stream perpendicular to the direction of flow.

Lower explosive limit (LEL): The minimum concentration of vapor in the air below which a flame does not occur in the presence of an ignition source.

Manifold: A common pipe chamber having several lateral outlets and constructed with a system of valves.

Operator: A device that uses manual, hydraulic, electrical, or pneumatic energy to open or close a valve.

Plug valves: Valves that have a closure element in the form of a tapered plug.

Pressure surge: A wave resulting when a fluid in motion is forced to suddenly stop or change directions. This commonly occurs when a valve is suddenly closed at the end of a pipeline system and a pressure wave propagates in the pipe.

Valves: Devices used to control the flow of fluid in a pipeline.

The devices that are installed at specific locations along liquid pipelines, providing a variety of functions, are called **valves**. These valves can be opened, partially opened, or closed to control and direct the flow of product. Locally operated valves must be opened or closed either manually, by an electrically controlled valve **actuator**, or by a portable valve **operator**. The failure of a valve to operate, or the improper operation of a valve, can be both costly and dangerous. The safe operation of valves is key to pipeline safety.

1.1.0 Local Valve Operation

Local valve operation begins by first identifying the valve to be operated and then positioning the valve manually or with the aid of a portable valve operator. The operation ends when the proper valve position has been established and verified by observing the valve position indicator.

An **abnormal operating condition (AOC)** is defined under federal regulations as a condition identified by the pipeline operator that may indicate (1) a malfunction of a component or (2) a deviation from normal operations that may indicate a condition exceeding design limits or result in a hazard to persons, property, or the environment.

The following steps outline the general procedure for locally operating valves on a liquid pipeline system.

Step 1 Identify potential abnormal operating conditions (AOCs) that may occur during performance of this CT, and know the appropriate actions to take in response to them.

Step 2 Utilize the appropriate personal protective equipment (PPE) according to relevant company procedures.

Step 3 Verify that flows, pressures, and product characteristics are within normal range.

Step 4 Before executing the sequence, identify the need to open or close a valve or start or stop a unit to control flow or pressure.

Step 5 Identify the valve(s) to be operated, and identify the intended flow path.

Step 6 Operate the valve in accordance with company policy and procedures.

Step 7 Check that the desired result was achieved. (If it was not, contact appropriate personnel.)

Step 8 Complete appropriate documentation as required by operator's procedures.

CAUTION

An individual's actions during the performance of a covered task could cause a safety device to activate. The activation represents a deviation from normal operations, resulting in an abnormal operating condition (AOC).

1.1.1 Tank Farm Valves

A valve is a mechanical unit that is used to control, open, or shut off a line. Some valves are used as automatic or semi-automatic safety devices. Those used extensively include check, gate, ball, plug, needle, and pressure relief valves. In the design of pipelines and terminals, valves are placed at specific locations for specific reasons or functions. In the terminal, a **manifold** is constructed with a system of valves so that, when in a particular alignment, the product can be diverted to the exact location desired by opening the proper series of valves in the correct sequence. A pipeline manifold is shown in *Figure 1*. Field personnel can choose a particular tank or group of tanks in the tank farm, processing equipment, or loading facilities, or may direct a shipment to another pipeline.

Figure 1 Pipeline manifold.

1.1.2 Valve Alignment

Opening and closing valves in the proper sequence, known as valve alignment, is a skill that must be mastered. Pipelines and terminal operations are based on the specific sequences in which valves are opened and closed (or aligned). Most valves have a position indicator, which is a shaft, rod, or pointer that extends from the valve stem or mechanism and indicates the exact position of the valve.

In order to start and stop the pipeline, valves must be opened and closed in a very specific sequence, which must be followed over many miles of the pipeline. Valves opened or closed unexpectedly can result in wrong deliveries, contamination, or loss of product, as well as lost time and work. In a worst-case scenario, a pipeline rupture or even an accident of catastrophic proportions could occur as a result of improper valve operation.

In the terminal, particular alignments are required in order for product to reach certain delivery points. Some valves are used to throttle the flow or divert flow to tankage in order to relieve pressure at another station (*Figure 2*). In emergencies, field personnel must be able to use all their skills and knowledge of valve locations to make quick decisions about which valves to open or close in order to resolve identified situations.

Terminals and stations have standard operating procedures covering the exact valve alignments or lineups necessary to accomplish tasks laid out on charts, as shown in *Figure 3*. Valves are designated by numbers and/or letters so that they can be exactly designated. If a problem occurs and the facility must be switched to local control, the location personnel must perform tank and line switches, custody transfers, and other duties.

Figure 2 Hand-operated valve on a storage tank.

1.1.3 Valve Operation

Many valves used in pipeline facilities are operated automatically or by remote control (*Figure 4*). However, every facility has valves that are operated by hand. It is necessary for the facilities personnel to be familiar with the types, functions, locations, and operations of these manually operated valves.

At strategic points, large manual gate valves that are only operated for maintenance or emergency situations are used to help contain product in case of a release. In times of power failure, many installations have backup generator units or battery-operated units to control the valves. In cases where there is no electrical power, portable valve operators may be used. These are handheld devices that engage and turn the shaft of the valve, thus speeding the process while putting less strain on the personnel involved. Portable valve operators may be powered pneumatically or hydraulically, or by electricity from a small generator.

WARNING!

Over-tightening a valve may cause serious damage to the valve, in some cases worsening a situation the valve was installed to prevent.

Although hand-cranking or hand-turning operations might be possible on some valves, the number of turns required to open other valves would make an attempt useless, if not prohibitive, in an emergency. An example of this is a large gate valve fitted with a hand wheel and gearbox. In low gear, this particular valve requires 10,000 turns of the wheel to fully open or close manually. If it were not for the geared drive mechanism, it might be impossible to turn it at all. In high speed, the same valve would still require 2,500 turns to fully operate the valve. The time consumed making hundreds or thousands of turns would seriously hamper operations. The requirement of a portable valve operator in such a case can readily be seen and is a practical necessity, even in terminals (*Figure 5*).

1.1.4 Failure to Operate

If a valve fails to operate, either a light in the control room will signal that the valve is not operating, or field personnel in the yard will notice that a valve indicator is not moving up and down as appropriate. If product is coming out of the tank, flow will be lost and the line will shut down because of low pressure. If a switch is being made, product may be lost. If the valve fails while product is trying to enter a tank, pressure will increase and the line will shut down because of high discharge. Especially in the receiving station, if one valve opens and another remains shut, the line will become over-pressured and a rupture will likely occur. If the valve supplying the product closes and will not reopen, the only solution is to shut down the line until personnel can operate the valve manually.

WARNING!

When a valve, manifold, or other pipeline equipment must be repaired, always remember to make the proper notifications per company policy, and to follow lockout/tagout procedures.

Total failure of electrical power requires decisive action on the part of field personnel. It is likely that all electronic valve status indicators will be lost. The established pumping and switching schedule may still be followed if the pipeline is not shut down.

In certain situations, it may be necessary to occupy some critical stations in remote locations. Call-outs may be initiated so that someone can operate strategic pipeline valves (Figure 6) manually until power can be restored and normal operating conditions are achieved again.

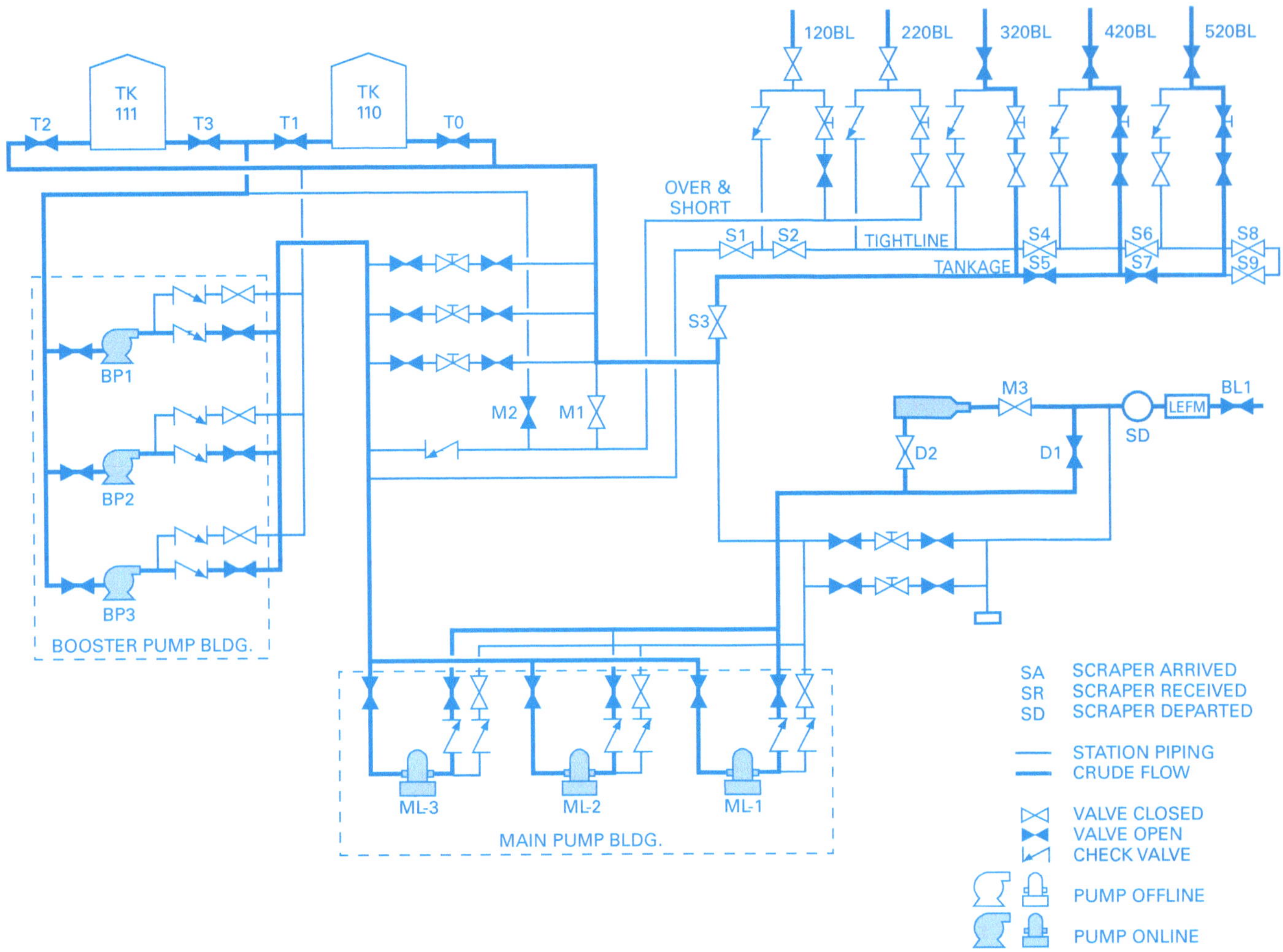

Figure 3 Valve alignment chart.

1.1.5 Valve Access

Valves are used in a pipeline system to control the flow of fluids and to isolate segments of pipe or piping systems for maintenance or emergency response activities. In order to perform these important functions, ready access to the valves is critical, particularly if they must be manually operated by field personnel in response to a release. In an emergency, mitigation of the hazards may be hampered by valves that are obscured by overgrown vegetation, or are difficult to access because of marginal roads or flooding.

The locations and functions of all valves should be clearly indicated on facility maps, pipeline alignment sheets, or otherwise linked with latitude-longitude coordinates so that they can be quickly and positively identified by field personnel. A potentially serious incident such as an inadvertent release can be minimized if location personnel are able to quickly locate and close the necessary valve.

Although valves must be accessible for qualified field personnel, they must also be locked or fenced to prevent unauthorized access by the general public. By regulation, pipeline operators are required to provide protection for each valve from unauthorized operation and from vandalism.

1.1.6 Excessive Vapors or Visible Product

Excessive vapors may not be visible but can be detected from the strong odor of product in the area. Meters may not have detected the **lower explosive limit (LEL)**, but such a level may be approaching if vapors are escaping from a leaking tank or line (*Figure 7*). Valve access must be restricted until the source of such vapor can be discovered and controlled, and vapor levels have dissipated.

> **WARNING!**
>
> Electrical switching causes arcing between switch contacts. For that reason all activity involving electrical equipment and switches must cease until vapor levels have subsided.

Figure 4 Valve control panel.

Figure 5 Large geared valve with both manual and electrical controls.

Figure 6 Pipeline valves at a remote station.

Figure 7 Flammable gas leak detector.

A visible vapor cloud is the most dangerous hindrance to valve access. Such clouds move with the slightest breeze. All personnel must remove themselves to a safe distance from any vapor cloud. At the same time, close observations of the vapor cloud must be maintained so that if it moves, personnel can be warned to move again.

WARNING!

Under no circumstances should any person attempt to reach a valve when a vapor cloud is observed in the area.

Frozen product in the form of a frost ball of liquid petroleum (LP) or possible gasoline such as butane on the ground near a release suggests a serious situation in which the product is frozen without a visible vapor cloud. It suggests a possible heavy accumulation of vapor around and near it. Danger of fire and explosion is critical in the area. Reaching a valve in such a location could be deadly. Consideration should be given to the possible presence of static electricity. All electrical activities in the area must cease until the release is controlled and the product is cleaned up.

If product is observed to be accumulating on the surface around a valve or manifold, the area must be avoided. It is obvious that the surface underneath is saturated to the point of supporting product. Wading in product without required PPE is never acceptable. Other than the danger of ignition, product saturates shoes and clothing. The fumes from some products could cause suffocation. If the product has a caustic character, skin and flesh can be burned. Such a valve must be isolated until the source of the release is controlled and the release is cleaned up. If personnel can determine the source and close a valve safely, containment can be greatly facilitated.

WARNING!

Wading in product is never an acceptable procedure unless proper PPE is worn and proper safety precautions are taken.

1.2.0 Personnel Knowledge and Skills

Personnel must possess a variety of skills in order to successfully operate valves on a liquid pipeline. The operation itself requires that the personnel locally operating valves must be competent in performing the operation. Personnel must also be aware of the particular AOCs that may occur during the procedure and know the proper responses to them.

1.2.1 Prerequisites and Competencies

Personnel operating pipeline valves must know how to determine the valve's position. Each valve, other than a **check valve**, has some means of indicating the valve's position. Examples may include the following:

- Rising stem
- Arrow
- Handle position
- Open/close flag or display

Items to be considered prior to valve operation include the following:

- Impact on the pipeline such as pressure, flow, and tank levels should be taken into account. A **pressure surge** and hydraulic shock can result from improper valve operation.
- Verify that the correct valve to be operated has been identified. Operation of an incorrect valve could cause an unsafe condition.
- Creation of a thermal trap by shutting-in segments of the pipeline system where it could be over-pressured because of an increase in product temperature.

Field personnel must be able to use the appropriate references to correctly identify the valve to be operated (*Figure 8*). Personnel must understand how to locally operate a valve, either manually or by the use of a motor-operated actuator. They must also know how to determine that the valve is in the intended position, and be able to communicate this information to field personnel and/or the control center.

Field personnel must know that for **plug valves** and **ball valves**, the wrench and/or position

Figure 8 DRA injection control valve.

indicator must be in line with the pipeline when the valve is open and transverse when the valve is closed. Valves fitted with a manual or powered actuator must be furnished with a visible indicator to show the open and closed positions parallel to the pipe in the fully open position and at right angles to the pipe in the fully closed position. An intermediate position indicates a partially opened or closed valve.

Field personnel must have a working understanding of pipeline hydraulics and the maximum operating pressure (MOP) requirements for the pipeline. Field personnel must also understand company emergency response procedures.

The control center has ultimate responsibility and authorization for actions that affect the safe operation of the pipeline. Personnel must understand what communications between local operators and/or the control center are required in association with valve operation.

1.2.2 Abnormal Operating Conditions

Personnel locally operating valves on a liquid pipeline system must be able to identify potential abnormal operating conditions (AOCs) and know the proper response to each, should they occur. The following are some potential AOCs, and the appropriate action personnel should take in response to them:

- *An unexplained pressure deviation occurs* – Personnel should make the condition safe according to the extent possible and the operator's procedures. Assess the situation for safety, and for environmental or physical damage. Reactions to an unexplained pressure deviation could include the following:
  - Shutting down the system (if qualified)
  - Returning the valve to its original position
  - Operating another valve to relieve the situation
  - Isolating damaged equipment
  - Making the proper notifications

  An unexplained pressure deviation may be indicated by pressure display(s), a sound or vibration, or an alarm or other pressure indicator (*Figure 9*) showing the unexplained pressure deviation.
- *An unexplained flow rate deviation occurs* – Personnel should make the condition safe to the extent possible, in accordance with the operator's procedures. Assess the situation for safety and for environmental or physical damage. Reactions to an unexplained flow rate deviation could include the following:

Figure 9 Electronic pressure transmitter.

  - Returning the valve to its original position
  - Operating another valve to relieve the situation
  - Isolating damaged equipment
  - Shutting down the system (if qualified)
  - Making the proper notifications

  An unexplained flow rate deviation may be indicated by flow gauge readings, alarms, improper tank levels, or other flow indicators.
- An *unexplained valve position indication occurs* – Personnel should first confirm the actual valve position. If the valve is not in the indicated position, investigate the cause and resolve the source of discrepancy between the actual valve position and the position indicator. Proper valve position is required before safe pipeline operation resumes. Make the proper notifications.
- *An inoperable valve occurs* – Personnel should make the condition safe to the extent possible and in accordance with operator's procedures. Assess the situation for safety, and for environmental or physical damage. Reactions could include the following:
  - Re-try operating the valve to see if the situation can be corrected
  - Relieve excessive differential pressure across the valve
  - Shutting down the system (if qualified and if necessary)
  - Make all appropriate notifications

A valve may not operate as intended, or may not open or close. An inoperable valve may be detected by observing that the valve position indicator does not show the intended position, or that valve operation results in unexpected flow or pressure outcomes. A valve may be stuck and unable to be operated either manually or by a valve operator or actuator. Failure to operate may also be due to excessive differential pressure across the valve.

- *An unexpected presence of hazardous liquid or vapor occurs* – Personnel should minimize the situation to the extent possible if it can be safely done, and make all appropriate notifications. Incorrect valve operation could lead to an unintended release of product that may be observed by sight, smell, sound, or alarms.
- *An exposed pipe or missing/damaged pipeline support occurs* – Personnel should make notifications as required by company policy.
- *Activation of a safety device, such as a pressure relief valve or a remotely controlled shutdown device (RCSD) occurs* – Personnel should investigate the cause of the AOC, take appropriate actions to mitigate the situation (if possible), and make any necessary notifications.
- *A failure of communication devices, SCADA, or electrical services occurs* – Personnel should follow troubleshooting procedures and make contacts as required by company policy.
- *A fire or explosion occurs* – Personnel should immediately evacuate the area and make proper notifications, including notifying 911 as appropriate.
- *Damage to the pipeline system occurs* – Personnel should minimize or eliminate pressures and/or flows as appropriate and make necessary notifications as required by company policy.
- *An unintended status change, such as a change in unit status or valve position occurs* – Personnel should investigate the cause of the AOC, take appropriate actions to mitigate the situation (if possible), and make any necessary notifications.

WARNING!

Proper PPE must be worn and proper safety precautions must be taken when dealing with an unexpected release of hazardous liquids or vapors.

Additional Resources

It is assumed that you will learn to perform these procedures under the direct supervision of your crew leader, foreman, or instructor. For independent study, refer to the recommended Additional Resources at the back of this module.

1.0.0 Section Review

1. A system of valves is called a(n) _____.
   a. gate
   b. array
   c. manifold
   d. cluster

2. If an unexplained pressure deviation occurs, _____.
   a. the line must be shut down
   b. try restoring the valve to its original position
   c. do not operate another valve to relieve the situation
   d. do not reverse the direction of the valve until the cause is determined

SUMMARY

Valves are installed and operated to provide a variety of functions in a pipeline system. Valve operation begins by identifying and positioning the valve, and ends when the valve position has been established and verified. Check, gate, plug, needle, and pressure relief valves are placed at specific locations for specific reasons. Manifolds are systems of valves that allow personnel to divert product to a particular location.

Valve alignment is the specific sequence in which valves are opened or closed. Valves opened or closed improperly can result in wrong delivery, contamination, loss of product, or loss of time and work. Some valves throttle or divert flow to relieve line pressure. Valves are identified by numbers or letters, and personnel must make quick decisions about which valve to open or close in an emergency.

While many valves are operated automatically or remotely, some must be manually operated, and personnel must know their types, functions, locations, and operations. Large manual gate valves for maintenance or emergency are located at strategic points and portable valve operators, which can quickly open a valve, may be needed in case of power failure.

Valve failure is indicated by a light in the control room or incorrect position indicator. Failure may result in product lost during a switch, loss of flow and pressure, an overpressure, or line rupture. If a valve closes and will not open, a line shutdown, manual operation, call-outs, and occupying remote locations may be necessary.

Valves should be conspicuous, clearly marked, and easy to locate on maps or alignment sheets. They must be protected from unauthorized operation. Valve access must also be restricted from personnel when hazardous vapors or conditions are present.

Personnel operating valves must be able to identify a valve's position. There are considerations to be taken before operating a valve, including its impact on pressure, flow, or tank levels; pressure surges and hydraulic shock; and identification of the appropriate valve(s) to be operated. Personnel must know how to operate a valve, determine its position, and communicate valve operations with the control center.

AOCs are conditions in a pipeline that exceed operating design limits specified in 49 *CFR* 195. These include unexplained pressure or flow rate deviations. Personnel should be aware of any potential AOCs, and be able to respond to them with the appropriate action. Responses include returning the valve to its original position, operating another valve to relieve the situation, isolating damaged equipment, or shutting down the system. Procedures should always be followed thoroughly, especially in the event of an AOC.

Review Questions

1. The opening and closing of valves in the proper sequence is called _____.
   a. valve actuating
   b. valve entrainment
   c. valve sequencing
   d. valve alignment

2. A shaft, rod, or pointer that shows a valve's position is called a(n) _____.
   a. manifold
   b. indicator
   c. marker
   d. flag

3. To resolve issues quickly in an emergency, it is important that operating personnel use their knowledge of valve _____.
   a. locations
   b. types
   c. indicators
   d. ratings

4. What type of manually operated valve can be placed at strategic points to help contain product in case of a release?
   a. Ball valve
   b. Plug valve
   c. Gate valve
   d. Actuator

5. If a valve position indicator shows an unexplained valve position, first ____.
   a. return the valve to its original position
   b. operate another valve to alleviate the situation
   c. reduce differential pressure and then recheck
   d. confirm the actual position

Trade Terms Quiz

Fill in the blank with the correct term that you learned from your study of this module.

1. A device used to remotely operate a valve using electrical, pneumatic, or hydraulic means is called a(n) ________.
2. A common pipe chamber having several lateral outlets is called a(n) ________.
3. A wave caused by a fluid in motion suddenly stopping or changing direction is called a(n) ________.
4. Valves that contain a closing element consisting of a wedge or slap, which moves perpendicular to the flow, are called ________.
5. A condition within a pipeline system that exceeds design limits is called a(n) ________.
6. Valves that have a closure element in the form of a tapered plug are called ________.
7. Devices used to control the flow of fluid in a pipeline are called ________.
8. Valves that use a spherical closure element that is rotated through 90° to open or close are called ________.
9. A device that uses manual, hydraulic, electrical, or pneumatic energy to open or close a valve is called a(n) ________.
10. The minimum concentration of vapor in air below which a flame does not occur in the presence of an ignition source is called the ________.
11. A valve that is held open by the fluid flow in one direction and closes when the flow is stopped or reversed is called a(n) ________.

Trade Terms

Abnormal operating condition (AOC)
Actuator
Ball valves
Check valve
Gate valves
Lower explosive limit (LEL)
Manifold
Operator
Plug valves
Pressure surge
Valves

Trade Terms Introduced in This Module

Abnormal operating condition (AOC): A condition within a pipeline system, identified by the operator, that may indicate (1) a malfunction of a component or (2) a deviation from normal operations that may indicate a condition exceeding design limits or result in a hazard to persons, property, or the environment.

Actuator: A device used to operate a valve using electrical, pneumatic, or hydraulic means. Actuators are often used for the remote control or sequencing of pipeline valves.

Ball valves: Valves that use a spherical closure element which is rotated through 90° to open or close the valve.

Check valve: A valve that is held open by the fluid flow in one direction and closes when the flow is stopped or reversed.

Gate valves: Valves that contain a closing element consisting of a wedge or slab positioned between two seating surfaces, which can be moved in or out of the flow stream perpendicular to the direction of flow.

Lower explosive limit (LEL): The minimum concentration of vapor in the air below which a flame does not occur in the presence of an ignition source.

Manifold: A common pipe chamber having several lateral outlets.

Operator: A device which uses manual, hydraulic, electrical, or pneumatic energy to open or close a valve.

Plug valves: Valves that have a closure element in the form of a tapered plug.

Pressure surge: A wave resulting when a fluid in motion is forced to suddenly stop or change direction. This commonly occurs when a valve is suddenly closed at the end of a pipeline system and a pressure wave propagates in the pipe.

Valves: Devices used to control the flow of fluid in a pipeline.

Additional Resources

The following recommended resources can provide additional helpful information related to the requirements for this covered task:

API Recommended Practice 1161, Third Edition, © 2014 American Petroleum Institute. API Publishing Services, 1220 L Street, NW, Washington, DC 20005. Distributed by Thompson Reuters (Scientific) LLC, **www.techstreet.com**.

NCCER Module AOCFG-17, *Abnormal Operating Conditions Field and Gas.*

NCCER Module AOCCC-17, *Abnormal Operating Conditions Control Center.*

NCCER Module 64106-02, *Pipeline Operations.*

NCCER Module 60103-15, *Pipeline Communications.*

NCCER Module 60102-15, *Basic Pipeline Hydraulics and Equipment.*

NCCER Module CT63_1-17, *Startup of a Liquid Pipeline.*

NCCER Module CT63_2-17, *Shutdown of a Liquid Pipeline.*

NCCER Module CT63_3-17, *Monitor Pressures, Flows, Communications, and Line Integrity and Maintain Them Within Allowable Limits on a Liquid Pipeline System (Field).*

NCCER Module CT21_2-17, *Disassembly/Re-assembly of a Valve.*

NCCER Module CT52_1-17, *Conduct a Vegetation Survey.*

NCCER Module CT52_2-17, *Conduct a Leak Survey with a GCD.*

NCCER Module CT52_3-17, *Conduct a Leak Survey with a Flame Ionization Unit.*

49 *CFR* 192, **www.ecfr.gov**.

49 *CFR* 195, **www.ecfr.gov**.

Figure Credits

© iStockphoto.com/Huyangshu, Module Opener
INSAGO/Shutterstock.com, Figure 1
Christian Lagerek/Shutterstock.com, Figure 2
Tim Mills, Figures 4, 8, 9
© iStockphoto.com/ekipaj, Figure 5
Jim Parkin/Shutterstock.com, Figure 6
General Monitors, Figure 7

Section Review Answer Key

Answer	Section Reference	Objective
Section One		
1. c	1.1.1	1a
2. b	1.2.2	1b

NCCER CURRICULA — USER UPDATE

NCCER makes every effort to keep its textbooks up-to-date and free of technical errors. We appreciate your help in this process. If you find an error, a typographical mistake, or an inaccuracy in NCCER's curricula, please fill out this form (or a photocopy), or complete the online form at **www.nccer.org/olf**. Be sure to include the exact module ID number, page number, a detailed description, and your recommended correction. Your input will be brought to the attention of the Authoring Team. Thank you for your assistance.

Instructors – If you have an idea for improving this textbook, or have found that additional materials were necessary to teach this module effectively, please let us know so that we may present your suggestions to the Authoring Team.

NCCER Product Development and Revision

13614 Progress Blvd., Alachua, FL 32615

Email: curriculum@nccer.org

Online: www.nccer.org/olf

❑ Trainee Guide ❑ Lesson Plans ❑ Exam ❑ PowerPoints Other ________________

Craft / Level: Copyright Date:

Module ID Number / Title:

Section Number(s):

Description:

Recommended Correction:

Your Name:

Address:

Email: Phone:

CT64_1-17

Start-Up of a Liquid Pipeline (Control Center)

Overview

The ultimate purpose of the control center is to move product through the system efficiently, economically, and safely. Control center personnel must use their training, experience, and judgment to make good decisions that support this goal. This module describes the routine but crucial task of starting up a pipeline. This task involves more than just flipping a few switches—pipeline controllers must understand scheduling and confirm that everything is ready before starting equipment.

Trainees with successful module completions may be eligible for credentialing through the NCCER Registry. To learn more, go to **www.nccer.org** or contact us at 1.888.622.3720. Our website has information on the latest product releases and training, as well as online versions of our *Cornerstone* magazine and Pearson's product catalog.

Your feedback is welcome. You may email your comments to **curriculum@nccer.org**, send general comments and inquiries to **info@nccer.org**, or fill in the User Update form at the back of this module.

This information is general in nature and intended for training purposes only. Actual performance of activities described in this manual requires compliance with all applicable operating, service, maintenance, and safety procedures under the direction of qualified personnel. References in this manual to patented or proprietary devices do not constitute a recommendation of their use.

CT64_1-17
Start-Up of a Liquid Pipeline (Control Center)

Objectives

When you have completed this module, you will be able to do the following:

1. Describe key liquid pipeline components and their operations.
   a. Describe tanks and the operations associated with them.
   b. Describe manifolds and the valve operations associated with them.
   c. Describe pumps and the operations associated with them.
2. Describe the requirements and procedures for starting up a liquid pipeline (CT64_1-17).
   a. Identify the prerequisites, competencies, and abnormal operating conditions associated with starting up a liquid pipeline.
   b. Describe the procedures used to start up a liquid pipeline.

Performance Task

Under the supervision of your instructor, you should be able to do the following:

1. Start up a liquid pipeline (CT64_1-17).
   - Identify potential abnormal operating conditions that may occur during performance of this CT, and know the appropriate actions to take in response to them.
   - Utilize the appropriate personal protective equipment according to relevant company procedures.
   - Verify that the flows, pressures, and product characteristics are within normal range.
   - Ensure that upstream and downstream stations are prepared to bring the pipeline up before beginning the start-up procedure.
   - Activate (start up) units beginning at the origination point and proceeding downstream to the delivery point.
   - Activate units at downstream stations upon arrival of a pressure rise from the upstream station.
   - Perform proper operation according to company policy and procedures.
   - Verify that the desired result was achieved.
   - Complete the appropriate documentation as required by operator's procedures.

Trade Terms

Abnormal operating conditions (AOCs)
Automatic overfill protection system (AOPS)
Automatic tank gauge (ATG)
Breakout tank
Computational pipeline monitoring (CPM)
Critical high (CH)
Floating
Gravitometer
High-high (HH)
Interface
Line fill
Maximum operating pressure (MOP)
Maximum working (MW)
Optical interface detector (OID)
Packed
Supervisory Control and Data Acquisition (SCADA)
Steady state
Stratification
Surge
Tank gauging system
Tightlining
Transmix
Valve interlock
Variable frequency drive (VFD)

Industry Recognized Credentials

If you are training through an NCCER-accredited sponsor, you may be eligible for credentials from NCCER's Registry. The ID number for this module is CT64_1-17. Note that this module may have been used in other NCCER curricula and may apply to other level completions. Contact NCCER's Registry at 888.622.3720 or go to **www.nccer.org** for more information.

Contents

Figures and Tables

Section One

1.0.0 Liquid Pipeline Components

Objective

Describe key liquid pipeline components and their operations.

a. Describe tanks and the operations associated with them.
b. Describe manifolds and the valve operations associated with them.
c. Describe pumps and the operations associated with them.

Trade Terms

Automatic overfill protection system (AOPS): A tank level monitoring system that acts automatically and independently of other systems to prevent tank overfilling. Also, an API-defined tank level at which the AOPS will activate.

Automatic tank gauge (ATG): A mechanical sensing system that uses a float to detect the product level within a tank and control an outside display.

Breakout tank: A temporary storage tank used for holding specific grades of product, product mixtures, or the output of a pressure relief system.

Critical high (CH): The API-defined tank level at which a tank begins to overflow.

Floating: The practice of using a tank as a buffer between pipelines by allowing product to enter and exit the tank at the same time, although not necessarily at the same rate.

Gravitometer: A pipeline instrument that determines the product's specific gravity.

High-high (HH): The API-defined tank level at which an alarm will activate and tank filling must halt within a specified time interval.

Interface: The region between batches of dissimilar products where both products combine to form a transmix.

Line fill: The fluid volume required to fill a pipeline section.

Maximum operating pressure (MOP): The highest pressure at which a liquid pipeline may be operated.

Maximum working (MW): The API-defined tank level considered the highest point to which liquid should rise during normal operations.

Optical interface detector (OID): A pipeline instrument that uses light absorption to detect the interface between batches of dissimilar product.

Supervisory Control and Data Acquisition (SCADA): A control system used in the pipeline industry that collects and assembles data, and also allows operators to control the pipeline remotely.

Steady state: The condition in which pipeline variables, such as pressure and flow, are relatively constant over a period of time.

Stratification: The tendency for liquid mixtures to separate over time, forming distinct layers according to density.

Tank gauging system: A system that measures and reports the liquid level within a tank.

Tightlining: The process of moving product directly from one pipeline to another without transferring it into a tank first.

Transmix: An unusable mixture of two different products, usually created by switching from one incompatible product to another (stands for "transportation mixture").

Valve interlock: A control system feature that prevents unsuitable valve alignments and ensures that valve sequencing occurs in the correct order.

Variable frequency drive (VFD): An electric motor paired with an electronic controller that adjusts motor speed by varying the power supply voltage and frequency.

Because pipelines and their associated facilities are now far more automated than they used to be, control room personnel have more responsibilities as a result. Their duties now include some tasks that used to be handled by field personnel. Their actions carry financial consequences as well; the careless contamination of product can be a costly mistake. Control center personnel must be thoroughly familiar with their responsibilities.

Operating a liquid pipeline involves interacting with and controlling many different pieces of equipment. Typical tasks like start-up, shutdown, and valve manipulation all involve many of these devices. For this reason, control center personnel must understand the functions of each piece of equipment, as well as its basic operation.

1.1.0 Tank Operations

Tanks are the pipeline's "warehouses". Using tanks safely and effectively requires training and experience. Tanks hold products as they move through the system, acting as buffers if one part cannot keep up with another. They assist in blending operations, scheduling, custody transfer, and holding **transmix** until it can be dealt with. Tanks support critical safety systems by dealing with excessive volume and providing a place for relief devices to transfer product.

Tankage falls into two broad categories. Storage tanks hold product inventory for various periods of time. A **breakout tank**, also known as a *relief tank*, provides temporary storage for relief devices when they open. Breakout tanks also hold product mixtures, and temporarily hold specific grades of product for mixing operations. In this way, they contribute to maintaining a **steady state** condition in the pipeline. The Department of Transportation Office of Pipeline Safety (DOT/OPS) exercises jurisdiction over breakout tanks.

Pipeline controllers view tanks primarily from a management perspective. They need to know if the product that they're about to move is ready, on-spec, and has a place to go when it is shipped. Scheduling is a major part of the controller's responsibilities. Operations personnel must manage tankage as efficiently as possible. This task includes allocating tanks to receive shipments, keeping inventories up to date, and knowing which products are available so shipments can be scheduled. If an off-specification product arrives, controllers must find the best way to store it without contaminating other products. Everything must be accurately documented at all times.

WARNING!

Because tanks have many associated hazards, there is no room for error in tank procedures. Personnel must always be cautious and alert, and follow company and industry procedures to the letter.

1.1.1 Tank Mixing

Due to the density differences of their components, products stored in tanks will eventually separate into distinct layers, as shown in *Figure 1*. This process is called **stratification**, and is usually undesirable. Each layer has its own properties, so essentially a stratified tank contains several different products. Because this condition is potentially hazardous, stratification must be prevented.

Figure 1 If undisturbed, liquid mixtures separate according to their density.

NOTE

In a limited number of situations, stratification can be helpful. For example, if a batch of crude oil contains excessive water, allowing it to stratify provides an easy way to remove the water, which forms a layer at the bottom.

Tank mixers (*Figure 2*) prevent stratification by keeping the contents of the tank in motion. This action maintains uniform quality and temperature. With crude oil, mixing keeps sediment suspended and water uniformly distributed. Mixers may run continuously or periodically as required by the particular product.

The control center usually manages tank mixers remotely, although mixers also have a manual switch for field use. Depending on their size, tanks may have more than one mixer. Mixers resemble a boat's outboard motor, with the propeller entering the tank from the side. The propeller angle may be fixed or variable. A variable-angle

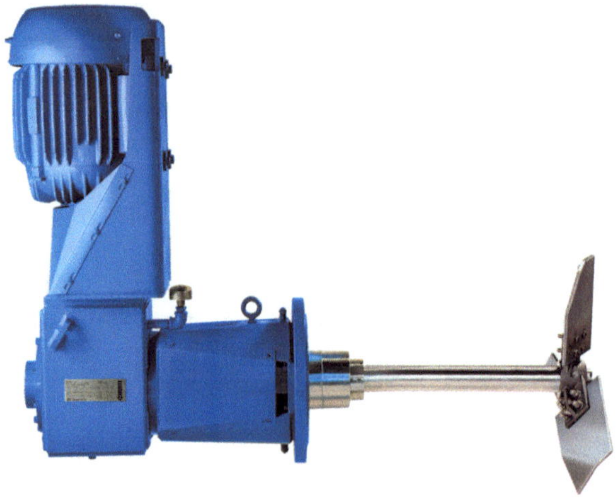

Figure 2 Tank mixer.

propeller can be adjusted to stir up the tank bottom to keep it clean from sediment.

A tank should be considered stratified if it has not been mixed in 24 hours or more. Never pump a stratified tank; the outcome could result in quality problems, and could be hazardous to the recipient. Instead, turn on the tank mixers and allow them to run long enough to mix the tank's contents. Four hours is a typical mixing time. Crude oil may require continuous mixing during pumping.

> **WARNING!**
>
> Always confirm that a tank's contents have been mixed sufficiently before pumping. Delivering stratified content could result in the recipient receiving an unexpectedly hazardous substance.

Before starting the tank mixers, confirm that the tank level isn't too low. Mixing when the level is low blends air into the product, which can create pumping problems. If the tank has a floating roof, the roof's flotation device may damage the mixer's propeller when the level is too low. Most companies have a standard policy that requires all mixers to be turned off, and possibly locked out, when the roof is below a specified safe level.

> **NOTE**
>
> Today, most tank roofs are of the cone or floating types. Floating roofs significantly reduce product evaporation and vapor release into the atmosphere.

1.1.2 Sumps

A sump is a holding tank, usually underground, with uses that vary depending on the installation. Companies often use sumps to collect and store contaminated product until it can be disposed of properly. Sumps can also act as breakout tanks for thermal relief valves from manifolds. Direct drain lines from dikes and firewalls may empty into sumps. Some sumps hold overflow, transmix, operational wastes, sludge, wash-downs, used lubricants, or even dirt and trash.

If a sump contains pure product, it may be acceptable to pump it into a tank or simply inject it back into the shipment. Mixed or contaminated product must be analyzed to determine its specifications. After testing, field personnel will determine the best way to blend, move, or dispose of the product. For example, a scheduled batch may support a carefully calculated injection, eliminating the need to discard the product.

If the sump contains transmix, the company may sell it to a processing firm. Some companies require field operations to pick up transmix and haul it away. Regardless of the ultimate destination, material from a sump must be accounted for and accurately measured. Sump volume must be integrated into calculations to avoid inventory problems.

Generally, sumps are managed by field personnel, who should measure and empty manually controlled sumps on a regular basis. Automatically pumped sumps should still be checked and measured periodically. Many automatic sump systems send an alarm to the control center when the high-level pump starts. When these alarms occur, personnel should inspect the sump visually to confirm that all is working properly. Sumps used for pipeline relief should receive routine, scheduled maintenance and have their communications system tested.

1.1.3 Receiving and Deliveries

When product arrives at the terminal or station, it must be stored until delivery to the customer. Receiving a shipment begins by diverting the incoming product stream out of the pipeline, and into a suitable tank. This action requires precision planning to ensure that the transfer is efficient and safe. Pipeline personnel must perform their jobs properly so each piece of equipment operates correctly. But beyond ensuring that all of the technology functions as expected, control personnel must be efficient in planning and executing the schedule.

Because tanks are dedicated to specific products, operations personnel decide which tanks fit the operation. Good tank management requires selecting tanks in advance, based on some or all of the following factors:

- Schedule
- Specific product involved
- Volume being pumped or delivered
- Available tank capacity
- Location efficiency
- Cargo requirements
- Flow rates
- Product compatibility

During receipt, operations personnel must open and close valves in exact sequence so the product takes the correct path from pipeline to tank. While control center personnel may perform these actions remotely, field personnel may have to confirm valve alignment. If a valve actuator fails, manual intervention will be required before operations personnel can open the pipeline delivery valve.

In most instances, the flow passes through strainers and metering runs before being routed into the selected tank (see *Figure 3*). Metering runs usually have straightening vanes upstream of the meter itself. These reduce flow turbulence and improve meter accuracy. In some cases, the flow will also go through pig traps. Personnel should verify that there is enough room in the tanks for the shipment.

Empty tanks require special precautions. Filling an empty tank can generate large amounts of static electricity. A flash can occur as the product begins to spread across the tank floor. (A flash is a sudden ignition of flammable vapor. In this case, it's the vapor in the storage tank that ignites when a static electric discharge accumulates and sparks during filling.) Beginning the fill very slowly reduces the potential for a flash. Each company has policies and procedures in place for filling empty tanks, which must always be followed.

CAUTION

Always follow your company's policies and procedures for filling empty tanks to prevent the occurrence of a flash.

If the tank has a floating roof (*Figure 4*), special precautions may be required. Filling too fast may result in the product flowing onto the roof, preventing it from floating. Not all tank-measuring systems give accurate level readings when the floating roof is resting on its legs. Until the roof starts floating, be particularly cautious about level readings.

Allowing product to flow in and out of a tank at the same time is a practice known as **floating**. The tank acts as a buffer if there is a difference in flow capacity between input and output lines. For example, if a large incoming stream must flow into

Figure 4 Floating roof.

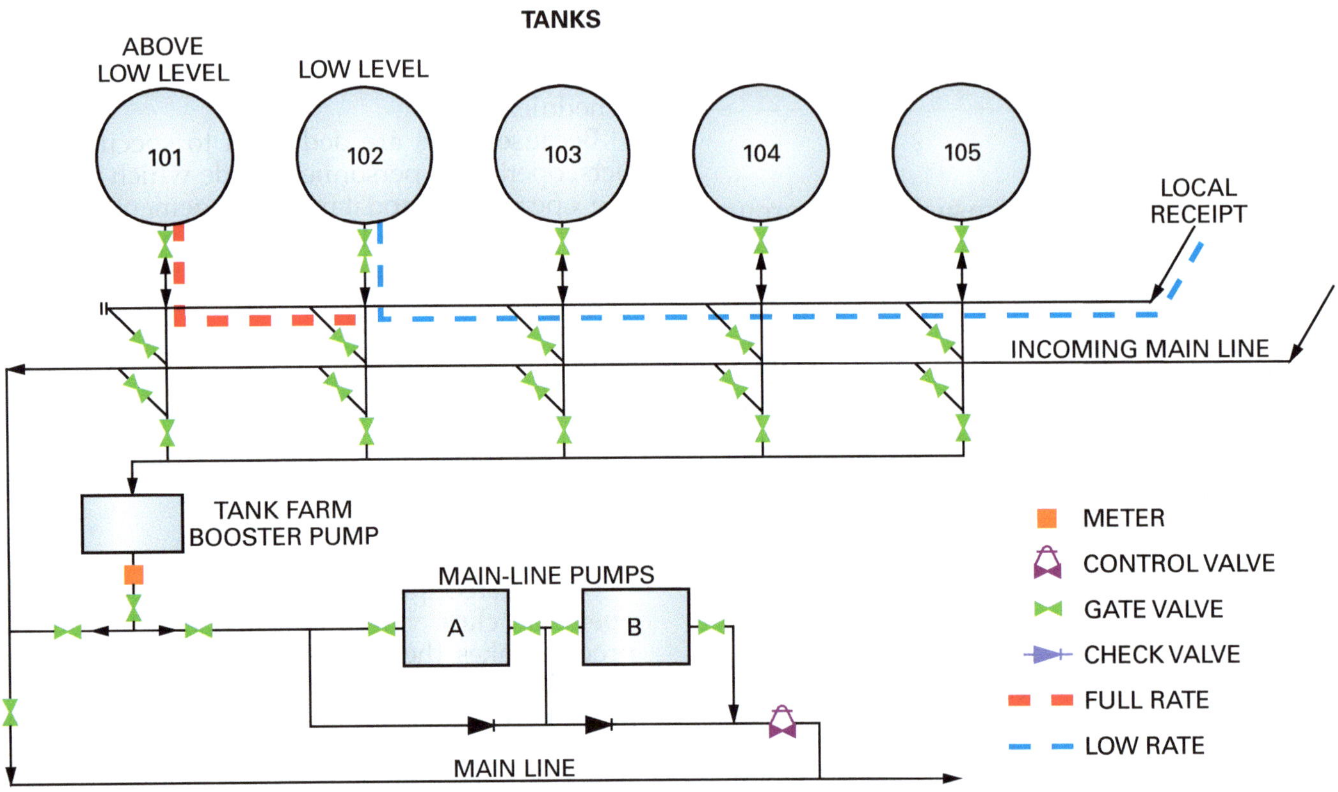

Figure 3 Selecting appropriate tankage.

a smaller pipeline, floating it with a tank allows the change without forcing the operator to reduce flow. The tank captures the large volume and holds it until the smaller pipeline can finish receiving it.

Operations personnel must be careful when floating a tank if the next batch is a different product. Unless the switch is made properly, an **interface** could wind up in the next batch.

1.1.4 Managing Inventory

Inventory fluctuates constantly as the terminal receives and delivers shipments, so operations personnel must stay aware of current inventory. The **Supervisory Control and Data Acquisition (SCADA)** system carries much of the load of this task by monitoring and calculating product quantity. Status reports give totals at the end of each operating day and confirm the terminal's inventory. These reports, along with SCADA tank level readings, help personnel know how much space is available so they can schedule receipts.

NOTE

The SCADA system actually updates its information every few seconds. It can provide hourly, cumulative, and on-demand readings. Pipeline controllers can get the latest values by clicking or otherwise selecting the appropriate indicator on their SCADA screens.

Every arriving shipment must go into a tank with enough reserve capacity to hold the incoming volume. Once the scheduling department has set up the shipment, operations personnel must prepare in advance to receive it. They will set up a plan that includes selecting appropriate tanks and confirming sufficient storage to accept the shipment. If the facility does not have enough capacity, operations personnel must notify the scheduling department immediately.

For example, imagine that a shipment of 125,000 barrels (19,873 m^3) of gasoline is expected. Operations personnel discover that there is space for 92,500 barrels (14,706 m^3). The terminal cannot find the extra space immediately. In this scenario, operations personnel should contact the scheduling department as quickly as possible to work out a solution.

Or consider a customer that is expecting to receive 200,000 barrels (31,797 m^3) of gasoline. Operations personnel discover that there is only 160,000 barrels (25,438 m^3) in the pipeline. They should contact the scheduling department as quickly as possible to work out a solution. The customer may have to be notified to use an alternative source.

SCADA uses data from meters to calculate product flow and other related information. This data appears on appropriate monitor screens (see *Figure 5*). The SCADA system calculates tank inventory using one of two methods: gross volume or net volume. Each must be calculated so that all volume can be accounted for.

Gross tank volume is the total liquid volume contained in a tank based on measurements before adjustments are applied. Net tank volume is a calculated value determined by applying adjustments to the gross volume. These adjustments include sediment and water (S&W), temperature, density, and any other applicable factors that may affect volume. Manual measuring may supplement the SCADA report.

During delivery, careful metering is crucial. The control room's key responsibility is providing a specific quantity of product. The receiver's responsibility is having sufficient space to absorb it as it arrives.

1.1.5 Tank Level Alarms

Maintaining the proper tank level may seem like a minor task, but it is actually extremely important. Tank overflows have caused some major pipeline disasters in the past. *Figure 6* shows the potential consequences of a tank overflow. API *Standard 2350* 4th Edition was created in response to a major overfill accident, known as the Buncefield fire.

Operations personnel must monitor and respond to tank level alarms. These will vary depending on the technological sophistication of the **tank gauging system** installed, as well as its compliance with API *Standard 2350*.

Tank gauges range anywhere from totally manual systems, such as a field technician with a weighted tape measure (*Figure 7*), to highly

Figure 5 Control room monitors.

Figure 6 An overflow can lead to a major disaster.

sophisticated systems. An **automatic tank gauge (ATG)** is a simple mechanical system that displays liquid level on a display mounted beside the tank (*Figure 8*). Some ATGs (but not all) can send level data to the control room electronically. Being mechanical, ATGs need regular maintenance and can malfunction fairly easily. A faulty ATG, among other factors, contributed to the Buncefield fire.

More sophisticated systems include tank gauges that determine level by bouncing energy off the liquid's surface and timing the reflections. These gauges use sonar, laser, or radar technology. A radar tank gauge is shown in *Figure 9*. All are highly accurate and reliable, and recent models offer full remote communications capabilities.

Control room personnel must be aware of several different tank levels. API *Standard 2350* 4th Edition defines four, although some installations may include others:

- **Maximum working (MW)** – This is the lowest high level and refers to the normal maximum height to which liquid should rise. Reaching this level will not generate an alarm, although some systems may notify the operator.
- **High-high (HH)** – If the liquid reaches this point, it will trigger an alarm and the flow should be terminated within a specific period of time.
- **Automatic overfill protection system (AOPS)** level – An AOPS is an automatic and independent system that takes corrective action to halt tank filling when it detects a set level (either at or above the HH level). The level at which the AOPS activates is called the *AOPS level*.
- **Critical high (CH)** – This is the level at which the tank starts to overflow. There are no alarms at this level since they should have been triggered below it. Control room personnel should know the CH for each tank, but liquid levels should never be allowed to get close to this point.

The Buncefield Fire

The explosion and subsequent fire at the Buncefield Oil Depot is one of the most memorable pipeline disasters in recent history. Located near Hemel Hempstead, a town northwest of London, the fuel depot was one of the largest in the United Kingdom.

The incident began with a tank overflow on December 11, 2005. Shortly afterwards, the first explosion occurred, followed by a rapidly spreading fire and additional explosions. Huge clouds of black smoke moved across the southeast of England, and many people had to evacuate their homes temporarily. Despite the efforts of firefighters, the fires continued to burn until December 15. Fortunately, no fatalities occurred.

The investigation that followed revealed numerous causes contributing to the disaster. A series of equipment malfunctions, poor site design, an overworked staff, and failure to follow procedure were the main causes. But like many disasters, the Buncefield fire stimulated changes in the industry that led to improved safety standards. Many of these are incorporated in API *Standard 2350* 4th Edition.

Figure 7 A handline used for manual tank gauging.

In addition to the four API-defined levels, some tank level sensors detect and report other levels:

- *High (H)* – This level is above the MW point, but below the HH point. It is optional under API *Standard 2350* 4th Edition.
- *Minimum working* – This is the lowest level the liquid should sink to during normal operation. Allowing the liquid level to drop below this point may cause equipment damage, floating roof damage, or other problems. Most systems send a notification when the liquid drops to this point.

Figure 10 summarizes all of these levels. The levels mandated by API *Standard 2350* 4th Edition are shown bold red type.

> **NOTE**
> Earlier versions of API *Standard 2350* used different names for some levels. It's important to be familiar with your particular facility and the terminology its tank gauging system uses to report levels.

Tank gauges that report to the SCADA system enable it to maintain up-to-date inventories and fill level status. During receiving operations, control room personnel should watch fill levels. If a value freezes, the situation must be investigated immediately. Some possible causes include hung gauges, communications failure, water in the gauging system, or a faulty tank gauge.

> **NOTE**
> Prior to the fire, an ATG at Buncefield got stuck and stopped reporting a level change. The AOPS system had become deactivated either by accident or through carelessness, and personnel were overloaded and not following consistent procedures.

If an alarm sounds, operations personnel must respond quickly, and company policy must be followed. The flow path may have to be redirected or the pipeline shut down. If an AOPS activates, it should take action to prevent the tank from overflowing, but control room personnel will have to perform additional steps to deal with the consequences.

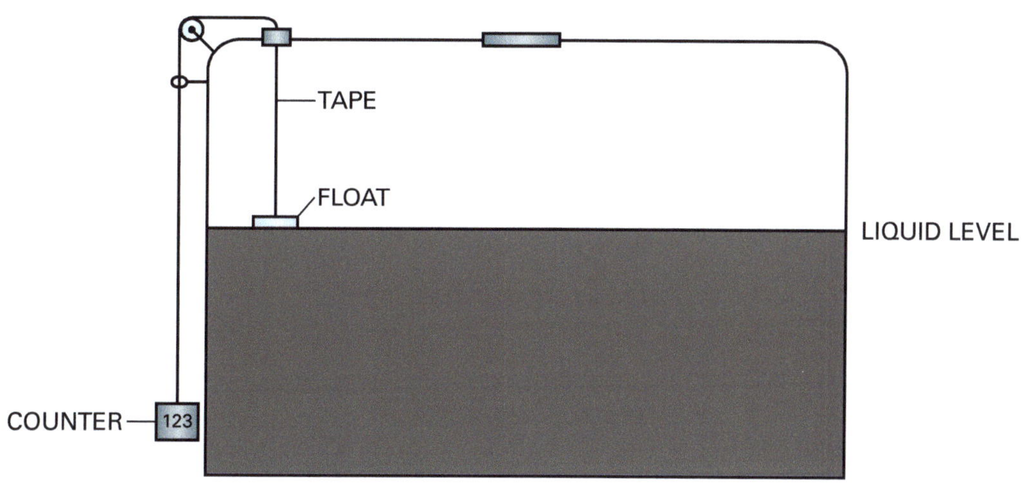

Figure 8 ATG diagram.

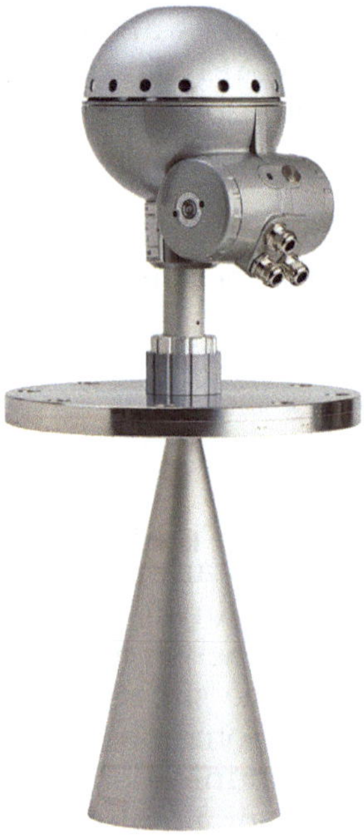

Figure 9 A radar tank gauge.

1.2.0 Manifolds and Valve Operations

Operations personnel control product movement by opening and closing terminal, station, and pipeline valves in the proper sequence. They may do this either by direct operation, or remotely through the SCADA system. Common tasks associated with these operations include making tank switches, verifying valve alignments, and making pipeline switches.

1.2.1 Valve Types

The pipeline and its associated systems contain numerous types of valves. Many of these perform specialized tasks, and all work together to make the pipeline both efficient and safe. At the most basic level, valves are control devices, opening and closing to start, stop, or regulate flow.

Hand valves (*Figure 11*) must be operated manually. They are common on meter runs and tanks. However, almost all valves can be operated by hand if necessary. Field personnel may have to use a portable power source to operate valves manually if they are very large.

Block valves operate in the fully open or fully closed positions only and offer minimal flow resistance. They are the main pipeline valves, and they can isolate pipeline segments as required. Block valves play an important role when the control center purges a pump station. The block valve closes while pump suction and discharge remain open. This configuration prevents a slug of the previous product from contaminating the new one.

Pipeline personnel operate block valves either at the local control point or remotely from the control center. While any valve can act as a block valve, gate, plug, butterfly, and ball valves are most commonly used. A gate valve is shown in *Figure 12*.

Control valves (*Figure 13*) function in the fully open, fully closed, or partially open (throttle) positions. For this reason, they are ideal for precision control, often being used to manage pressure and flow. For example, control valves manage both the station suction and station discharge ends of pumps. Their action protects both the upstream and downstream sides. In delivery facilities, they control the delivery rate. Globe and angle valves are the most common control valves, although butterfly valves are also used. They can be operated by hand, although most are managed by pneumatic, hydraulic, or electric actuators.

Control valves are usually smaller than the pipeline itself to provide some built-in throttling action. This reduces wear on the valve seat as well as the stem motion required to control the flow. Because partially open valves reduce pressure downstream, they must be used with care to avoid damaging other components, such as

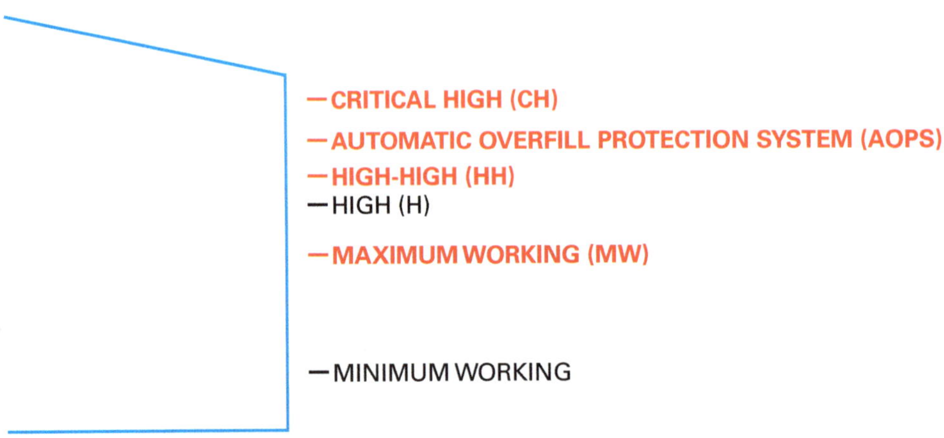

Figure 10 Tank levels.

pumps. Constant-speed centrifugal pumps are often used with control valves.

Pipeline personnel use control valves in any situation requiring flow or pressure adjustment. These valves limit flow until conditions normalize in cases of low suction, during pipeline start-up, or after a main-line pump loss.

Figure 11 Hand-operated valve.

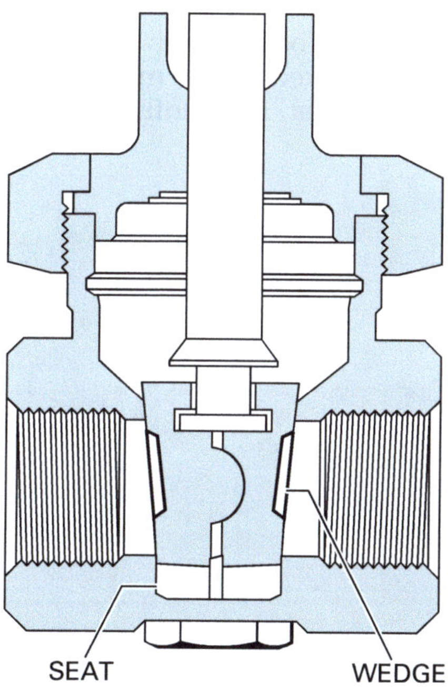

Figure 12 Gate valve.

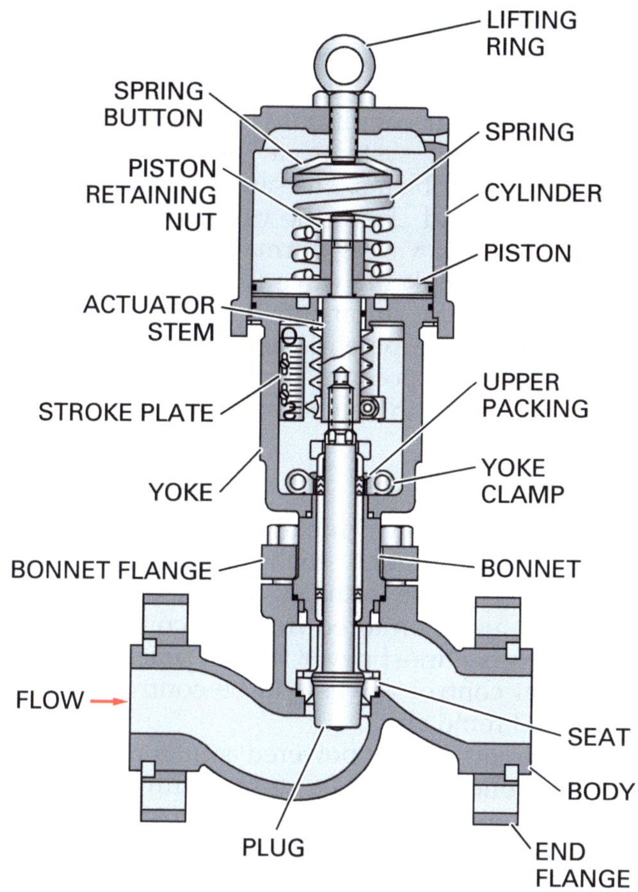

Figure 13 Control valve.

> **NOTE**
> Pipeline controllers may change some valve setpoints through the SCADA system. Other setpoints are dictated by engineering standards and design and may not be changed without authorization. However, pipeline controllers should know all setpoint values, regardless of whether or not they can alter them.

A backpressure valve is a special type of control valve, which intentionally creates backpressure through throttling. The valve raises pipeline pressure upstream, preventing a slack line. Backpressure valves reduce pipeline vibration and protect low-pressure equipment, such as metering installations. Normally, these valves come preset to generate a specific amount of backpressure and are not remotely adjustable.

Pressure relief valves protect the pipeline during an overpressure condition by opening to release product into a breakout tank. Once the pressure has dropped back to safe levels, the valve closes. Relief valves use a number of different technologies to determine when to open. The most common type uses a spring whose force is equal to the pressure at which the device should open. Relief valves may also be built around electronic or hydraulic technologies.

Safety valves protect the pipeline by activating during abnormal conditions. For example, with electrical control systems, a power loss could cause serious problems. The safety system detects the flow loss and closes the valve to protect the manifold. Safety valves normally appear in the line below the takeoff valve (the last valve that opens to start the shipment), but may appear in other parts of the pipeline as well. In all cases, however, the purpose of the safety valve is to place the pipeline in a safe condition.

1.2.2 Valve Control

Remote-control valves make the pipeline and its associated equipment much more manageable. Ideally, every valve in the system should be both controlled and monitored from the control room. While this is almost never true in practice, most important control valves will be controlled and monitored remotely.

Valves managed by powered actuators are usually controlled by a PLC (programmable logic controller) or an RTU (remote terminal unit), which are computer-like devices that talk to the SCADA system. Examples of PLCs are shown in *Figure 14*. Commands to valves are issued by control room personnel through the SCADA system, which also provides valve status information (*Figure 15*) and indicates when valves are traveling between positions. This approach allows control decisions to be made hundreds of miles away from the actual valves.

Moving product into or out of a terminal requires operations personnel to set the proper valve alignments to direct the product correctly. The valve sequence opens a path through which the product flows. Using appropriate commands, the control room can direct product into or out of tanks, operate metering and prover systems, or bypass a station altogether.

1.2.3 Interlocks

Control systems may include **valve interlock** logic designed to prevent dangerous or problematic valve combinations. For example, the control system's interlock programming might not allow a main-line valve to close until the corresponding pumps are shut down. Such automation prevents many careless accidents, protecting both the station and the pipeline.

Some companies use interlocks on booster stations to prevent the line from becoming blocked. Interlock program logic monitors all major valves at the stations, including the control valves. If a valve begins to close unexpectedly, the control program will send an Open command to reopen the valve. If the control valve closes by more than 65 percent, the program will send the main-line block valve an Open command to ensure that the line is not blocked.

Unstaffed booster stations may have interlock logic that prevents the main-line block valve and the booster pump suction and discharge valves from closing at the same time. They may also have interlock logic designed to open the main-line block valve if the control valve closes by more than 65 percent.

Interlocks can also shut down the station in response to a line block command. This action requires less time than shutting down all units individually.

While the interlock system can fail, human error causes blocked lines more often. These failures occur most frequently with deliveries

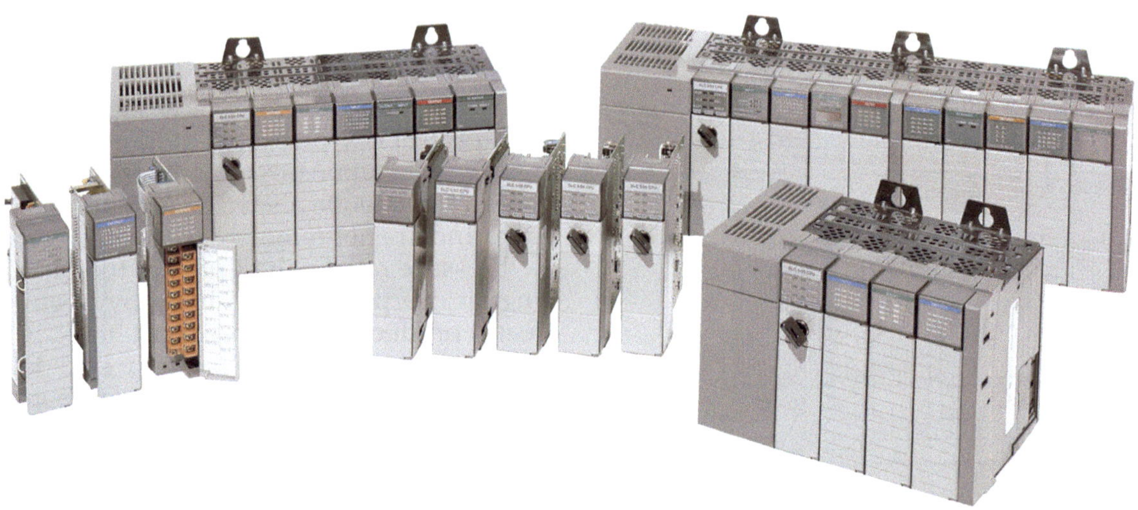

Figure 14 PLCs are often found in control systems.

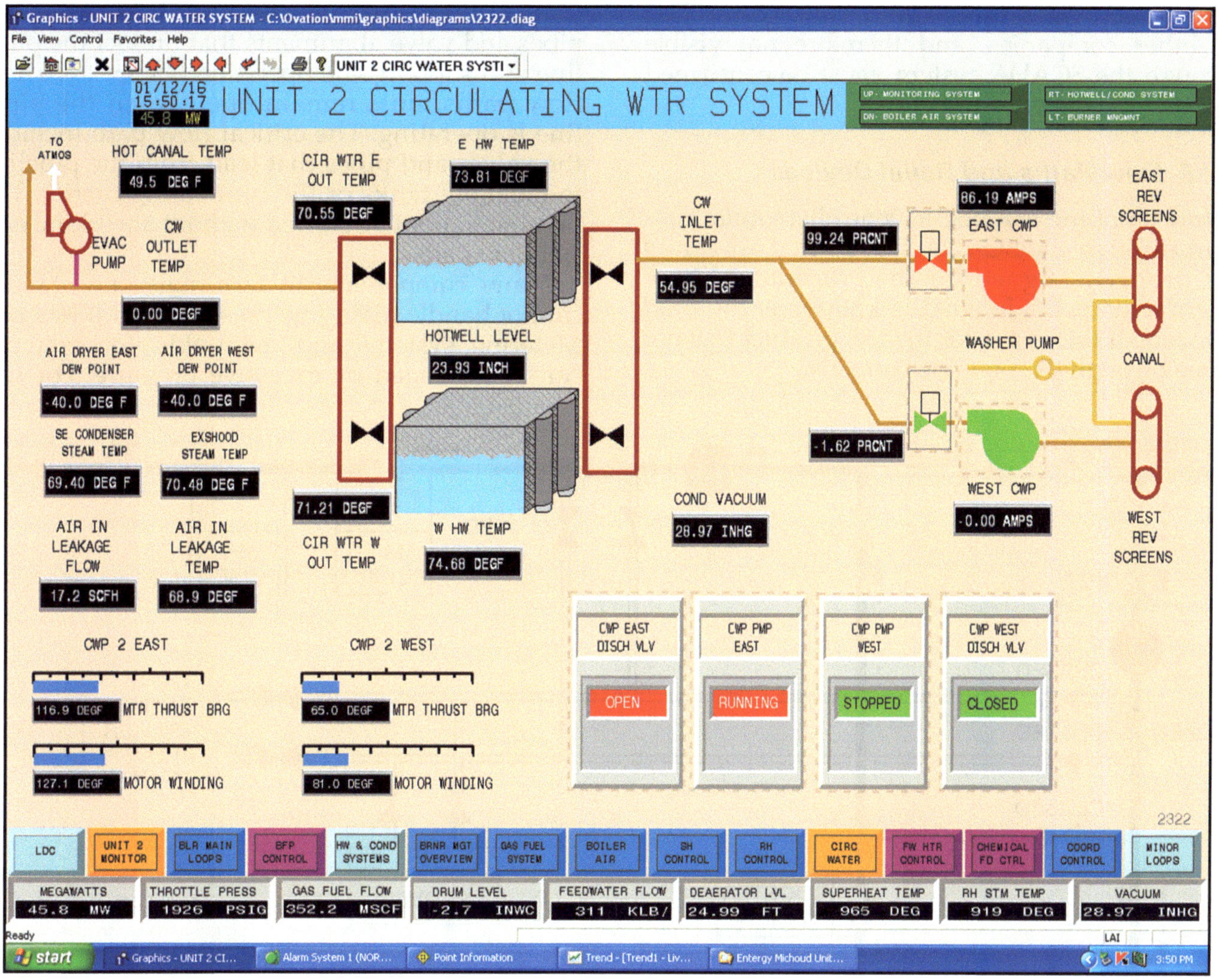

Figure 15 Valve status display.

into staffed facilities. Regardless of the cause, the controller must shut down the line, following company procedures. Interlock failures should be handled by following the specified procedure. In most cases, the time to respond to and resolve overpressure situations is short. Control room personnel must be quick to recognize problems and know the correct response so they can act immediately.

1.2.4 Valve Alignment

Control center personnel must verify valve alignment for every shipment or change in operations. Once the scheduling department issues shipping orders, the control center sets up the procedures to fulfill those orders. This task includes determining the proper valve alignment needed to get the product from point A to point B safely and efficiently. Operators will look at piping layouts to pick the best path, and then prepare a list of valves to be opened or closed. They will then use the SCADA system to operate the valves in the correct sequence.

Most valve operations occur by remote control. Similarly, SCADA screens show valve status and indicate whether valves are open, closed, or traveling. Control center personnel must keep an eye on status indicators to confirm that everything is working. The operation may not begin until all valve positions have been confirmed. As previously discussed, many control systems include valve interlock logic that prevents inappropriate or unsafe valve combinations or sequencing.

If a valve does not respond correctly, it may be necessary to have field personnel go to the location and check or operate the valve manually. Standard operating policies require control center personnel to verify that valves are properly set before any product enters the system. Some companies dispatch field personnel to check each valve individually, although this practice is becoming uncommon due to unstaffed facilities. Verification is important, especially when several companies are involved. Valves controlled by other companies, and therefore not visible through the SCADA system, must be confirmed by other means prior to delivery.

1.2.5 Flow Paths and Relief Devices

Before shipping can begin, controllers must confirm through the SCADA system that all valves are properly aligned for the procedure. These alignments provide what is known as the *valid flow path*. If this is not confirmed, product will not move through the pipeline correctly.

Critical flow path, on the other hand, refers to pipes and valve alignments that protect the pipeline and support safe operation (*Figure 16*). These flow paths must remain open when the pipeline is operating. The critical flow path includes the valves and pipes that lead from the pipeline to breakout tanks or the connecting carrier. The pipeline cannot be started without a working critical flow path.

Many components in terminals and stations cannot handle main-line pressure and flow rates. Metering systems and manifolds, for example, can be damaged by excessive pressure. For this

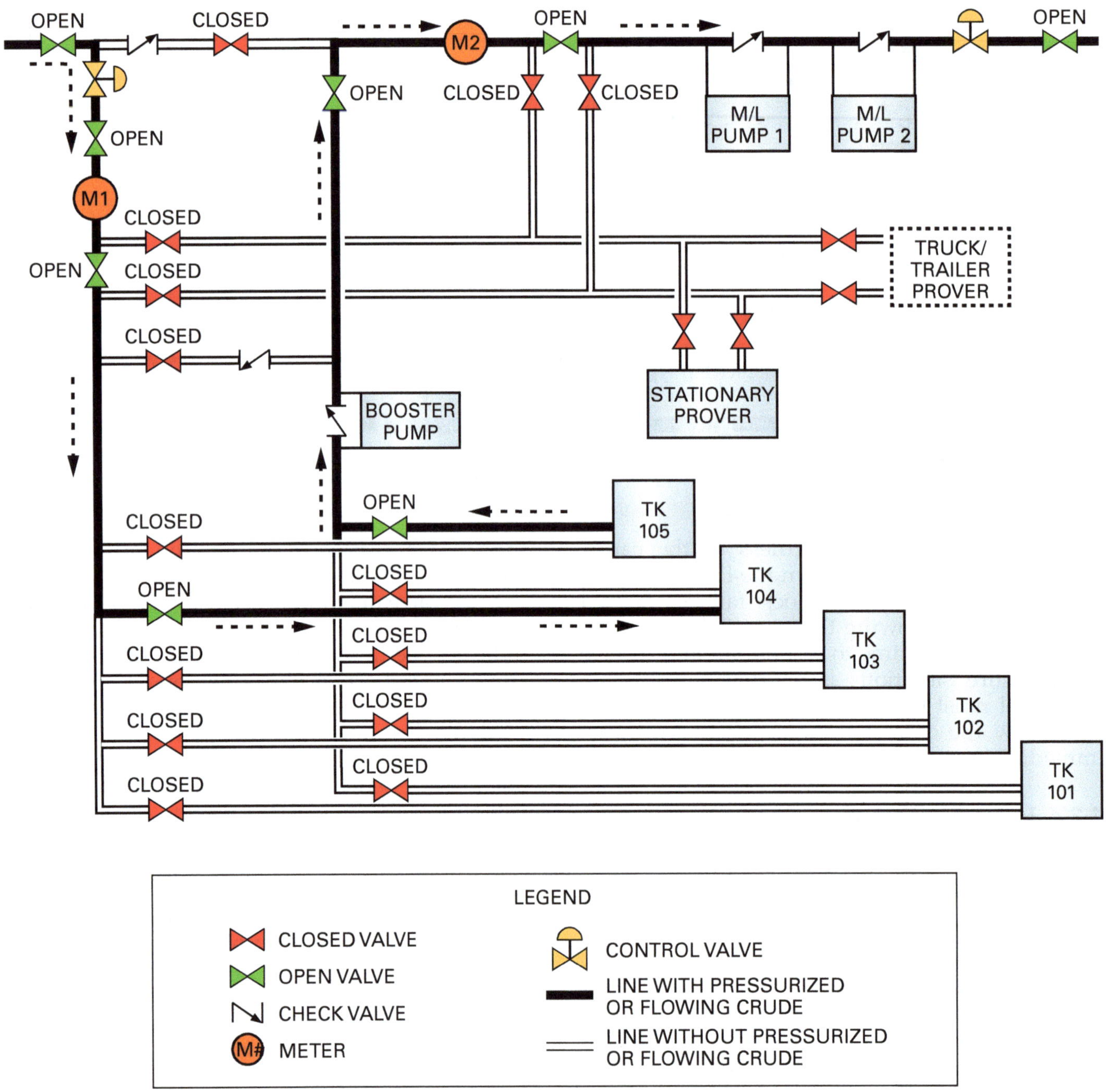

Figure 16 Critical flow path diagram.

reason, pressure relief and pressure control devices must be present and correctly set to protect these vulnerable components.

Pressure relief valves (*Figure 17*) open when pressure becomes excessive, and release product into a breakout tank. Pressure control devices manipulate valves or divert flow to prevent pressure from rising to dangerous levels. Thermal relief devices are similar to pressure relief valves. They relieve pressure caused by thermal expansion—the tendency of the product to expand when it heats up.

Generally, the control center will not be able to change protective device setpoints. In some cases, these may be fixed by the manufacturer and cannot be changed. Other devices may have to be changed in the field either through a mechanical adjustment or by reprogramming a control device. Regardless, the control center must know protective device setpoints, and be alert to pipeline pressure and flow conditions. Controllers must also monitor relief devices to ensure that they're performing properly.

1.2.6 Dealing with Incompatible Products

There are many products that cannot be mixed because the combination might be dangerous or simply fail to meet standard requirements. For example, mixing gasoline with a shipment of fuel oil

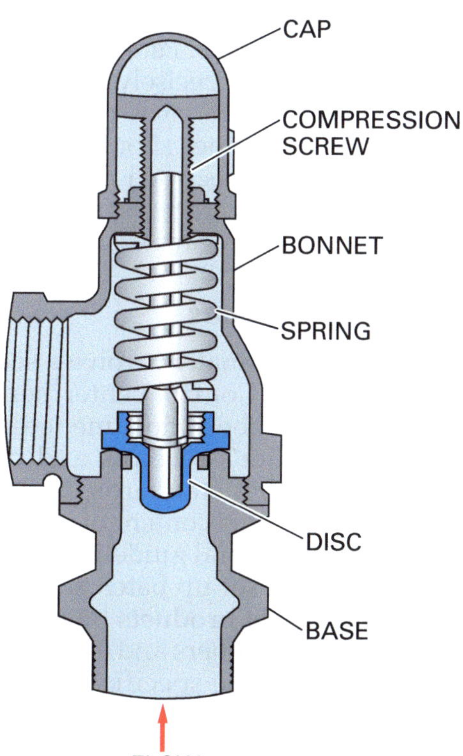

Figure 17 Pressure relief valve.

is dangerous because it has a lower flash point than fuel oil should have. Mixing regular gasoline with a shipment of premium gasoline is not dangerous, but the mixture doesn't have the required quality level. If ultra-low-sulfur diesel becomes contaminated with other grades of diesel, it violates the strict quality standards for diesel fuels. For these reasons, switching between incompatible products requires one or more special procedures.

Purging (also known as *line wash* or *line flush*) involves using another product to push out the remaining product still in the line or manifold. Operations personnel must manage the **line fill** between each manifold as well as other volumes of product residue remaining in equipment. In most installations, line fill is known or can be calculated for various valve alignments and delivery points. Control center personnel should use this information when purging. For example, if the line fill is 300 barrels (48 m^3), that amount must be removed and diverted when purging the line.

The line should always be purged with a product that is compatible with the upcoming shipment. Personnel must calculate the correct amount and purge the line, diverting the output into an appropriate tank. Then the second shipment can begin. Knowing or calculating line fills ensures that the line gets purged without using excess product.

Measuring the product's API gravity is one way to identify and track batches as they flow through the pipeline. Gravity testing also reveals changes in the product that may occur in shipment. For this reason, many pipelines have some type of **gravitometer** (*Figure 18*) installed upstream of the terminal. The SCADA system receives data from the unit and relays it to the control room.

> **NOTE**
>
> Gravitometers measure product density or specific gravity. Calculations then convert this information to API gravity. Gravitometers use one of several different methods to measure density. One method involves vibrating an element in the product stream and measuring the oscillation frequency. As the product becomes denser, the element oscillates more slowly. Gravitometers are usually checked periodically against laboratory density measurements.

The SCADA system calculates, records, and displays all gravity information. Control center personnel should monitor the API gravity patterns, and any variations should be investigated.

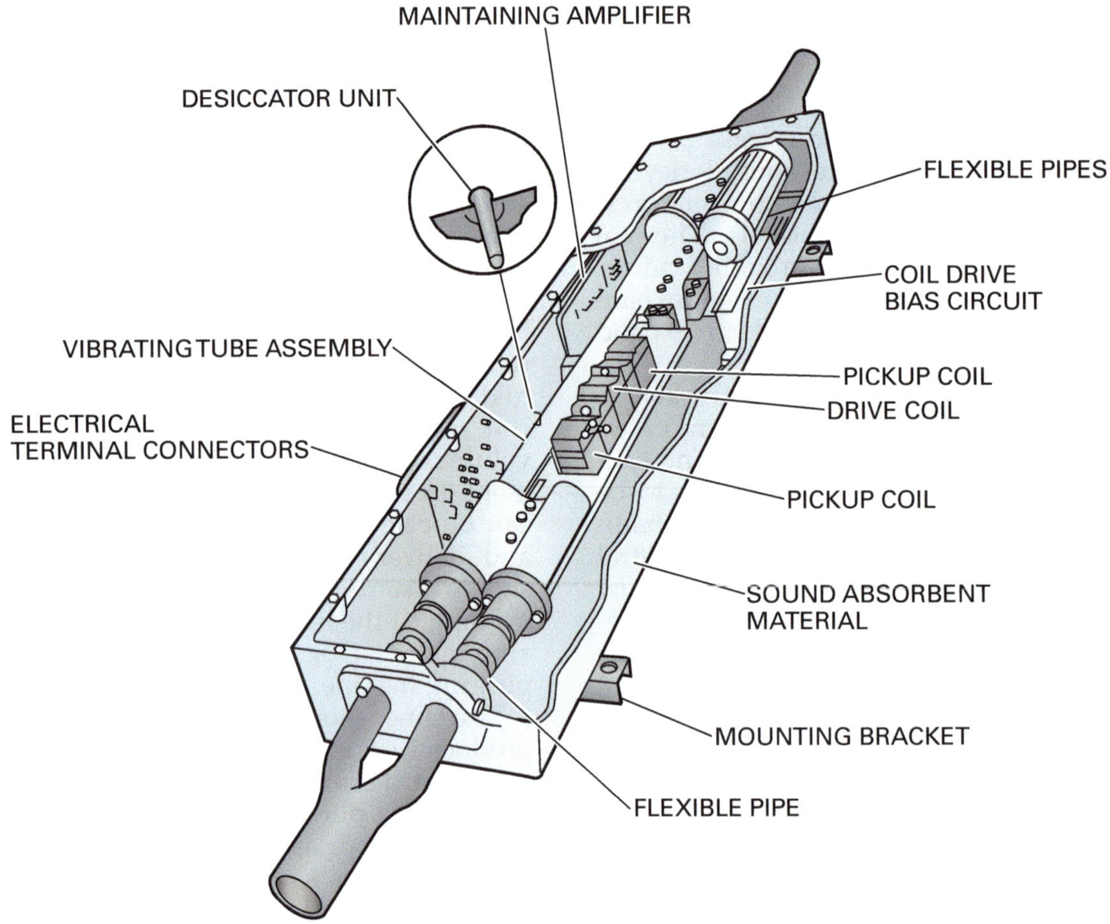

Figure 18 Gravitometer.

Significant variations must be documented and addressed according to company specifications.

API gravity measurements alone are not always useful for identifying the interface between batches. Some pipelines use a second instrument, called an optical interface detector (OID), to supplement the gravitometer's readings. An OID passes light through the product and reads the absorption pattern. This method is especially useful with products having similar API gravity values.

Before a scheduled change arrives, control center personnel should begin monitoring the mainline stream and documenting the API gravity according to company policy. If the pipeline is not using an OID, the company-specified method for detecting the interface should be used. When the first break occurs, operations personnel should begin constant monitoring and perform the switch according to company policy.

Some facilities rely on time calculations to determine when to switch. Using information such as the pumping rate and the volume to be delivered, they can determine when the switches should occur. Some valve and tank switches are made based on the length of time that product flows through the pipe. This method is common in systems with hand valves. Personnel must perform manual switches as precisely as possible.

Because schedules may vary over long distances, operations personnel should supplement timing with measurements, such as API gravity, to confirm interfaces. Switches should occur at the correct gravity point regardless of time.

1.2.7 Batching Operations

Batching operations normally involve a sequence of different products. Control center personnel should be meticulous about documentation and handle every customer's receipts and deliveries accurately. Strict EPA standards, particularly those involving the sulfur content of diesel fuels, have led to many procedural guidelines and rules. Scheduling personnel set up batch sequences to minimize incompatible products being run one after the other. Code numbers and/or letters identify products with unique specifications. Charts may be set up to illustrate which products can be pumped directly in front of and behind other products. *Table 1* is an example of such a chart.

> **NOTE**
> In addition to being skillful at scheduling, controllers must be expert at maintaining product quality.

API gravity ranges vary somewhat in similar products. A deviation of one or two degrees API from the maximum or minimum shown on a chart is not considered significant. However, products with unique specifications (such as jet fuel) are handled more carefully in batch arrangements than are variations in products (such as fuel oils).

Testing for the interface and making a correct switch during an interface protects the product integrity. Personnel must sample and document each shipment's product as it arrives at the terminal. Doing so determines the switch point, at which time operations personnel open the pipeline valve, switching the product into the terminal. Transmix goes into separate tankage, and the pure product goes into storage.

1.2.8 Measuring Volume

The term *barrel count* refers to the total number of barrels measured during a given operation or time interval. SCADA flow computers (*Figure 19*)

Table 1 Sample Batch Arrangement Chart

Grade	Type of Product	60°F API Gravity Range	Normal Pump Sequence Head End	Normal Pump Sequence Tail End
A	Premium Unleaded Gasoline 91/93 Undyed	45-75	N, M	N, M
B	Commercial Aviation Gasoline 100/103 Octane – Dyed Blue	66-72	N, M with HP-7 Buffer	N, M with HP-7 Buffer
C	Refinery Charging Stock Crude Oil	20-55	X5	X5
D	Ultra-Low-Sulfur Diesel Fuel Oil – Normally Undyed 47.5 Cetane Range	36-41	X, X5, Y	X, X5, Y
E	Denatured Fuel Ethanol		Not Pumped	Not pumped
F	#2 Fuel Oil Distillate Cenex Proprietary – Dyed Red	30–37	D, Y, Q, X, X5, N	D, Y, X, X5, N
G		Naptha	Special Delivery	Special Delivery
H	Normal Butane	0.580-0.588	N, L	N, L
I	Isobutane	0.560-0.570	H	H
J	Military Jet Fuel JP8 or J8	41-48	N, V	N, V, D, X
L	Propane	0.500-0.510	H	H
M	Mid-Grade Unleaded Gasoline 89 Octane Range	48-75	A, D, N, X	A, D, N, X
N	Regular Unleaded Gasoline 87 Octane Range	48-75	A, F, J, M, Q, V, X, X5, Y	A, F, J, M, Q, V, X, X5, Y
O	Water	1	Special Delivery	Special Delivery
P	Refinery Blendstock		A, N, X	A, N, X
Q	Commercial Jet Fuel	41-48	N, V	N, V, D, X
R	Leaded Racing Gasoline	60-80	N, V	N, V
T	Transmix Commercial Purchases	35-65	N, X, X5	N, X, X5
V	Sub-Octane Unleaded Gasoline, 84 Octane Range	48-75	F, J, M, N, Q, V, X, X5, Y	F, J, M, N, Q, V, X, X5, Y
W	Natural Gasoline Liquid	79-86	D, X, X5, N, H	D, X, X5, N, H
X5	#2 Fuel Oil High Sulfur (HSFO) Normally Dyed Red 0.50 sulfur maximum (The 5 in X5 may not be present when listed in specific grade)	30-40	D, F, N, V, X, Y	D, F, N, V, X, Y
Y	#1 Fuel Oil GrDistillate	40-45	D, F, N, V, X, X5	D, F, N, V, X, X5

Figure 19 Flow computer.

calculate the barrel count using the pipeline metering system. The barrel count may indicate the total volume flowing through the pipeline, the total volume of a shipment, or the volume being moved, blended, or injected.

Control center personnel should verify the totals, calculations, and all parameters that affect volume for each operation or time frame. The temperature and pressure are confirmed at the meter's manual thermometer and pressure gauge. The true total barrel count comes from a calculation involving the meter reading and all compensating factors. This value can be compared to the physical totals in tankage to verify inventory and operations.

Barrel counts give control center personnel a way to account for the day's operations, calculate the over-and-short, verify receipts and deliveries, and track operations. They also can act as a type of leak detection system, since significant discrepancies may indicate leakage.

1.2.9 Pipeline Switching

Various pipeline companies serve different geographical areas. Pipelines often run parallel to each other and sometimes cross (see *Figure 20*). Transferring product from one pipeline to another has numerous advantages to customers, so many companies have developed methods to make it possible. Control center personnel must understand the geography, facilities, and installations of other companies to schedule shipments between pipelines.

During transfers between companies, all involved control centers must closely coordinate with each other. Close communications are vital after confirming schedules. The same testing and sampling methods used for any regular custody transfer still apply. When changing carriers, extra care should be taken to verify and document batches. Confirm readiness with each pipeline before beginning the shipment.

Figure 20 Abbreviated map of the main US pipelines.

Sending a shipment to a tank before changing carriers makes measurement and verification easier. If the scheduling department specifies that the shipment be tightlined into the other pipeline, control center personnel should take extra care with documenting specifications and measurements. (**Tightlining** is the process of moving a product from one pipeline to another without transferring it into a tank first.) If a question comes up, this information will be needed. Always document valve alignments and verification records so product movement can be easily traced.

1.2.10 Procedures

Each company has standard operating procedures for receipts, deliveries, and custody transfers. In order to manage these transfers efficiently, control center personnel should review scheduled shipments. They should understand the exact shipment type, specifications, time, and flow rate.

> **NOTE**
>
> Always know and follow your company's procedures for all product transfers.

When possible, prepare and document terminal piping and valves as far in advance of shipments as possible. Make appropriate plans for switching local line fill. Confirm that receiving equipment is operating properly and that there is sufficient tankage, if needed. If tankage is a problem, contact the scheduling and dispatching departments immediately.

Although consecutive batches may be compatible, the characteristics or additives of one batch may be different from others. Because delivery points may be different for each batch, switching will also be different. Operations personnel must plan the switch so each batch reaches the appropriate destination. The following tools may help with this planning:

- Company schedule
- Company product change report form
- API gravity conversion chart
- Software batch cut programs

To signal the change between two batches when there is little or no difference between them, some companies inject a dye flag. If the facility is using an OID, it can detect these batches easily, making switching simple. If the pipeline is using a gravitometer only, the batch is not switched when the dye flag arrives. Rather, the switch occurs at the mid-point of the gravity change. Personnel should verify dye flags or color markers before the shipment begins. Dye flags may be color-coded for various products.

1.3.0 Pump Operations

Pumps move product from one point in the pipeline to another. Centrifugal pumps (*Figure 21*) act as main-line and booster pumps on most pipelines. With this kind of pump, fluid enters the case near the axis of a rotating impeller. The impeller accelerates the fluid and flings it outwards at high velocity. The surrounding casing captures the fluid, which then exits the pump at high pressure. Some pipelines also use positive displacement (PD) pumps for auxiliary or special purpose applications.

All pumps use a driver of some kind to supply power. Electric motors, internal combustion engines, and turbines are examples of typical drivers. Power is expressed in horsepower (hp). Moving a heavy product at a high flow rate requires more power than moving a lighter one at a low flow rate. Control center personnel plan and calculate shipping procedures to use the lowest horsepower possible to accomplish the job.

> **NOTE**
> In countries that use the metric system, power is expressed in watts (W) or kilowatts (kW), rather than horsepower (1 hp = 746 W, or 0.746 kW).

Pipeline pumps fall into three main categories:

- Main-line
- Booster
- Sump

1.3.1 Main-Line Pumps

Main-line pumps move the product through the pipeline, generating the pressure and flow required for meeting shipment schedules. Most main-line units are large centrifugal pumps (*Figure 22*) with high horsepower drivers. Each location will likely have a number of these. Main-line units operate individually, in parallel, or in series. In many installations, main-line pumps receive suction from a common line supplied by a tank manifold and supply flow to the pipeline through a common manifold. The manifold allows several tanks to supply the pumps during an operation.

The control center manages pumps, starting and stopping them as required. It also sets and adjusts suction and discharge setpoints, depending on how the pumps are working. When multiple units work together, control center personnel

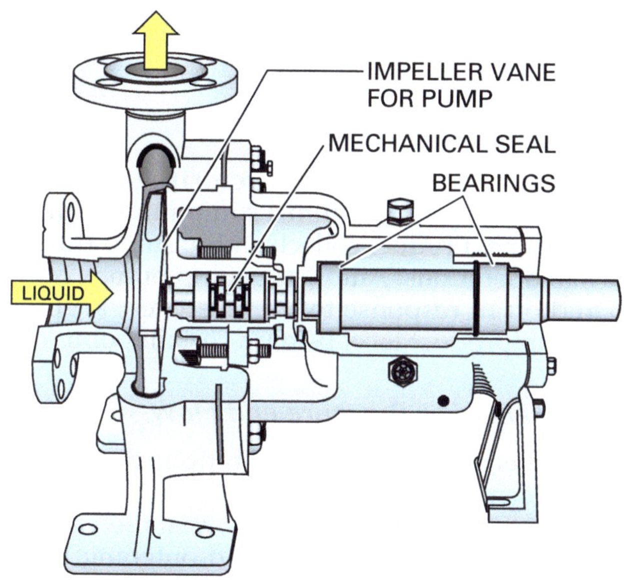

Figure 21 Centrifugal pump.

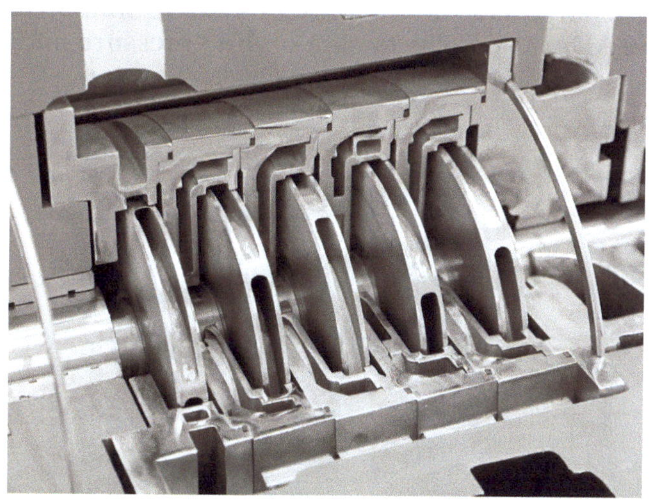

Figure 22 Multi-impeller centrifugal pump.

must ensure compatible flow rates. Check valves prevent the pumps from recirculating.

Pumps do not have to be of the same size or capacity. When operating in parallel, they must be adjusted so their head settings match, because unequal pressure will cause a shutdown. Pumps operating in series complement each other, so it's not necessary to match head settings. Total discharge is the sum of each pump's head.

1.3.2 Booster Pumps

Booster pumps (*Figure 23*) are high-volume, low-head units that pump product from a tank into the suction side of a main-line unit. When operating in series, they can also act as injection pumps. Booster pumps provide the flow rate needed by main-line pumps running at high settings. Product level in a tank may supply sufficient suction pressure for the main-line pump, but if not, booster pumps come on line to supply the needed pressure.

1.3.3 Sump Pumps

Sump pumps are low-head centrifugal pumps used to move product from a sump into a tank or the pipeline. Product collects in the sump during normal operations. Automatic sump pumps come on and off in response to a level switch, keeping the sump product level below a certain point.

1.3.4 Managing Pressure and Flow

Safe and efficient pipeline operations depend on control center personnel monitoring and managing pressure. Control valves, installed in the pumping station's incoming and outgoing lines, adjust the pressure and flow. Transducers monitor and report pressure at the pump's suction and discharge sides to the SCADA system. Similarly, SCADA also collects data on line pressure and

Figure 23 Booster pumps on a manifold.

case pressure. Control center personnel use this information to adjust control valves to sustain steady state operating conditions on the outgoing line.

Control valves provide crucial pressure and flow regulation in systems with constant-speed main-line pumps. Valve throttling action at the suction and discharge sides keeps pressure and flow within appropriate ranges. Control center personnel determine the appropriate setpoints and adjust the valves from the control center. Station design may allow personnel to adjust pressure and flow at several different locations (*Figure 24*). Generally, the control center can manage valves at origin and delivery points, but not at booster stations.

Backpressure valves maintain line pressure above the product vapor pressure point in case of problems or operational changes. They also protect terminal and station low-pressure piping. These devices are preset and will operate as required.

1.3.5 Pressure Points

Suction pressure is generated on the pump's upstream side, while discharge pressure is generated on its downstream (pipeline) side. The pump controller, usually a PLC, manages both of these pressures by monitoring them and comparing their current values to programmed setpoints. Control valves throttle the flow to maintain safe and appropriate operating conditions.

Increasing discharge pressure increases flow, but it decreases suction pressure because of the greater demand on the supply. Suction must also be increased, or the pump may shut down. Decreasing discharge pressure increases suction pressure, but decreases flow because the pump is ejecting less product.

Product density can significantly affect pressure and flow as well. A pump will experience a density change when an interface passes through it. A density increase will cause discharge pressure to also increase. The pump control PLC may make adjustments to maintain appropriate suction and discharge. The control room may also modify settings through the SCADA system.

In a large pipeline, different segments may operate at different flow rates and pressures. Some sections may supply low-pressure equipment, such as metering and manifolds. On older pipelines, segments may be de-rated to a lower **maximum operating pressure (MOP)**. These must be handled separately. When starting the pipeline, control operators will probably open segments in a specific sequence rather than all at once. They

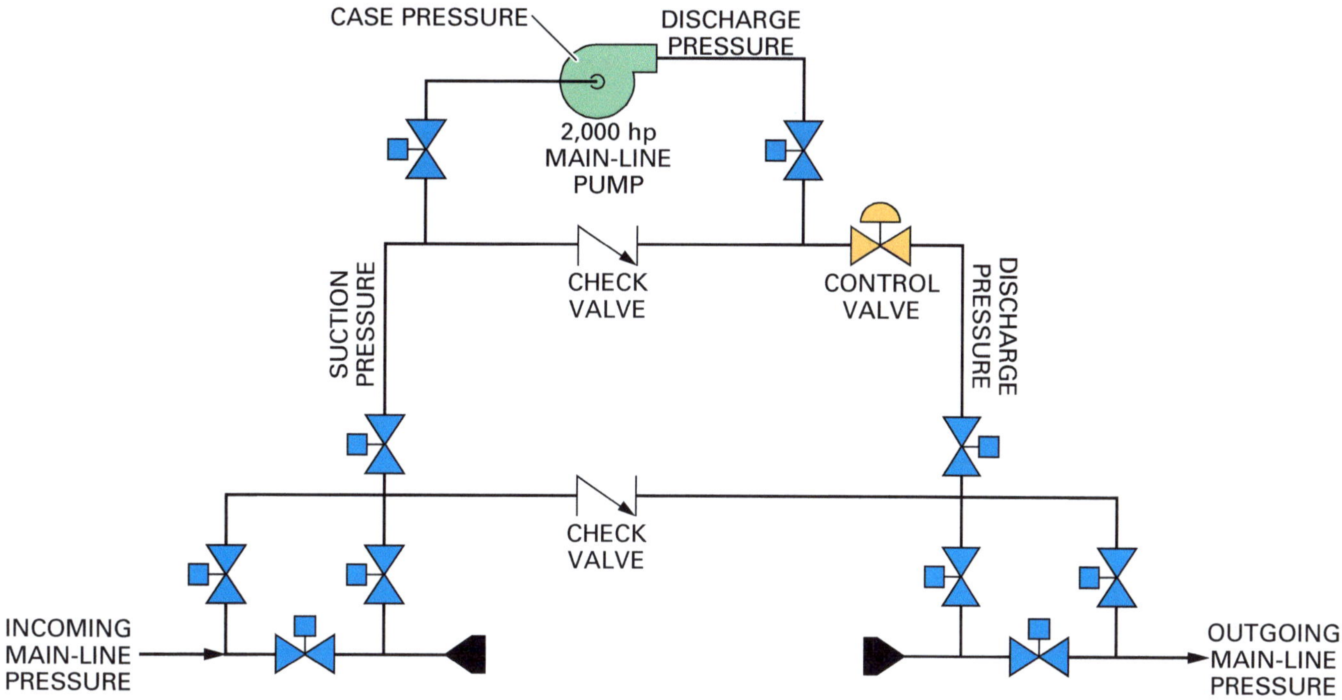

Figure 24 Simplified diagram of a pump station.

will allow the initially opened segments to build up pressure before opening others.

Control center personnel must note pressure differences from one segment to the next, and be careful not to confuse them. Errors could cause an overpressure situation or a slack line condition. If there is a great difference in segment sizes, pressure management may be difficult. One section may require throttling while another may not. Pressure transducers provide full information through the SCADA system so operators can make decisions.

A pump's discharge side setting determines the pressure within the pump case. It will always be higher than the pressure past the discharge side. Pump-case pressure can cause damage over time, but it might not be noticed immediately. As the condition worsens, high pump-case pressure may cause unstable pump controls, piping vibration, and high noise levels. The control center can deal with this problem by making pressure adjustments upstream and/or downstream. Pumps should be inspected periodically for signs of pressure-related damage.

1.3.6 Pumping Applications

Pipeline engineers select pumps based on product type and throughput requirements. Increasing or decreasing horsepower increases or decreases flow

Valves and Pigging

Pump stations may include technology designed to handle pipeline pigs. Pigs are cylindrical devices that fit snugly in the pipeline and are forced through it by the flowing product. They are used for one of several reasons:

- Cleaning or decontaminating the pipeline
- Inspecting the pipeline
- Pushing out gases or water
- Separating batches of product

Pigs come in a number of types and styles, depending on their purpose. Some are very simple, while others, such as "smart" pigs, contain sophisticated electronics that monitor and record conditions within the pipeline.

When pigs are used, appropriate technology is required to handle them. This includes valves for introducing and removing them, as well as detectors to sense their presence. The station shown in *Figure 24* includes valves to accommodate pigging.

rate. Similarly, adding or removing pumps has the same effect. Control center personnel determine the best way to utilize the pumps under their control to make the shipment move on schedule.

If the pipeline climbs to a higher elevation, extra horsepower will be needed beyond that required to handle a level grade. Calculations determine the extra horsepower necessary to do the work. A station may need to start several additional main-line units to supply enough horsepower to go uphill (*Figure 25*).

If pumps come on or go off line unexpectedly, pressure and flow will probably change. The possibility of such occurrences increases with the shipment distance. Pressure and flow rates must be watched closely and adjusted to compensate for these changes.

1.3.7 Pump Timing

Systems made up of multiple stations, terminals, and tank farms require a specific pump start-up sequence. For example, a booster pump may have to be started before a main-line pump in order to provide sufficient suction. The booster pump delivers product to the suction side of the main-line unit. Failing to start the booster pump first could cause damage and force the main-line unit to shut down due to low suction (*Figure 26*).

NOTE

Raising suction pressure and discharge pressure also raises pump-case pressure.

On the other hand, if discharge or main-line pressure rises too high, the control center may shut down a booster pump or a main-line pump. These actions would reduce suction pressure, causing the pump-case and discharge pressures to drop in response.

The control center brings up an idle pipeline by starting the pumps at the originating station first and gradually adding downstream stations as the product stream and pressure allow. Prior to start-up, main-line block, terminal, and tank valves must be opened for flow to increase. Booster pumps come on line first. Main-line units then follow. After the initial start-up, each successive pump downstream comes on line. Pressure builds until the system reaches a steady state condition.

Figure 25 Increased horsepower required for elevation.

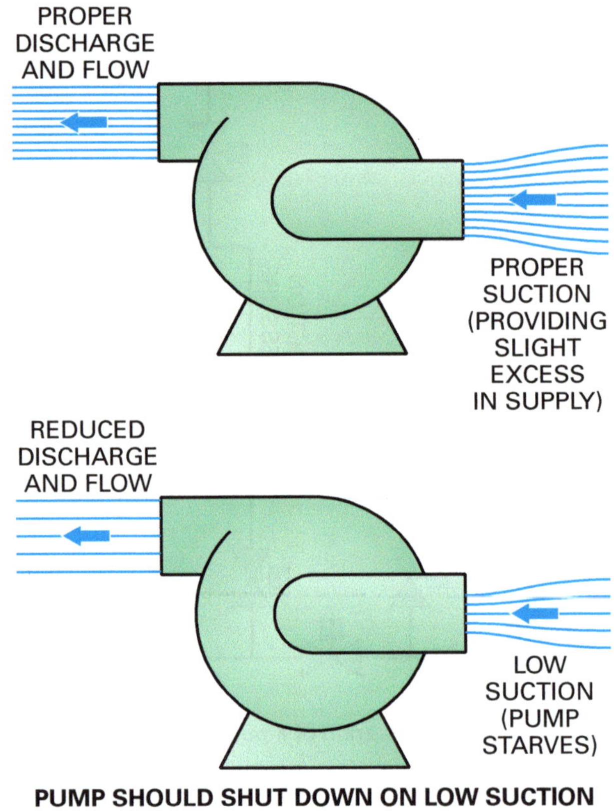

Figure 26 Effects of normal and low suction.

1.3.8 Pump Management and Power Optimization

No matter the type, power costs money. Almost every pipeline activity consumes power, so control center personnel work hard to manage power carefully and optimize its use. By using training, experience, and good judgment, personnel set up shipments to make the most economical use of the equipment. Some facilities compare utility rates when planning a delivery. They then schedule pumping routes to take advantage of the best rates and times of day.

By being thoroughly familiar with pump condition indicators, such as flow rate, differential pressure, and consumption of current (measured in amperes, or amps), operators can estimate efficiency. For example, a constant-speed pump's efficiency drops when throttle pressure increases. With a variable-speed pump, some speeds are more efficient than others for a given set of conditions. Any unexplained change requires investigation immediately. By staying on top of problems, the control center can reduce costs and keep operational efficiency high.

Control center personnel can use natural power sources, such as gravity, to accomplish some parts of their work. For example, a tank with a high level can deliver high flow rates to pumps without booster pump assistance. Gravity can also transfer product between tanks if they are at different heights. A booster pump may have to be used for part of the transfer, however.

For variable-speed pumps, a **variable frequency drive (VFD)** can control the pump for maximum efficiency. VFDs manage their electric motor's speed by adjusting the voltage and frequency supplied to it. This technique controls pressure and flow without resorting to throttling, which wastes more power. It also offers an efficient way to reduce power when full power is not needed.

Power consumption can be calculated if a few key factors are known. The following formula is one such example based on the pipeline units used in the United States:

$$\text{power} = (\text{flow} \times \text{head}) \div (2{,}450 \times \text{efficiency})$$

Where

power = the power consumed in hp
flow = the pump's flow rate in bph
head = the pump's head in psi
2,450 = a conversion constant
efficiency = the pump's efficiency as a percentage

Another way to manage power is by switching pumps, an option available in installations that have multiple units. Pumps can be alternated based on power requirements in order to meet rate conditions or to perform maintenance, depending on company policy and procedure. One pumping unit can be brought on line and another unit taken off in order to improve pipeline hydraulics or to achieve line balance. Control center personnel manage suction and discharge at the origin station. They must also sequence valves and handle switching. Booster stations have their own control systems.

1.3.9 Drag Reducing Agents

Drag reducing agents (DRAs) are long-chain polymers that enhance flow by reducing pipeline friction. Personnel introduce DRAs into the pipeline through annular ring injectors in order to coat the inner surface (*Figure 27*). Pipelines treated with DRAs experience about 20 percent greater throughput and a corresponding 20 percent reduction in pressure. Consequently, DRA injections greatly reduce the amount of power required to pump certain products, especially crude oil. However, pipeline personnel must calculate the DRA cost and the injection expense to be sure that the energy savings are large enough to justify using a DRA.

Figure 27 DRA injection control valve.

NOTE

DRAs are also used during maintenance when a station must be bypassed.

DRA injections are planned by the scheduling department. By reducing friction, a DRA affects the pumping rate in positive ways. Adding a DRA results in a shorter pumping schedule and creates a negligible change in volume. Since DRAs are expensive and have a strong effect on operations, use care in following schedule orders.

Cautions and guidelines associated with using DRAs include the following:

- DRAs may only be used with compatible products, such as crude oil.
- DRAs may not be used with refined highly volatile liquid (HVL) products.
- DRAs injection must be stopped at least two hours before pumping specialty products, such as jet fuel.

Control center personnel can manage some injections from the control center. Others may require field personnel to be at the site. Control center personnel should consult their company's DRA product compatibility chart and the product schedule to verify that a planned DRA injection is compatible with the product. DRAs should not be injected if any of the following situations exist:

- An incompatible shipment will occur within a specific time period after the injection.
- An incompatible product interface has passed through the injection site less than a specific time before the injection is scheduled to start.
- A residue of product or DRA could cause some type of contamination.

Additional Resources

It is assumed that you will learn to perform this procedure under the direct supervision of your crew leader, foreman, or instructor. For independent study, refer to the recommended Additional Resources at the back of this module.

1.0.0 Section Review

1. A tank mixer helps prevent _____.
   a. evaporation
   b. stratification
   c. explosions
   d. contamination

2. Tank volume based on measurements that have been adjusted for factors such as temperature and density is known as _____.
   a. net volume
   b. API volume
   c. SCADA volume
   d. gross volume

3. Which of the following valves can operate in the open, closed, and throttle positions?
   a. Block valves
   b. Safety valves
   c. Gate valves
   d. Control valves

4. When main-line pumps run in parallel, their head settings _____.
   a. add together
   b. must match
   c. can be different
   d. subtract from each other

Section Two

2.0.0 Starting Up a Liquid Pipeline

Objective

Describe the requirements and procedures for starting up a liquid pipeline (CT64_1-17).

a. Identify the prerequisites, competencies, and abnormal operating conditions associated with starting up a liquid pipeline.
b. Describe the procedures used to start up a liquid pipeline.

Performance Task

1. Start up a liquid pipeline—control center (CT64_1-17).
   - Identify potential abnormal operating conditions that may occur during performance of this CT, and know the appropriate actions to take in response to them.
   - Utilize the appropriate personal protective equipment according to relevant company procedures.
   - Verify that the flows, pressures, and product characteristics are within normal range.
   - Ensure that upstream and downstream stations are prepared to bring the pipeline up before beginning the start-up procedure.
   - Activate (start up) units beginning at the origination point and proceeding downstream to the delivery point.
   - Activate units at downstream stations upon arrival of a pressure rise from the upstream station.
   - Perform proper operation according to company policy and procedures.
   - Verify that the desired result was achieved.
   - Complete the appropriate documentation as required by operator's procedures.

Trade Terms

Abnormal operating conditions (AOCs): Conditions within a pipeline system, identified by the operator, that may indicate (1) a malfunction of a component or (2) a deviation from normal operations that may indicate a condition exceeding design limits or result in a hazard to persons, property, or the environment.

Computational pipeline monitoring (CPM): A computer-based pipeline monitoring system that creates a real-time hydraulic model of the entire pipeline and also detects leaks and breaks.

Packed: The condition in which the pipeline is kept under pressure so its contents do not vaporize.

Surge: A wave of high pressure that occurs when fluid starts or stops moving suddenly. Surges occur both at start-up and shutdown, although from different causes.

Starting a liquid pipeline requires careful planning and coordination between the originating terminal, the receiving terminal, and all other stations along the pipeline. Normal start-up procedures begin when the schedule arrives from the scheduling department. They end when the line is packed and flowing along its entire length to the receiving terminal.

Successful start-up requires precise timing and careful monitoring. Each station or piece of equipment needs to start at the right moment so the line eventually reaches a steady state condition. Many different pieces of equipment contribute to this goal, including pumps, valves, tanks, and monitoring technology. Once the product is flowing, control center personnel make adjustments to maintain safe and efficient operation.

The fact that most stations are unstaffed makes this task even more demanding. Consequently, the control room relies heavily on remote monitoring and control through the SCADA system. Personnel must be proficient with this system and be able to use it quickly and efficiently.

2.1.0 Prerequisites, Competencies, and Abnormal Operating Conditions

Control room personnel must possess a variety of skills in order to start-up a liquid pipeline. They must also be aware of the particular abnormal operating conditions (AOCs) that may occur as they carry out these tasks.

2.1.1 Prerequisites and Competencies

Control room personnel must possess the following basic knowledge and skills prior to starting up a liquid pipeline:

- Be familiar with SCADA displays and know how to interpret their informational pages as well as their alarm and alert pages.
- Know how to use the SCADA system to change or adjust pipeline settings.
- Understand how to interpret leak detection system displays—computational pipeline monitoring (CPM) and non-CPM—and use them to localize a leak.
- Know the maximum discharge pressure for each station as well as the pressure and flow values typical for a steady state condition. Know the MOP values for key points in the pipeline if this information is available.
- Have a basic understanding of pipeline hydraulics and the factors that influence it.
- Be thoroughly familiar with abnormal operating conditions and emergencies, understanding both their entry conditions and the proper responses to each.
- Understand and be skilled at communicating both within the control room and with field personnel.
- Understand and be skilled at documenting all activities using the proper methods and formats.

2.1.2 Abnormal Operating Conditions

Control room personnel starting up a liquid pipeline must be able to identify the following potential abnormal operating conditions and know the proper response to each, should they occur. The following are potential AOCs, and the appropriate action the controller should take in response to them:

- *Exposed pipe, missing or damaged pipeline supports* – It is likely that these conditions would be noticed if the operator were physically at the location but not if the operator were in a remote control center. If the line is operated from a central control center, contact the control center and any additional persons as required by company policy. In any event, as a minimum, notification should be made in accordance with company policy.
- *Activation of a safety device, such as an overpressure device or overfill device* – Never ignore these devices because each is typically designed to reduce or eliminate a hazardous situation. If a safety device activates, minimize the situation and verify the operation of the safety device. Notifications should be made per company policy.
- *An alarm associated with a safety device activates* – Identify the specific device involved and verify its operation. Minimize the situation and make necessary notifications per company policy.
- *Communications, control system or power interruption or failure, such as loss of radio, telephone, SCADA, or electrical services on all or part of the pipeline* – Follow troubleshooting procedures and make contacts as required by company policy.
- *Fire/explosion, or ignition of flammable liquids or gases* – If the situation cannot be safely controlled immediately by hand, immediately evacuate the area, limit access, and make proper notifications. Notify local 911 as appropriate.
- *Damaged pipeline system components (broken, bent, crushed, gouged, or burned)* – Minimize or eliminate pressures and/or flows as appropriate for the severity of the situation and make necessary notifications according to company policy.
- *Unexpected hazardous liquid or carbon dioxide encountered* – If the leak cannot be isolated safely at the site, the control center should be contacted to isolate or shut down the line. Evacuate the area.
- *Unexplained flow rate deviation, such as high flow, low flow, and no flow* – With a high, low, or no flow, verify that the proper valves are aligned and opened. Always work to minimize any potential hazardous situation. Verify equipment readings. Verify that any units and/or valves operated prior to this deviation are still in the desired position.
- *Unexplained pressure deviation, such as a pressure increase, pressure decrease, or lack of pressure reading* – Assume the reading is correct and work to identify the cause. Always minimize the severity of the situation. If you are not immediately aware of the cause, you may need to shut down the line until the situation can be identified and corrected. Verify that any units and/or valves operated prior to this deviation are still in the desired position.
- *Unexplained status change, such as unexplained changes in units, valves, gravity, tank levels, temperature, and flash* – Determine the cause and minimize the situation.

2.2.0 Pipeline Start-Up

Starting up the liquid pipeline involves numerous tasks, many of which must happen in a specific and well-timed sequence. The importance of timing should never be underestimated. Personnel must be able to perform all required start-up tasks confidently and without hesitation.

2.2.1 Start-Up Overview

Pipeline start-up is governed by 49 *CFR* 195.402. The central goal is to keep all components within their design limits and safe operating conditions. Control center personnel carry the responsibility for ensuring a proper start up.

Start-ups fall into three categories: normal, maintenance, and abnormal. Normal start-ups come after a routine pipeline shutdown, such as one that follows an uneventful delivery. Maintenance start-ups come after a pipeline shutdown and a maintenance task. Abnormal start-ups come after a pipeline shutdown caused by an AOC.

Basic procedures are essentially the same for all three types of start-ups, but for maintenance and abnormal start-ups, additional precautions are required. These precautions will be discussed after the standard start-up preliminaries and procedures are explained.

2.2.2 Preliminary Tasks

Control center personnel must receive authorization in the form of scheduling orders to start-up the pipeline. The scheduling department plans the product shipment, then sends the orders to the control center. The control center manages the pipeline equipment needed to fulfill those orders, while adhering to company policies and procedures. Its tasks include selecting pumps, establishing setpoints and valve alignments, using the SCADA system to control operations, and monitoring key parameters during delivery to prevent abnormal operating conditions.

Control center personnel should be skilled at reading schedules. Although each company's scheduling system and schedule format will vary to some degree, all schedules contain common elements. These include the following:

- Product type
- Product source
- Product destination
- Product volume to be transferred
- Desired flow rate
- Delivery times

Personnel must also understand the batch arrangement (*Figure 28*) that the schedule dictates. Recall that the batch sequence has a significant influence on procedures and handling. If compatible products run in sequence, operations are relatively simple. With incompatible products, interfaces must be detected and transmix must be transferred to suitable breakout tanks for eventual disposal. Understanding compatibility is a key skill that runs in parallel with understanding how to read the schedule.

Inventory consists of the product in storage and available for shipment. Control center personnel must be familiar with inventory status and review the daily inventory report. Tanks should be gauged before shipment begins. If there is not enough of the specified product to fulfill the shipment, the scheduling department and the customer should be informed immediately.

Control center personnel must contact the receiving terminal before shipment begins in order to verify sufficient tankage for the delivery. If there is insufficient space, the scheduling and dispatching departments should be contacted immediately and other arrangements made. Another shipment will have to be selected.

CAUTION

Never begin a shipment without confirming sufficient tankage.

Always notify all appropriate personnel before start-up, including those at staffed facilities. Confirm the readiness of unstaffed facilities through the SCADA system. Verify that all required equipment is available and functioning prior to start-up. In some terminals, several shipments may be coming in simultaneously, so metering and proving facilities may be limited. Space in transmix tanks or relief tanks could also be limited.

Never start a pipeline that has integrity problems or abnormalities. Review the status and maintenance reports to confirm a normal

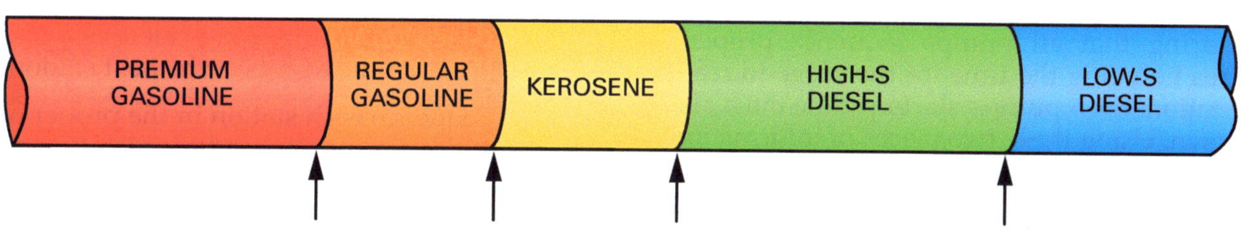

Figure 28 Visual representation of a batch sequence.

condition. Question any situation that could be problematic. Be aware of any maintenance activities going on along the line and confirm that they will not interfere with the planned operation. Always verify valve positions to avoid the risk of overpressure conditions or product contamination.

The critical flow path ensures pipeline safety by providing a pipe and valve arrangement that leads to relief tanks. These flow paths must remain open when the pipeline is operating. If there is a problem with the valves, select an alternative alignment that guarantees emergency relief when it's required. Control center personnel should verify that all critical flow paths are ready before starting the pipeline.

Personnel need to determine the number of barrels required to pressurize the line. The CPM leak detection system cannot start functioning until the line is under pressure. If the CPM system does not start operating after the requisite number of barrels has been pumped, there may be a leak in the system. This should be investigated without delay to avoid unnecessary product loss.

2.2.3 Timing

Starting a pipeline successfully depends on good timing. From a start-up perspective, think of the pipeline as a series of pumps spread out over hundreds of miles. Main-line pumps move the product from station to station, and booster pumps provide extra suction as needed. But product doesn't move instantly from the source to the destination. Rather, the origin station's pumps start and the product travels through the pipeline toward the next station. The initial product **surge** requires a certain amount of time to arrive. This time depends on distance, pipeline hydraulics, pressure, and flow rate.

Ideally, the main-line pump at the next stage comes on line exactly as the product surge arrives. The product then moves toward the next station, whose pumps should come on line as it arrives. This process continues until the product reaches the receiving station. At this point, all pumps will be operating, and the pipeline should reach the steady state condition.

Ensuring that all pumps sequence properly requires knowing the time for the surge to reach each station. Each pump's start-up time must also be known. Using these two pieces of information, the control center can start pumps at exactly the right moment.

2.2.4 Normal Start-Up Procedures

Although every pipeline system is unique, start-up procedures tend to have a common pattern. Control center personnel must protect life, the environment, and the pipeline's safety, and must always follow company policy and procedures.

> **CAUTION**
>
> Personnel working in the control room must be alert to the possibility of abnormal operating conditions and be prepared to initiate the proper responses from memory.

The following steps summarize the procedures involved in starting up the pipeline:

Step 1 Confirm that the origin and delivery facilities have been notified, and that all preliminary steps have been completed.

Step 2 Confirm that the proper flow path has been established and that all valve alignments are in place.

Step 3 Determine all essential delivery parameters such as pressure, flow rate, and line fills.

Step 4 Determine which pumps will be used. These include both booster and main-line units.

Step 5 Determine the appropriate start-up times for the downstream main-line units. Calculate the time at which the product surge will arrive at each station from the one immediately upstream of it. Be sure to include each pump's start-up time in the calculations.

Step 6 Start the suction booster pumps at the supply source.

Step 7 Start the main-line unit at the originating station.

Step 8 Lower suction and relief setpoints at the stations downstream.

Step 9 Confirm that the next station is ready for loading when the surge arrives.

Step 10 Based on your timing calculations, start the main-line units at the next station.

Step 11 Sequence each station in the proper order so its pumps are ready for loading when the surge arrives from upstream. Communicate with the delivery station so it will know when to expect its units to come on line.

Step 12 Increase the speed of the main-line unit to the desired flow rate while watching for the rise in suction pressures downstream.

Step 13 Confirm that all downstream stations come up to the desired main-line flow rate.

> **NOTE**
> Slow pressure and/or flow rise could be caused by pipeline topography or a leak somewhere along the line. If a leak is suspected, it must be investigated without delay to avoid unnecessary product loss.

Step 14 As soon as flow has stabilized and all units are on line, set all alarms to the required setpoints.

Step 15 Start any required DRA injections.

2.2.5 Monitoring

During and after start-up, the control center must track pressure and flow rates carefully to ensure that the maximum discharge pressure stays within appropriate limits. These rates will likely be different for each location along the pipeline. The SCADA system should contain the proper limits for each and will raise an alarm if any are exceeded. Remember that some pipeline sections may have been de-rated due to age; they cannot withstand the same discharge pressures as other segments.

> **NOTE**
> Throughout the shipment, the control center must maintain communications with the receiving terminal. Personnel must communicate any changes in flow. They must also complete all required documentation according to company procedure.

2.2.6 Maintenance and Abnormal Start-Ups

Sometimes pipelines must be shut down for extended maintenance and repair work. Following repair, inspection, and approval, the pipeline can be restarted. The first start-up after a maintenance cycle follows the same procedure as a normal start-up. However, control center personnel must be especially diligent to confirm that all systems are ready prior to start-up. They must also wait for notification and approval from the control center manager or lead manager before beginning.

Monitoring the pipeline during and after start-up is especially important since problems may occur in the repaired sections. Watch SCADA monitors particularly closely but also be aware of early warning signs that may appear before SCADA raises an alarm. Experience makes it easier to spot these. Catching an early warning sign may prevent any need to halt pipeline operations.

Pipeline shutdowns caused by AOCs require an official investigation to determine the exact cause. Trained and qualified personnel may have to survey the suspected area. Once they have identified the problem, a maintenance team will make repairs or corrections.

Before restarting the pipeline, the control center must receive appropriate reports from maintenance personnel confirming that the problem has been fixed. This confirmation will include test results, inspections, and approvals that meet company guidelines. Finally, official confirmation and approval will be received from the control center manager or lead manager to restart the pipeline. At this point, the control center should verify that all systems show as operational and begin the restart procedures.

While the pipeline start-up follows the standard procedure, control center personnel should be especially vigilant and monitor the pipeline carefully. They must watch systems associated with the AOC to verify that they are operating normally without a hint of a continuing problem.

Additional Resources

It is assumed that you will learn to perform this procedure under the direct supervision of your crew leader, foreman, or instructor. For independent study, refer to the recommended Additional Resources at the back of this module.

2.0.0 Section Review

1. Control room personnel must know how to interpret SCADA pages as well as how to use the SCADA system to _____.
   a. reset the CPM system
   b. adjust pipeline settings
   c. complete required documentation
   d. monitor system security

2. If an unexpected change in flow occurs, the controller must first _____.
   a. record the alarm in the control center logbook
   b. check SCADA for related problems
   c. alert higher management
   d. verify that the proper valves are aligned and opened

3. Pipeline start-up largely depends on _____.
   a. proper pump start-up timing
   b. idling pumps correctly
   c. communicating with booster station personnel
   d. compensating for pressure changes

SUMMARY

A liquid pipeline is comparable to a huge machine, made up of many different parts. Pumps, valves, tanks, and pipes are its main components. However, this machine spans hundreds of miles and must perform reliably with minimal human presence; operating it requires more than just equipment—it requires control. Pipelines, therefore, also include sophisticated systems that monitor the equipment and allow personnel to manage almost everything from a central location. It is through this control system that personnel start up the pipeline.

In order to start the pipeline, operators must understand its key components. Tanks store product, keeping it in good condition until it's ready to move through the pipeline. Tanks also receive deliveries and provide a place for relief devices to vent. Managing tanks and keeping them from overflowing are major control center responsibilities.

Valves control product as it moves through tanks, manifolds, and the pipeline itself. The pipeline contains numerous types and sizes of valves, many of which can be controlled remotely. Personnel must know how to operate valves and check their status through the SCADA system. They must also know how to manage them to maintain product quality and prevent contamination.

Pumps move product from one part of the pipeline to another. Most pipeline pumps are centrifugal. Properly controlling them leads to efficient energy use and lower operating costs. Similarly, using drag reducing agents can save money by lowering pipeline friction.

Starting the pipeline requires precision timing; pumps must come on line at just the right moment. But control center personnel must also watch for abnormal operating conditions and other problems as they bring the pipeline to a steady state condition. In addition, start-up after maintenance or an abnormal shutdown requires special attention to ensure that there are no lingering issues.

Review Questions

1. Filling an empty tank can create a large amount of _____.
   a. sediment and water
   b. transmix
   c. static electricity
   d. stratification

2. The process of allowing product to flow into and out of a tank at the same time is known as _____.
   a. floating the tank
   b. mixing the tank
   c. gauging the tank
   d. dewatering the tank

3. Valves that help prevent a slack line condition are called _____.
   a. hand valves
   b. safety valves
   c. block valves
   d. backpressure valves

4. What control system feature prevents inappropriate valve alignments and ensures that valve sequences happen in the correct order?
   a. SCADA
   b. Valve interlocks
   c. Safety valves
   d. Fieldbus networks

5. What kind of pressure is generated on the downstream side of a pump?
   a. Discharge pressure
   b. Mainstream pressure
   c. Pump-case pressure
   d. Suction pressure

6. Before special products can be run through the pipeline, a DRA injection must have been stopped for at least _____.
   a. 1 hour
   b. 2 hours
   c. 5 hours
   d. 10 hours

7. In addition to knowing the pressure and flow values for a steady state condition, control center personnel must know the _____.
   a. critical high level for the main manifold
   b. AOC trip point for the main-line pumps
   c. MOP for the entire pipeline
   d. maximum discharge pressure for each station

8. If a tank level is outside of safe limits, the first thing that the pipeline controller must do is _____.
   a. halt the current operation
   b. determine the cause of the problem
   c. notify relevant personnel
   d. activate the AOPS

9. The control center receives authorization to start the pipeline through _____.
   a. the SCADA system
   b. the scheduling department
   c. the control manager's office
   d. the receiving station

10. Calculating when to bring a main-line pump online requires control center personnel to know the surge arrival time and the pump's _____.
    a. operating speed
    b. maximum flow rate
    c. minimum flow rate
    d. start-up time

Trade Terms Quiz

Fill in the blank with the correct term that you learned from your study of this module.

1. A pipeline instrument that determines the product's specific gravity is called a(n) ________.
2. The tendency for liquid mixtures to separate over time, forming distinct layers according to density is known as ________.
3. A temporary storage tank used for holding specific grades of product, product mixtures, or the output of a pressure relief system is a(n) ________.
4. A control system used in the pipeline industry that collects and assembles data and also allows operators to control the pipeline remotely is called a(n) ________.
5. A control system feature that prevents unsuitable valve alignments and ensures that valve sequencing occurs in the correct order is referred to as ________.
6. The highest pressure at which a liquid pipeline may be operated is the ________.
7. The process of moving product directly from one pipeline to another without transferring it into a tank first is known as ________.
8. A tank level monitoring system that acts automatically and independently of other systems to prevent tank overfilling is called a(n) ________.
9. The region between batches of dissimilar product in which both products combine to form a transmix is called the ________.
10. A system that measures and reports the liquid level within a tank is called a(n) ________.
11. The condition in which pipeline variables are relatively constant over a period of time is called ________.
12. The API-defined tank level at which an alarm will activate and tank filling must halt within a specified time interval is known as the ________.
13. A pipeline instrument that uses light absorption to detect the interface between batches of dissimilar product is called a(n) ________.
14. A wave of high pressure that occurs when fluid starts or stops moving suddenly is known as a(n) ________.
15. The condition in which the pipeline is kept under pressure so its contents do not vaporize is known as ________.
16. A mechanical sensing system that uses a float to detect the product level within a tank and control an outside display is called a(n) ________.
17. The practice of using a tank as a buffer between pipelines by allowing product to enter and exit the tank at the same time is known as ________.
18. The fluid volume required to fill a pipeline section is called the ________.
19. The API-defined tank level at which a tank begins to overflow is the ________.
20. An electric motor paired with an electronic controller that adjusts motor speed by varying the power supply voltage and frequency is called a(n) ________.
21. An unusable mixture of two different products, usually created by switching from one incompatible product to another, is referred to as ________.
22. The API-defined tank level considered the highest point to which liquid should rise during normal operations is called the ________.

23. Conditions within a pipeline system that may indicate (1) a malfunction of a component or (2) a condition exceeding design limits or resulting in a hazard to persons, property, or the environment are called ________.

24. A computer-based pipeline monitoring system that creates a real-time hydraulic model of the entire pipeline and also detects leaks and breaks is called ________.

Trade Terms

- Abnormal operating conditions (AOCs)
- Automatic overfill protection system (AOPS)
- Automatic tank gauge (ATG)
- Breakout tank
- Computational pipeline monitoring (CPM)
- Critical high (CH)
- Floating
- Gravitometer
- High-high (HH)
- Interface
- Line fill
- Maximum operating pressure (MOP)
- Maximum working (MW)
- Optical interface detector (OID)
- Packed
- Supervisory Control and Data Acquisition (SCADA)
- Steady state
- Stratification
- Surge
- Tank gauging system
- Tightlining
- Transmix
- Valve interlock
- Variable frequency drive (VFD)

Trade Terms Introduced in This Module

Abnormal operating conditions (AOCs): Conditions within a pipeline system, identified by the operator, that may indicate (1) a malfunction of a component or (2) a deviation from normal operations that may indicate a condition exceeding design limits or result in a hazard to persons, property, or the environment.

Automatic overfill protection system (AOPS): A tank level monitoring system that acts automatically and independently of other systems to prevent tank overfilling. Also, an API-defined tank level at which the AOPS will activate.

Automatic tank gauge (ATG): A mechanical sensing system that uses a float to detect the product level within a tank and control an outside display.

Breakout tank: A temporary storage tank used for holding specific grades of product, product mixtures, or the output of a pressure relief system.

Computational pipeline monitoring (CPM): A computer-based pipeline monitoring system that creates a real-time hydraulic model of the entire pipeline and also detects leaks and breaks.

Critical high (CH): The API-defined tank level at which a tank begins to overflow.

Floating: The practice of using a tank as a buffer between pipelines by allowing product to enter and exit the tank at the same time, although not necessarily at the same rate.

Gravitometer: A pipeline instrument that determines the product's specific gravity.

High-high (HH): The API-defined tank level at which an alarm will activate and tank filling must halt within a specified time interval.

Interface: The region between batches of dissimilar products where both products combine to form a transmix.

Line fill: The fluid volume required to fill a pipeline section.

Maximum operating pressure (MOP): The highest pressure at which a liquid pipeline may be operated.

Maximum working (MW): The API-defined tank level considered the highest point to which liquid should rise during normal operations.

Optical interface detector (OID): A pipeline instrument that uses light absorption to detect the interface between batches of dissimilar product.

Packed: The condition in which the pipeline is kept under pressure so its contents do not vaporize.

Supervisory Control and Data Acquisition (SCADA): A control system used in the pipeline industry that collects and assembles data, and also allows operators to control the pipeline remotely

Steady state: The condition in which pipeline variables, such as pressure and flow, are relatively constant over a period of time.

Stratification: The tendency for liquid mixtures to separate over time, forming distinct layers according to density.

Surge: A wave of high pressure that occurs when fluid starts or stops moving suddenly. Surges occur both at start-up and shutdown, although from different causes.

Tank gauging system: A system that measures and reports the liquid level within a tank.

Tightlining: The process of moving product directly from one pipeline to another without transferring it into a tank first.

Transmix: An unusable mixture of two different products, usually created by switching from one incompatible product to another (stands for "transportation mixture").

Valve interlock: A control system feature that prevents unsuitable valve alignments and ensures that valve sequencing occurs in the correct order.

Variable frequency drive (VFD): An electric motor paired with an electronic controller that adjusts motor speed by varying the power supply voltage and frequency.

Additional Resources

The following recommended resources can provide additional helpful information related to the requirements for this covered task:

NCCER Module AOCCC-17, *Abnormal Operating Conditions Control Center*

NCCER Module AOCFG-17, *Abnormal Operating Conditions Field and Gas*

NCCER Module 12406-16, *Programmable Logic Controllers*, from *Instrumentation Level Four*

NCCER Module 64205-02, *Supervisory Control Systems*

NCCER Module CT23_1-17, *Maintain/Repair Relief Valves*

NCCER Module CT24_1-17, *Inspect, Test, and Calibrate Pressure Limiting Devices*

NCCER Module CT44_3-17, *Inspect, Test, and Maintain a Liquid Pipeline Leak Detection Flow Computer*

NCCER Module CT44_4-17, *Inspect, Test, and Perform Corrective and Preventative Maintenance of Tank Gauging for Leak Detection*

NCCER Module CT44_7-17, *Inspect, Test, and Maintain Gravitometers / Densitometers for Hazardous Liquid Leak Detection*

The following websites offer resources for products and training:

American Petroleum Institute, **www.api.org**

Pipeline and Hazardous Materials Safety Administration, **www.phmsa.dot.gov**

Figure Credits

ABB, Module Opener, Figure 5

© Ginton/Dreamstime.com, Figure 1

Chemineer HS agitator. Photo provided by National Oilwell Varco., Figure 2

© iStockphoto.com/Olivier Lantzendorffer, Figure 4

© Burnstuff2003/Dreamstime.com, Figure 6

U.S. Tape Company, Inc., Figure 7

Emerson Process Management, Figure 9

Apollo Valves and Elkhart Products Corporation, Figure 11

Rockwell Automation, Inc., Figure 14

Photos courtesy of Entergy New Orleans, Inc. © 2016 All Rights Reserved, Figure 15

OMNI Flow Computers, Inc., Figure 19

© Stepan Popov/Dreamstime.com, Figure 22

Satam, Figure 23

Tim Mills, Figure 27

Section Review Answer Key

Answer	Section Reference	Objective
Section One		
1. b	1.1.1	1a
2. a	1.1.4	1a
3. d	1.2.1	1b
4. b	1.3.1	1c
Section Two		
1. b	2.1.1	2a
2. d	2.1.2	2a
3. a	2.2.3	2b

NCCER CURRICULA — USER UPDATE

NCCER makes every effort to keep its textbooks up-to-date and free of technical errors. We appreciate your help in this process. If you find an error, a typographical mistake, or an inaccuracy in NCCER's curricula, please fill out this form (or a photocopy), or complete the online form at **www.nccer.org/olf**. Be sure to include the exact module ID number, page number, a detailed description, and your recommended correction. Your input will be brought to the attention of the Authoring Team. Thank you for your assistance.

Instructors – If you have an idea for improving this textbook, or have found that additional materials were necessary to teach this module effectively, please let us know so that we may present your suggestions to the Authoring Team.

NCCER Product Development and Revision

13614 Progress Blvd., Alachua, FL 32615

Email: curriculum@nccer.org
Online: www.nccer.org/olf

❑ Trainee Guide ❑ Lesson Plans ❑ Exam ❑ PowerPoints Other ____________

Craft / Level: Copyright Date:

Module ID Number / Title:

Section Number(s):

Description:

Recommended Correction:

Your Name:

Address:

Email: Phone:

CT64_2-17

Shutdown of a Liquid Pipeline (Control Center)

OVERVIEW

The ultimate purpose of the control center is to move product through the system efficiently, economically, and safely. Control center personnel must use their training, experience, and judgment to make good decisions that support this goal. This module explores the crucial task of shutting down a pipeline, something that happens for a number of reasons. Though it may sound simple, this task involves planning, coordinating, and multiple steps. Pipeline controllers must know how to bring the pipeline to a halt without damaging equipment, creating unsafe conditions, or losing valuable product.

Trainees with successful module completions may be eligible for credentialing through the NCCER Registry. To learn more, go to **www.nccer.org** or contact us at 1.888.622.3720. Our website has information on the latest product releases and training, as well as online versions of our *Cornerstone* magazine and Pearson's product catalog.

Your feedback is welcome. You may email your comments to **curriculum@nccer.org**, send general comments and inquiries to **info@nccer.org**, or fill in the User Update form at the back of this module.

This information is general in nature and intended for training purposes only. Actual performance of activities described in this manual requires compliance with all applicable operating, service, maintenance, and safety procedures under the direction of qualified personnel. References in this manual to patented or proprietary devices do not constitute a recommendation of their use.

CT64_2-17
SHUTDOWN OF A LIQUID PIPELINE (CONTROL CENTER)

Objectives

When you have completed this module, you will be able to do the following:

1. Describe key liquid pipeline components and their operations.
   a. Describe tanks and the operations associated with them.
   b. Describe manifolds and the valve operations associated with them.
   c. Describe pumps and the operations associated with them.
2. Describe the requirements and procedures for shutting down a liquid pipeline (CT64_2-17).
   a. Identify the prerequisites, competencies, and abnormal operating conditions associated with shutting down a liquid pipeline.
   b. Describe the procedures used to shut down a liquid pipeline.

Performance Task

Under the supervision of your instructor, you should be able to do the following:

1. Shut down a liquid pipeline (control center) (CT64_2-17).
   - Identify potential abnormal operating conditions that may occur during performance of this CT, and know the appropriate actions to take in response to them.
   - Utilize the appropriate personal protective equipment according to relevant company procedures.
   - In accordance with company policies, be familiar with the desired pressure to keep on a non-running line and the desired pressure to keep on a static line.
   - Shut down pump units or compressor units beginning at the origination point and proceeding downstream to the destination. Depending on the line profile, sufficient units at the source station and at the other stations may be kept operational. Then shut down in conjunction with valve closure at the delivery terminals or stations to satisfy the policies on maintaining minimum pressures on shutdown of the particular line.
   - The closure of isolation valves at intermediate delivery points should occur in conjunction with the shutdown of that station's units unless otherwise specified in company policy or procedure.
   - Perform proper operation according to company policy and procedures.
   - Verify that the desired result was achieved.
   - Complete the appropriate documentation as required by operator's procedures.

Trade Terms

Abnormal operating conditions (AOCs)
Automatic overfill protection system (AOPS)
Automatic tank gauge (ATG)
Breakout tank
Computational pipeline monitoring (CPM)
Critical high (CH)
Floating
Gravitometer
High-high (HH)
Interface
Line fill
Maximum operating pressure (MOP)
Maximum working (MW)
Optical interface detector (OID)
Packed
Supervisory Control and Data Acquisition (SCADA)
Static
Steady state
Stratification
Surge
Tank gauging system
Tightlining
Transmix
Valve interlock
Variable frequency drive (VFD)

Industry Recognized Credentials

If you are training through an NCCER-accredited sponsor, you may be eligible for credentials from NCCER's Registry. The ID number for this module is CT64_2-17. Note that this module may have been used in other NCCER curricula and may apply to other level completions. Contact NCCER's Registry at 888.622.3720 or go to **www.nccer.org** for more information.

Contents

Figures and Tables

Section One

1.0.0 Liquid Pipeline Components

Objective

Describe key liquid pipeline components and their operations.

a. Describe tanks and the operations associated with them.
b. Describe manifolds and the valve operations associated with them.
c. Describe pumps and the operations associated with them.

Trade Terms

Automatic overfill protection system (AOPS): A tank level monitoring system that acts automatically and independently of other systems to prevent tank overfilling. Also, an API-defined tank level at which the AOPS will activate.

Automatic tank gauge (ATG): A mechanical sensing system that uses a float to detect the product level within a tank and control an outside display.

Breakout tank: A temporary storage tank used for holding specific grades of product, product mixtures, or the output of a pressure relief system.

Critical high (CH): The API-defined tank level at which a tank begins to overflow.

Floating: The practice of using a tank as a buffer between pipelines by allowing product to enter and exit the tank at the same time, although not necessarily at the same rate.

Gravitometer: A pipeline instrument that determines the product's specific gravity.

High-high (HH): The API-defined tank level at which an alarm will activate and tank filling must halt within a specified time interval.

Interface: The region between batches of dissimilar products where both products combine to form a transmix.

Line fill: The fluid volume required to fill a pipeline section.

Maximum operating pressure (MOP): The highest pressure at which a liquid pipeline may be operated.

Maximum working (MW): The API-defined tank level considered the highest point to which liquid should rise during normal operations.

Optical interface detector (OID): A pipeline instrument that uses light absorption to detect the interface between batches of dissimilar product.

Supervisory Control and Data Acquisition (SCADA): A control system used in the pipeline industry that collects and assembles data, and also allows operators to control the pipeline remotely.

Steady state: The condition in which pipeline variables, such as pressure and flow, are relatively constant over a period of time.

Stratification: The tendency for liquid mixtures to separate over time, forming distinct layers according to density.

Tank gauging system: A system that measures and reports the liquid level within a tank.

Tightlining: The process of moving product directly from one pipeline to another without transferring it into a tank first.

Transmix: An unusable mixture of two different products, usually created by switching from one incompatible product to another (stands for "transportation mixture").

Valve interlock: A control system feature that prevents unsuitable valve alignments and ensures that valve sequencing occurs in the correct order.

Variable frequency drive (VFD): An electric motor paired with an electronic controller that adjusts motor speed by varying the power supply voltage and frequency.

Since pipelines and their associated facilities are far more automated than in former days, control room personnel carry wider responsibilities. Their duties now include some that used to be handled by field personnel. Similarly, their actions carry financial consequences. Carelessly contaminating product can be a costly mistake. For these reasons, control center personnel must know their duties thoroughly.

Operating a liquid pipeline involves interacting with and controlling many different pieces of equipment. Typical tasks like starting the pipeline, shutting it down, and manipulating valves all involve many of these devices. For this reason, control center personnel must understand the main ideas associated with each piece of equipment as well as its basic operation.

1.1.0 Tank Operations

Tanks are the pipeline's "warehouses". Using tanks safely and effectively requires training and experience. Tanks hold products as they move through the system, acting as buffers if one part cannot keep up with another. They assist in blending operations, scheduling, custody transfer, and holding **transmix** until it can be dealt with. Tanks support critical safety systems by dealing with excessive volume and providing a place for relief devices to transfer product.

Tankage falls into two broad categories. Storage tanks hold product inventory for various periods of time. A **breakout tank**, also known as a *relief tank*, provides temporary storage for relief devices when they open. Breakout tanks also hold product mixtures, and temporarily hold specific grades of product for mixing operations. In this way, they contribute to maintaining a **steady state** condition in the pipeline. The Department of Transportation Office of Pipeline Safety (DOT/OPS) exercises jurisdiction over breakout tanks.

Pipeline controllers view tanks primarily from a management perspective. They need to know if the product that they're about to move is ready, on-spec, and has a place to go when it is shipped. Scheduling is a major part of the controller's responsibilities. Operations personnel must manage tankage as efficiently as possible. This task includes allocating tanks to receive shipments, keeping inventories up to date, and knowing which products are available so shipments can be scheduled. If an off-specification product arrives, controllers must find the best way to store it without contaminating other products. Everything must be accurately documented at all times.

> **WARNING!** Because tanks have many associated hazards, there is no room for error in tank procedures. Personnel must always be cautious and alert, and follow company and industry procedures to the letter.

1.1.1 Tank Mixing

Due to the density differences of their components, products stored in tanks will eventually separate into distinct layers, as shown in *Figure 1*. This process is called **stratification**, and is usually undesirable. Each layer has its own properties, so essentially a stratified tank contains several different products. Because this condition is potentially hazardous, stratification must be prevented.

Figure 1 If undisturbed, liquid mixtures separate according to their density.

> **NOTE** In a limited number of situations, stratification can be helpful. For example, if a batch of crude oil contains excessive water, allowing it to stratify provides an easy way to remove the water, which forms a layer at the bottom.

Tank mixers (*Figure 2*) prevent stratification by keeping the contents of the tank in motion. This action maintains uniform quality and temperature. With crude oil, mixing keeps sediment suspended and water uniformly distributed. Mixers may run continuously or periodically as required by the particular product.

The control center usually manages tank mixers remotely, although mixers also have a manual switch for field use. Depending on their size, tanks may have more than one mixer. Mixers resemble a boat's outboard motor, with the propeller entering the tank from the side. The propeller angle may be fixed or variable. A variable-angle propeller can be adjusted to stir up the tank bottom to keep it clean from sediment.

A tank should be considered stratified if it has not been mixed in 24 hours or more. Never pump a stratified tank; the outcome could result in quality problems, and could be hazardous to the recipient. Instead, turn on the tank mixers and allow them to run long enough to mix the tank's contents. Four hours is a typical mixing time. Crude oil may require continuous mixing during pumping.

> **WARNING!** Always confirm that a tank's contents have been mixed sufficiently before pumping. Delivering stratified content could result in the recipient receiving an unexpectedly hazardous substance.

Figure 2 Tank mixer.

Before starting the tank mixers, confirm that the tank level isn't too low. Mixing when the level is low blends air into the product, which can create pumping problems. If the tank has a floating roof, the roof's flotation device may damage the mixer's propeller when the level is too low. Most companies have a standard policy that requires all mixers to be turned off, and possibly locked out, when the roof is below a specified safe level.

> **NOTE**
> Today, most tank roofs are of the cone or floating types. Floating roofs significantly reduce product evaporation and vapor release into the atmosphere.

1.1.2 Sumps

A sump is a holding tank, usually underground, with uses that vary depending on the installation. Companies often use sumps to collect and store contaminated product until it can be disposed of properly. Sumps can also act as breakout tanks for thermal relief valves from manifolds. Direct drain lines from dikes and firewalls may empty into sumps. Some sumps hold overflow, transmix, operational wastes, sludge, wash-downs, used lubricants, or even dirt and trash.

If a sump contains pure product, it may be acceptable to pump it into a tank or simply inject it back into the shipment. Mixed or contaminated product must be analyzed to determine its specifications. After testing, field personnel will determine the best way to blend, move, or dispose of the product. For example, a scheduled batch may support a carefully calculated injection, eliminating the need to discard the product.

If the sump contains transmix, the company may sell it to a processing firm. Some companies require field operations to pick up transmix and haul it away. Regardless of the ultimate destination, material from a sump must be accounted for and accurately measured. Sump volume must be integrated into calculations to avoid inventory problems.

Generally, sumps are managed by field personnel, who should measure and empty manually controlled sumps on a regular basis. Automatically pumped sumps should still be checked and measured periodically. Many automatic sump systems send an alarm to the control center when the high-level pump starts. When these alarms occur, personnel should inspect the sump visually to confirm that all is working properly. Sumps used for pipeline relief should receive routine, scheduled maintenance and have their communications system tested.

1.1.3 Receiving and Deliveries

When product arrives at the terminal or station, it must be stored until delivery to the customer. Receiving a shipment begins by diverting the incoming product stream out of the pipeline, and into a suitable tank. This action requires precision planning to ensure that the transfer is efficient and safe. Pipeline personnel must perform their jobs properly so each piece of equipment operates correctly. But beyond ensuring that all of the technology functions as expected, control personnel must be efficient in planning and executing the schedule.

Because tanks are dedicated to specific products, operations personnel decide which tanks fit the operation. Good tank management requires selecting tanks in advance, based on some or all of the following factors:

- Schedule
- Specific product involved
- Volume being pumped or delivered
- Available tank capacity
- Location efficiency
- Cargo requirements
- Flow rates
- Product compatibility

During receipt, operations personnel must open and close valves in exact sequence so the product takes the correct path from pipeline to tank. While control center personnel may perform these actions remotely, field personnel may have to confirm valve alignment. If a valve actuator fails, manual intervention will be required before operations personnel can open the pipeline delivery valve.

In most instances, the flow passes through strainers and metering runs before being routed into the selected tank (see *Figure 3*). Metering runs usually have straightening vanes upstream of the meter itself. These reduce flow turbulence and improve meter accuracy. In some cases, the flow will also go through pig traps. Personnel should verify that there is enough room in the tanks for the shipment.

Empty tanks require special precautions. Filling an empty tank can generate large amounts of static electricity. A flash can occur as the product begins to spread across the tank floor. (A flash is a sudden ignition of flammable vapor. In this case, it's the vapor in the storage tank that ignites when a static electric discharge accumulates and sparks during filling.) Beginning the fill very slowly reduces the potential for a flash. Each company has policies and procedures in place for filling empty tanks, which must always be followed.

> **CAUTION**
>
> Always follow your company's policies and procedures for filling empty tanks to prevent the occurrence of a flash.

If the tank has a floating roof (*Figure 4*), special precautions may be required. Filling too fast may result in the product flowing onto the roof, preventing it from floating. Not all tank-measuring systems give accurate level readings when the floating roof is resting on its legs. Until the roof starts floating, be particularly cautious about level readings.

Allowing product to flow in and out of a tank at the same time is a practice known as **floating**. The tank acts as a buffer if there is a difference in flow capacity between input and output lines. For example, if a large incoming stream must flow into a smaller pipeline, floating it with a tank allows the change without forcing the operator to reduce flow. The tank captures the large volume and holds it until the smaller pipeline can finish receiving it.

Operations personnel must be careful when floating a tank if the next batch is a different product. Unless the switch is made properly, an **interface** could wind up in the next batch.

1.1.4 Managing Inventory

Inventory fluctuates constantly as the terminal receives and delivers shipments, so operations personnel must stay aware of current inventory. The **Supervisory Control and Data Acquisition (SCADA)** system carries much of the load of this task by monitoring and calculating product quantity. Status reports give totals at the end of each operating day and confirm the terminal's inventory. These reports, along with SCADA tank level readings, help personnel know how much space is available so they can schedule receipts.

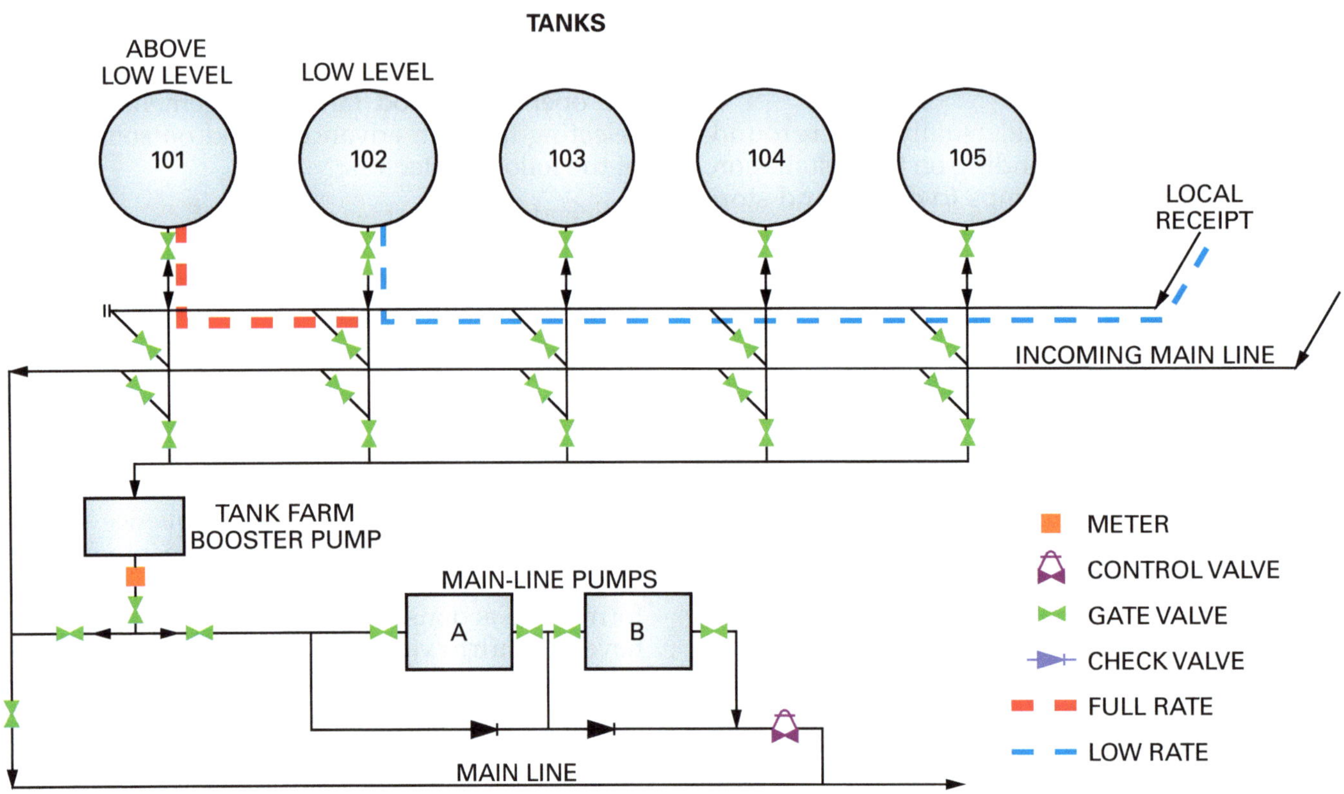

Figure 3 Selecting appropriate tankage.

Figure 4 Floating roof.

Figure 5 Control room monitors.

NOTE

The SCADA system actually updates its information every few seconds. It can provide hourly, cumulative, and on-demand readings. Pipeline controllers can get the latest values by clicking or otherwise selecting the appropriate indicator on their SCADA screens.

Every arriving shipment must go into a tank with enough reserve capacity to hold the incoming volume. Once the scheduling department has set up the shipment, operations personnel must prepare in advance to receive it. They will set up a plan that includes selecting appropriate tanks and confirming sufficient storage to accept the shipment. If the facility does not have enough capacity, operations personnel must notify the scheduling department immediately.

For example, imagine that a shipment of 125,000 barrels (19,873 m^3) of gasoline is expected. Operations personnel discover that there is space for 92,500 barrels (14,706 m^3). The terminal cannot find the extra space immediately. In this scenario, operations personnel should contact the scheduling department as quickly as possible to work out a solution.

Or consider a customer that is expecting to receive 200,000 barrels (31,797 m^3) of gasoline. Operations personnel discover that there is only 160,000 barrels (25,438 m^3) in the pipeline. They should contact the scheduling department as quickly as possible to work out a solution. The customer may have to be notified to use an alternative source.

SCADA uses data from meters to calculate product flow and other related information. This data appears on appropriate monitor screens (see *Figure 5*). The SCADA system calculates tank inventory using one of two methods: gross volume or net volume. Each must be calculated so that all volume can be accounted for.

Gross tank volume is the total liquid volume contained in a tank based on measurements before adjustments are applied. Net tank volume is a calculated value determined by applying adjustments to the gross volume. These adjustments include sediment and water (S&W), temperature, density, and any other applicable factors that may affect volume. Manual measuring may supplement the SCADA report.

During delivery, careful metering is crucial. The control room's key responsibility is providing a specific quantity of product. The receiver's responsibility is having sufficient space to absorb it as it arrives.

1.1.5 Tank Level Alarms

Maintaining the proper tank level may seem like a minor task, but it is actually extremely important. Tank overflows have caused some major pipeline disasters in the past. *Figure 6* shows the potential consequences of a tank overflow. API *Standard 2350* 4th Edition was created in response to a major overfill accident, known as the Buncefield fire.

Operations personnel must monitor and respond to tank level alarms. These will vary depending on the technological sophistication of the **tank gauging system** installed, as well as its compliance with API *Standard 2350*.

Tank gauges range anywhere from totally manual systems, such as a field technician with a weighted tape measure (*Figure 7*), to highly sophisticated systems. An **automatic tank gauge (ATG)** is a simple mechanical system that displays liquid level on a display mounted beside the tank (*Figure 8*). Some ATGs (but not all) can send level data to the control room electronically.

Figure 6 An overflow can lead to a major disaster.

Being mechanical, ATGs need regular maintenance and can malfunction fairly easily. A faulty ATG, among other factors, contributed to the Buncefield fire.

More sophisticated systems include tank gauges that determine level by bouncing energy off the liquid's surface and timing the reflections. These gauges use sonar, laser, or radar technology. A radar tank gauge is shown in *Figure 9*. All are highly accurate and reliable, and recent models offer full remote communications capabilities.

Figure 7 A handline used for manual tank gauging.

Control room personnel must be aware of several different tank levels. API *Standard 2350* 4th Edition defines four, although some installations may include others:

- **Maximum working (MW)** – This is the lowest high level and refers to the normal maximum height to which liquid should rise. Reaching this level will not generate an alarm, although some systems may notify the operator.
- **High-high (HH)** – If the liquid reaches this point, it will trigger an alarm and the flow should be terminated within a specific period of time.
- **Automatic overfill protection system (AOPS)** level – An AOPS is an automatic and independent system that takes corrective action to halt tank filling when it detects a set level (either at or above the HH level). The level at which the AOPS activates is called the *AOPS level*.
- **Critical high (CH)** – This is the level at which the tank starts to overflow. There are no alarms at this level since they should have been triggered below it. Control room personnel should know the CH for each tank, but liquid levels should never be allowed to get close to this point.

In addition to the four API-defined levels, some tank level sensors detect and report other levels:

- *High (H)* – This level is above the MW point, but below the HH point. It is optional under API *Standard 2350* 4th Edition.
- *Minimum working* – This is the lowest level the liquid should sink to during normal operation.

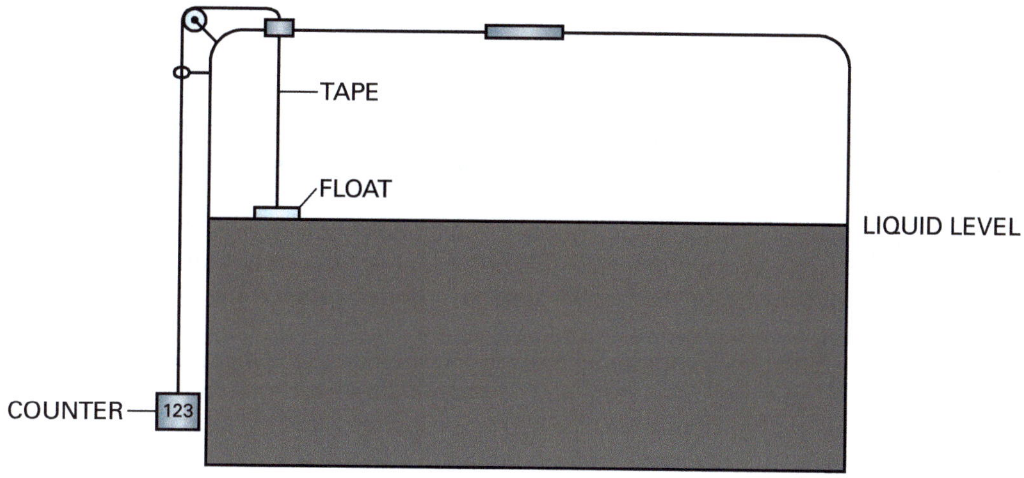

Figure 8 ATG diagram.

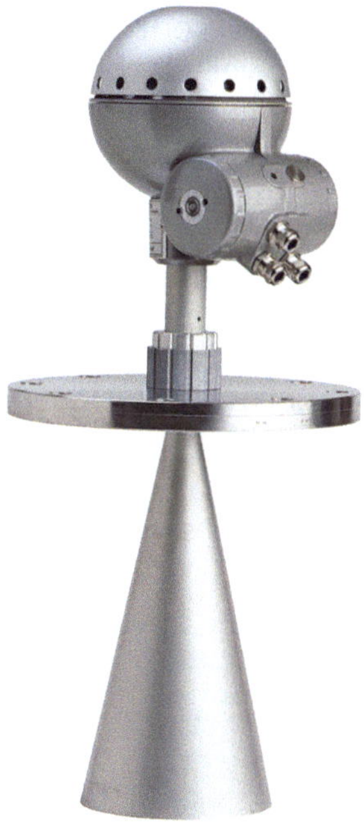

Figure 9 A radar tank gauge.

Allowing the liquid level to drop below this point may cause equipment damage, floating roof damage, or other problems. Most systems send a notification when the liquid drops to this point.

Figure 10 summarizes all of these levels. The levels mandated by API *Standard 2350* 4th Edition are shown bold red type.

> **NOTE:** Earlier versions of API *Standard 2350* used different names for some levels. It's important to be familiar with your particular facility and the terminology its tank gauging system uses to report levels.

Tank gauges that report to the SCADA system enable it to maintain up-to-date inventories and fill level status. During receiving operations, control room personnel should watch fill levels. If a value freezes, the situation must be investigated immediately. Some possible causes include hung gauges, communications failure, water in the gauging system, or a faulty tank gauge.

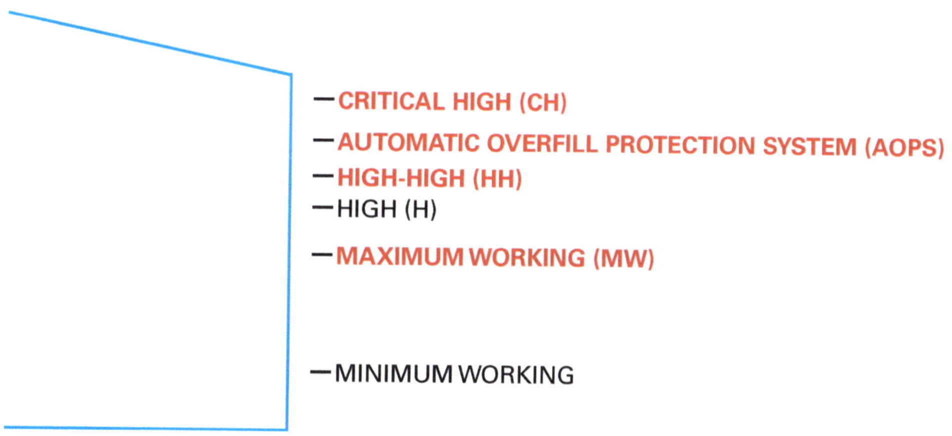

Figure 10 Tank levels.

The Buncefield Fire

The explosion and subsequent fire at the Buncefield Oil Depot is one of the most memorable pipeline disasters in recent history. Located near Hemel Hempstead, a town northwest of London, the fuel depot was one of the largest in the United Kingdom.

The incident began with a tank overflow on December 11, 2005. Shortly afterwards, the first explosion occurred, followed by a rapidly spreading fire and additional explosions. Huge clouds of black smoke moved across the southeast of England, and many people had to evacuate their homes temporarily. Despite the efforts of firefighters, the fires continued to burn until December 15. Fortunately, no fatalities occurred.

The investigation that followed revealed numerous causes contributing to the disaster. A series of equipment malfunctions, poor site design, an overworked staff, and failure to follow procedure were the main causes. But like many disasters, the Buncefield fire stimulated changes in the industry that led to improved safety standards. Many of these are incorporated in API *Standard 2350* 4th Edition.

NOTE: Prior to the fire, an ATG at Buncefield got stuck and stopped reporting a level change. The AOPS system had become deactivated either by accident or through carelessness, and personnel were overloaded and not following consistent procedures.

If an alarm sounds, operations personnel must respond quickly, and company policy must be followed. The flow path may have to be redirected or the pipeline shut down. If an AOPS activates, it should take action to prevent the tank from overflowing, but control room personnel will have to perform additional steps to deal with the consequences.

1.2.0 Manifolds and Valve Operations

Operations personnel control product movement by opening and closing terminal, station, and pipeline valves in the proper sequence. They may do this either by direct operation, or remotely through the SCADA system. Common tasks associated with these operations include making tank switches, verifying valve alignments, and making pipeline switches.

1.2.1 Valve Types

The pipeline and its associated systems contain numerous types of valves. Many of these perform specialized tasks, and all work together to make the pipeline both efficient and safe. At the most basic level, valves are control devices, opening and closing to start, stop, or regulate flow.

Hand valves (*Figure 11*) must be operated manually. They are common on meter runs and tanks. However, almost all valves can be operated by hand if necessary. Field personnel may have to use a portable power source to operate valves manually if they are very large.

Block valves operate in the fully open or fully closed positions only and offer minimal flow resistance. They are the main pipeline valves, and they can isolate pipeline segments as required. Block valves play an important role when the control center purges a pump station. The block valve closes while pump suction and discharge remain open. This configuration prevents a slug of the previous product from contaminating the new one.

Figure 11 Hand-operated valve.

Pipeline personnel operate block valves either at the local control point or remotely from the control center. While any valve can act as a block valve, gate, plug, butterfly, and ball valves are most commonly used. A gate valve is shown in *Figure 12*.

Control valves (*Figure 13*) function in the fully open, fully closed, or partially open (throttle) positions. For this reason, they are ideal for precision control, often being used to manage pressure and flow. For example, control valves manage both the station suction and station discharge ends of pumps. Their action protects both the upstream and downstream sides. In delivery facilities, they control the delivery rate. Globe and angle valves are the most common control valves, although butterfly valves are also used. They can be operated by hand, although most are managed by pneumatic, hydraulic, or electric actuators.

Control valves are usually smaller than the pipeline itself to provide some built-in throttling action. This reduces wear on the valve seat as well as the stem motion required to control the flow. Because partially open valves reduce pressure downstream, they must be used with care to avoid damaging other components, such as pumps. Constant-speed centrifugal pumps are often used with control valves.

Pipeline personnel use control valves in any situation requiring flow or pressure adjustment. These valves limit flow until conditions normalize in cases of low suction, during pipeline start-up, or after a main-line pump loss.

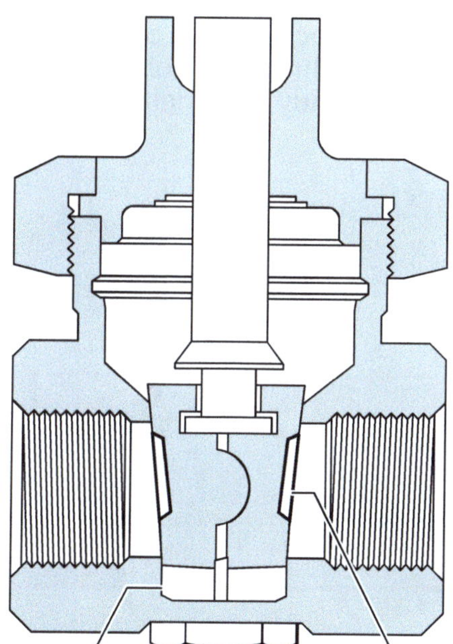

Figure 12 Gate valve.

NOTE: Pipeline controllers may change some valve setpoints through the SCADA system. Other setpoints are dictated by engineering standards and design and may not be changed without authorization. However, pipeline controllers should know all setpoint values, regardless of whether or not they can alter them.

A backpressure valve is a special type of control valve, which intentionally creates backpressure through throttling. The valve raises pipeline pressure upstream, preventing a slack line. Backpressure valves reduce pipeline vibration and protect low-pressure equipment, such as metering installations. Normally, these valves come preset to generate a specific amount of backpressure and are not remotely adjustable.

Pressure relief valves protect the pipeline during an overpressure condition by opening to release product into a breakout tank. Once the pressure has dropped back to safe levels, the valve closes. Relief valves use a number of different technologies to determine when to open. The most common type uses a spring whose force is equal to the pressure at which the device should open.

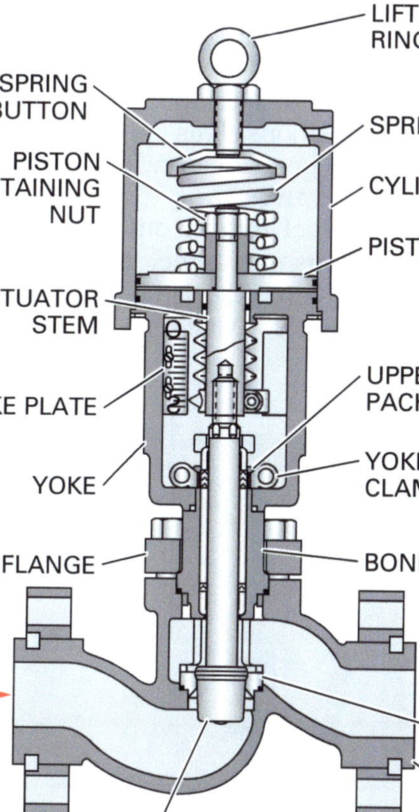

Figure 13 Control valve.

Relief valves may also be built around electronic or hydraulic technologies.

Safety valves protect the pipeline by activating during abnormal conditions. For example, with electrical control systems, a power loss could cause serious problems. The safety system detects the flow loss and closes the valve to protect the manifold. Safety valves normally appear in the line below the takeoff valve (the last valve that opens to start the shipment), but may appear in other parts of the pipeline as well. In all cases, however, the purpose of the safety valve is to place the pipeline in a safe condition.

1.2.2 Valve Control

Remote-control valves make the pipeline and its associated equipment much more manageable. Ideally, every valve in the system should be both controlled and monitored from the control room. While this is almost never true in practice, most important control valves will be controlled and monitored remotely.

Valves managed by powered actuators are usually controlled by a PLC (programmable logic controller) or an RTU (remote terminal unit), which are computer-like devices that talk to the SCADA system. Examples of PLCs are shown in *Figure 14*. Commands to valves are issued by control room personnel through the SCADA system, which also provides valve status information (*Figure 15*) and indicates when valves are traveling between positions. This approach allows control decisions to be made hundreds of miles away from the actual valves.

Moving product into or out of a terminal requires operations personnel to set the proper valve alignments to direct the product correctly. The valve sequence opens a path through which the product flows. Using appropriate commands, the control room can direct product into or out of tanks, operate metering and prover systems, or bypass a station altogether.

1.2.3 Interlocks

Control systems may include **valve interlock** logic designed to prevent dangerous or problematic valve combinations. For example, the control system's interlock programming might not allow a main-line valve to close until the corresponding pumps are shut down. Such automation prevents many careless accidents, protecting both the station and the pipeline.

Some companies use interlocks on booster stations to prevent the line from becoming blocked. Interlock program logic monitors all major valves at the stations, including the control valves. If a valve begins to close unexpectedly, the control program will send an Open command to reopen the valve. If the control valve closes by more than 65 percent, the program will send the main-line block valve an Open command to ensure that the line is not blocked.

Unstaffed booster stations may have interlock logic that prevents the main-line block valve and the booster pump suction and discharge valves from closing at the same time. They may also have interlock logic designed to open the main-line block valve if the control valve closes by more than 65 percent.

Interlocks can also shut down the station in response to a line block command. This action requires less time than shutting down all units individually.

While the interlock system can fail, human error causes blocked lines more often. These failures occur most frequently with deliveries into staffed

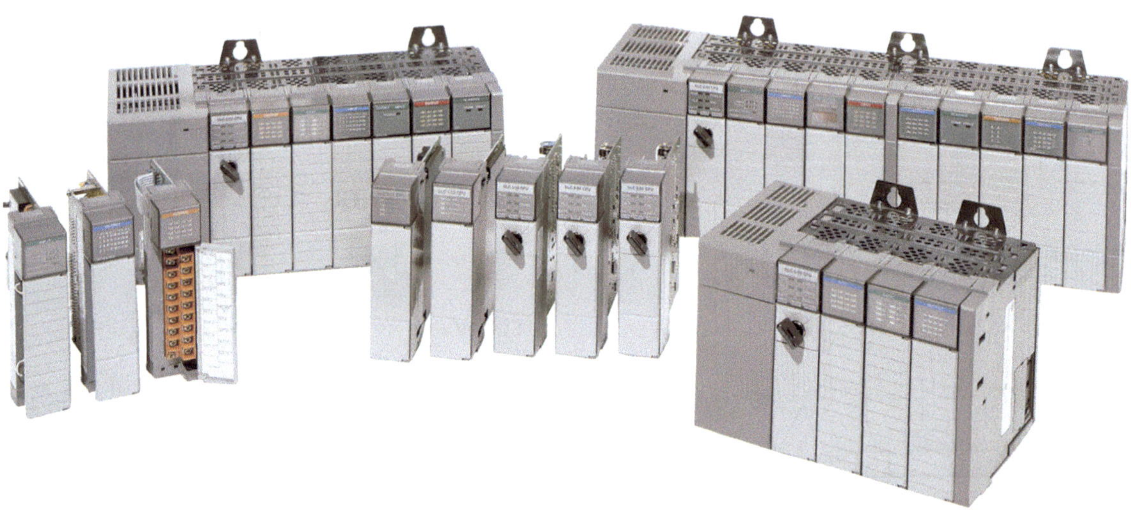

Figure 14 PLCs are often found in control systems.

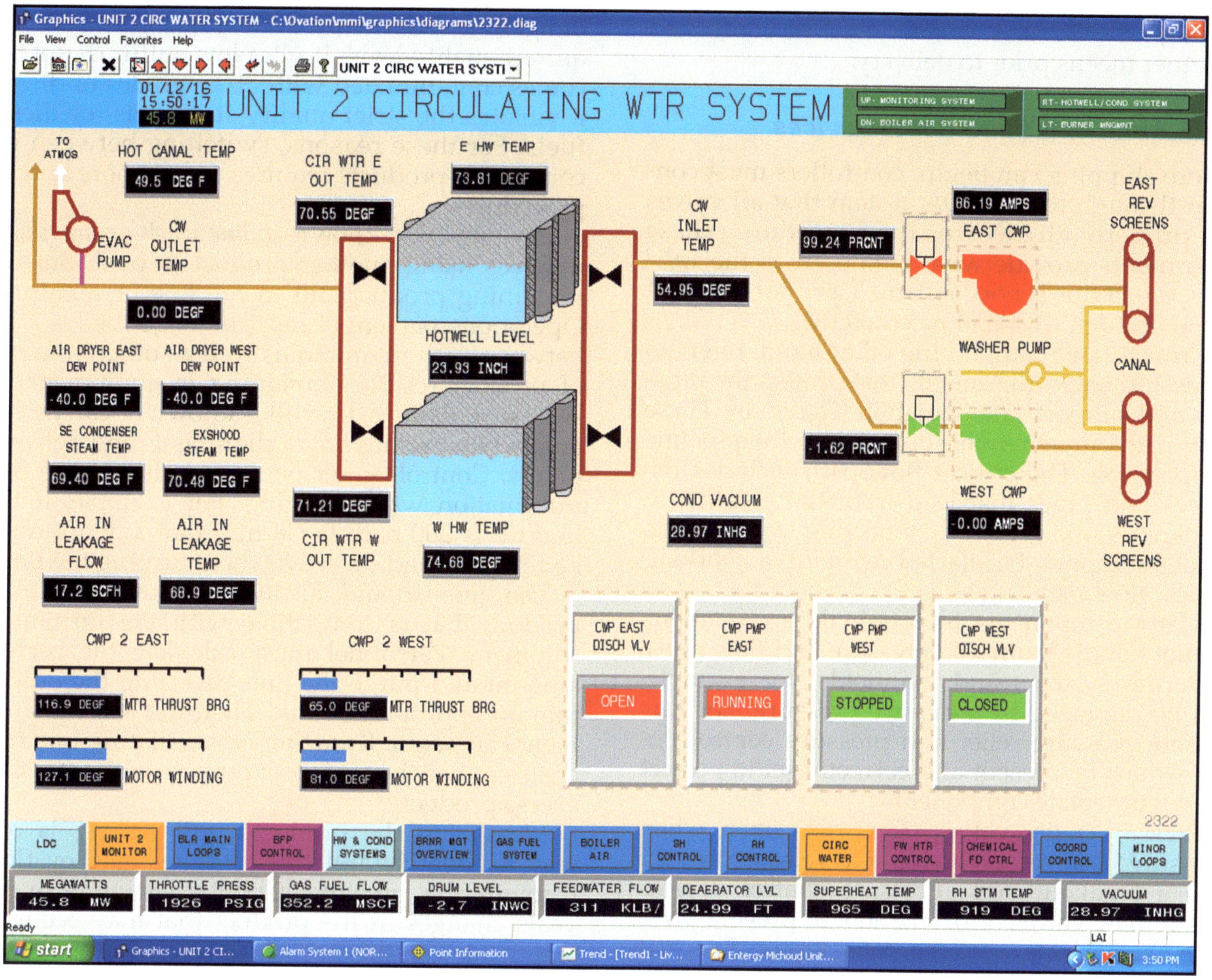

Figure 15 Valve status display.

facilities. Regardless of the cause, the controller must shut down the line, following company procedures. Interlock failures should be handled by following the specified procedure. In most cases, the time to respond to and resolve overpressure situations is short. Control room personnel must be quick to recognize problems and know the correct response so they can act immediately.

1.2.4 Valve Alignment

Control center personnel must verify valve alignment for every shipment or change in operations. Once the scheduling department issues shipping orders, the control center sets up the procedures to fulfill those orders. This task includes determining the proper valve alignment needed to get the product from point A to point B safely and efficiently. Operators will look at piping layouts to pick the best path, and then prepare a list of valves to be opened or closed. They will then use the SCADA system to operate the valves in the correct sequence.

Most valve operations occur by remote control. Similarly, SCADA screens show valve status and indicate whether valves are open, closed, or traveling. Control center personnel must keep an eye on status indicators to confirm that everything is working. The operation may not begin until all valve positions have been confirmed. As previously discussed, many control systems include valve interlock logic that prevents inappropriate or unsafe valve combinations or sequencing.

If a valve does not respond correctly, it may be necessary to have field personnel go to the location and check or operate the valve manually. Standard operating policies require control center personnel to verify that valves are properly set before any product enters the system. Some companies dispatch field personnel to check each valve individually, although this practice is becoming uncommon due to unstaffed facilities. Verification is important, especially when several companies are involved. Valves controlled by other companies, and therefore not visible through the SCADA system, must be confirmed by other means prior to delivery.

1.2.5 Flow Paths and Relief Devices

Before shipping can begin, controllers must confirm through the SCADA system that all valves are properly aligned for the procedure. These alignments provide what is known as the *valid flow path*. If this is not confirmed, product will not move through the pipeline correctly.

Critical flow path, on the other hand, refers to pipes and valve alignments that protect the pipeline and support safe operation (*Figure 16*). These flow paths must remain open when the pipeline is operating. The critical flow path includes the valves and pipes that lead from the pipeline to breakout tanks or the connecting carrier. The pipeline cannot be started without a working critical flow path.

Many components in terminals and stations cannot handle main-line pressure and flow rates. Metering systems and manifolds, for example, can be damaged by excessive pressure. For this reason, pressure relief and pressure control devices must be present and correctly set to protect these vulnerable components.

Pressure relief valves (*Figure 17*) open when pressure becomes excessive, and release product into a breakout tank. Pressure control devices manipulate valves or divert flow to prevent pressure from rising to dangerous levels. Thermal relief devices are similar to pressure relief valves. They relieve pressure caused by thermal expansion—the tendency of the product to expand when it heats up.

Generally, the control center will not be able to change protective device setpoints. In some cases, these may be fixed by the manufacturer and cannot be changed. Other devices may have to be changed in the field either through a mechanical adjustment or by reprogramming a control device. Regardless, the control center must know protective device setpoints, and be alert to pipeline pressure and flow conditions. Controllers must also monitor relief devices to ensure that they're performing properly.

1.2.6 Dealing with Incompatible Products

There are many products that cannot be mixed because the combination might be dangerous or simply fail to meet standard requirements. For example, mixing gasoline with a shipment of fuel oil is dangerous because it has a lower flash point than fuel oil should have. Mixing regular gasoline with a shipment of premium gasoline is not dangerous, but the mixture doesn't have the required quality level. If ultra-low-sulfur diesel becomes contaminated with other grades of diesel, it violates the strict quality standards for diesel fuels. For these reasons, switching between incompatible products requires one or more special procedures.

Purging (also known as *line wash* or *line flush*) involves using another product to push out the remaining product still in the line or manifold. Operations personnel must manage the **line fill** between each manifold as well as other volumes of product residue remaining in equipment. In most installations, line fill is known or can be calculated for various valve alignments and delivery points. Control center personnel should use this information when purging. For example, if the line fill is 300 barrels (48 m^3), that amount must be removed and diverted when purging the line.

The line should always be purged with a product that is compatible with the upcoming shipment. Personnel must calculate the correct amount and purge the line, diverting the output into an appropriate tank. Then the second shipment can begin. Knowing or calculating line fills ensures that the line gets purged without using excess product.

Measuring the product's API gravity is one way to identify and track batches as they flow through the pipeline. Gravity testing also reveals changes in the product that may occur in shipment. For this reason, many pipelines have some type of **gravitometer** (*Figure 18*) installed upstream of the terminal. The SCADA system receives data from the unit and relays it to the control room.

> **NOTE**
> Gravitometers measure product density or specific gravity. Calculations then convert this information to API gravity. Gravitometers use one of several different methods to measure density. One method involves vibrating an element in the product stream and measuring the oscillation frequency. As the product becomes denser, the element oscillates more slowly. Gravitometers are usually checked periodically against laboratory density measurements.

The SCADA system calculates, records, and displays all gravity information. Control center personnel should monitor the API gravity patterns, and any variations should be investigated. Significant variations must be documented and addressed according to company specifications.

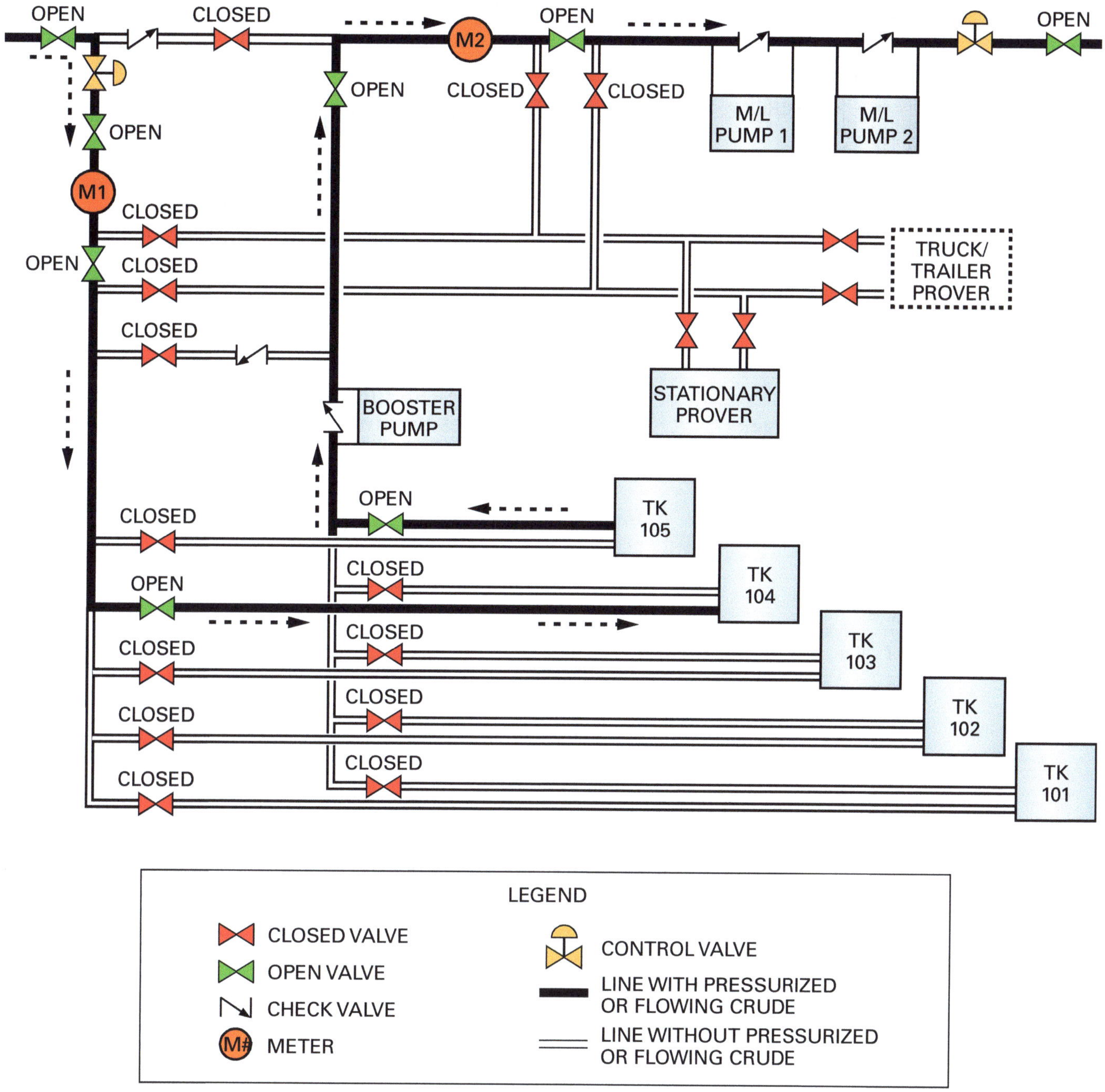

Figure 16 Critical flow path diagram.

API gravity measurements alone are not always useful for identifying the interface between batches. Some pipelines use a second instrument, called an **optical interface detector (OID)**, to supplement the gravitometer's readings. An OID passes light through the product and reads the absorption pattern. This method is especially useful with products having similar API gravity values.

Before a scheduled change arrives, control center personnel should begin monitoring the mainline stream and documenting the API gravity according to company policy. If the pipeline is not using an OID, the company-specified method for detecting the interface should be used. When the first break occurs, operations personnel should begin constant monitoring and perform the switch according to company policy.

Some facilities rely on time calculations to determine when to switch. Using information such as the pumping rate and the volume to be delivered, they can determine when the switches should occur. Some valve and tank switches are made based on the length of time that product flows through the pipe. This method is common in systems with hand valves. Personnel must perform manual switches as precisely as possible.

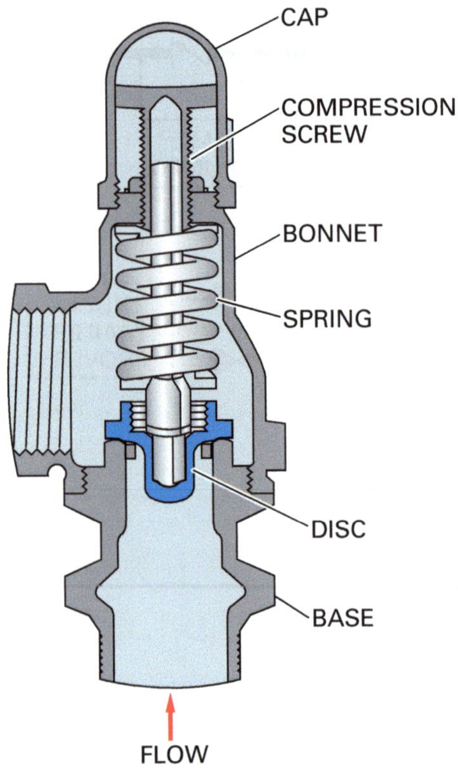

Figure 17 Pressure relief valve.

Because schedules may vary over long distances, operations personnel should supplement timing with measurements, such as API gravity, to confirm interfaces. Switches should occur at the correct gravity point regardless of time.

1.2.7 Batching Operations

Batching operations normally involve a sequence of different products. Control center personnel should be meticulous about documentation and handle every customer's receipts and deliveries accurately. Strict EPA standards, particularly those involving the sulfur content of diesel fuels, have led to many procedural guidelines and rules. Scheduling personnel set up batch sequences to minimize incompatible products being run one after the other. Code numbers and/or letters identify products with unique specifications. Charts may be set up to illustrate which products can be pumped directly in front of and behind other products. *Table 1* is an example of such a chart.

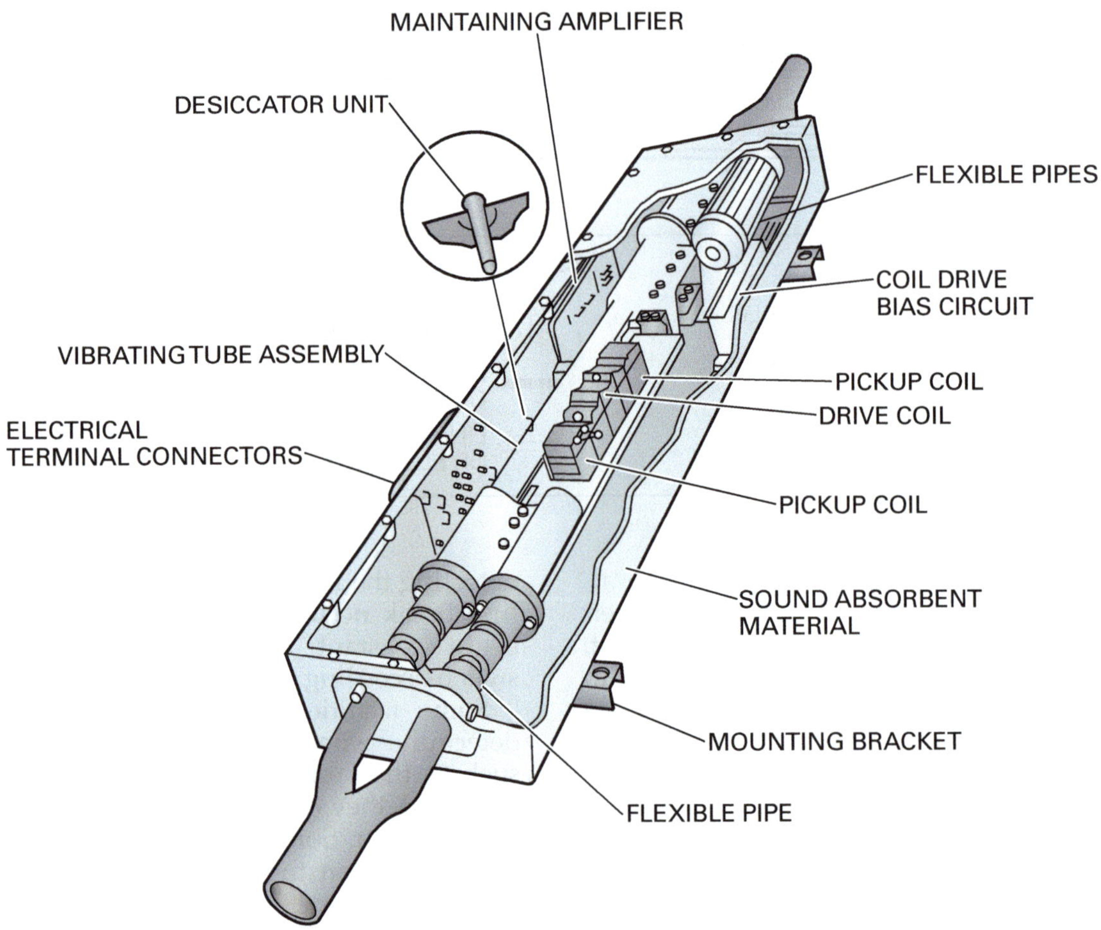

Figure 18 Gravitometer.

NOTE: In addition to being skillful at scheduling, controllers must be expert at maintaining product quality.

API gravity ranges vary somewhat in similar products. A deviation of one or two degrees API from the maximum or minimum shown on a chart is not considered significant. However, products with unique specifications (such as jet fuel) are handled more carefully in batch arrangements than are variations in products (such as fuel oils).

Testing for the interface and making a correct switch during an interface protects the product integrity. Personnel must sample and document each shipment's product as it arrives at the terminal. Doing so determines the switch point, at which time operations personnel open the pipeline valve, switching the product into the terminal. Transmix goes into separate tankage, and the pure product goes into storage.

1.2.8 Measuring Volume

The term *barrel count* refers to the total number of barrels measured during a given operation or time interval. SCADA flow computers (*Figure 19*) calculate the barrel count using the pipeline metering system. The barrel count may indicate the total volume flowing through the pipeline, the total volume of a shipment, or the volume being moved, blended, or injected.

Table 1 Sample Batch Arrangement Chart

Grade	Type of Product	60°F API Gravity Range	Normal Pump Sequence Head End	Normal Pump Sequence Tail End
A	Premium Unleaded Gasoline 91/93 Undyed	45-75	N, M	N, M
B	Commercial Aviation Gasoline 100/103 Octane – Dyed Blue	66-72	N, M with HP-7 Buffer	N, M with HP-7 Buffer
C	Refinery Charging Stock Crude Oil	20-55	X5	X5
D	Ultra-Low-Sulfur Diesel Fuel Oil – Normally Undyed 47.5 Cetane Range	36-41	X, X5, Y	X, X5, Y
E	Denatured Fuel Ethanol		Not Pumped	Not pumped
F	#2 Fuel Oil Distillate Cenex Proprietary – Dyed Red	30–37	D, Y, Q, X, X5, N	D, Y, X, X5, N
G		Naptha	Special Delivery	Special Delivery
H	Normal Butane	0.580-0.588	N, L	N, L
I	Isobutane	0.560-0.570	H	H
J	Military Jet Fuel JP8 or J8	41-48	N, V	N, V, D, X
L	Propane	0.500-0.510	H	H
M	Mid-Grade Unleaded Gasoline 89 Octane Range	48-75	A, D, N, X	A, D, N, X
N	Regular Unleaded Gasoline 87 Octane Range	48-75	A, F, J, M, Q, V, X, X5, Y	A, F, J, M, Q, V, X, X5, Y
O	Water	1	Special Delivery	Special Delivery
P	Refinery Blendstock		A, N, X	A, N, X
Q	Commercial Jet Fuel	41-48	N, V	N, V, D, X
R	Leaded Racing Gasoline	60-80	N, V	N, V
T	Transmix Commercial Purchases	35-65	N, X, X5	N, X, X5
V	Sub-Octane Unleaded Gasoline, 84 Octane Range	48-75	F, J, M, N, Q, V, X, X5, Y	F, J, M, N, Q, V, X, X5, Y
W	Natural Gasoline Liquid	79-86	D, X, X5, N, H	D, X, X5, N, H
X5	#2 Fuel Oil High Sulfur (HSFO) Normally Dyed Red 0.50 sulfur maximum (The 5 in X5 may not be present when listed in specific grade)	30-40	D, F, N, V, X, Y	D, F, N, V, X, Y
Y	#1 Fuel Oil GrDistillate	40-45	D, F, N, V, X, X5	D, F, N, V, X, X5

Figure 19 Flow computer.

Control center personnel should verify the totals, calculations, and all parameters that affect volume for each operation or time frame. The temperature and pressure are confirmed at the meter's manual thermometer and pressure gauge. The true total barrel count comes from a calculation involving the meter reading and all compensating factors. This value can be compared to the physical totals in tankage to verify inventory and operations.

Barrel counts give control center personnel a way to account for the day's operations, calculate the over-and-short, verify receipts and deliveries, and track operations. They also can act as a type of leak detection system, since significant discrepancies may indicate leakage.

1.2.9 Pipeline Switching

Various pipeline companies serve different geographical areas. Pipelines often run parallel to each other and sometimes cross (see *Figure 20*). Transferring product from one pipeline to another has numerous advantages to customers, so many companies have developed methods to make it possible. Control center personnel must understand the geography, facilities, and installations of other companies to schedule shipments between pipelines.

Figure 20 Abbreviated map of the main US pipelines.

During transfers between companies, all involved control centers must closely coordinate with each other. Close communications are vital after confirming schedules. The same testing and sampling methods used for any regular custody transfer still apply. When changing carriers, extra care should be taken to verify and document batches. Confirm readiness with each pipeline before beginning the shipment.

Sending a shipment to a tank before changing carriers makes measurement and verification easier. If the scheduling department specifies that the shipment be tightlined into the other pipeline, control center personnel should take extra care with documenting specifications and measurements. (**Tightlining** is the process of moving a product from one pipeline to another without transferring it into a tank first.) If a question comes up, this information will be needed. Always document valve alignments and verification records so product movement can be easily traced.

1.2.10 Procedures

Each company has standard operating procedures for receipts, deliveries, and custody transfers. In order to manage these transfers efficiently, control center personnel should review scheduled shipments. They should understand the exact shipment type, specifications, time, and flow rate.

> **NOTE**
> Always know and follow your company's procedures for all product transfers.

When possible, prepare and document terminal piping and valves as far in advance of shipments as possible. Make appropriate plans for switching local line fill. Confirm that receiving equipment is operating properly and that there is sufficient tankage, if needed. If tankage is a problem, contact the scheduling and dispatching departments immediately.

Although consecutive batches may be compatible, the characteristics or additives of one batch may be different from others. Because delivery points may be different for each batch, switching will also be different. Operations personnel must plan the switch so each batch reaches the appropriate destination. The following tools may help with this planning:

- Company schedule
- Company product change report form
- API gravity conversion chart
- Software batch cut programs

To signal the change between two batches when there is little or no difference between them, some companies inject a dye flag. If the facility is using an OID, it can detect these batches easily, making switching simple. If the pipeline is using a gravitometer only, the batch is not switched when the dye flag arrives. Rather, the switch occurs at the mid-point of the gravity change. Personnel should verify dye flags or color markers before the shipment begins. Dye flags may be color-coded for various products.

1.3.0 Pump Operations

Pumps move product from one point in the pipeline to another. Centrifugal pumps (*Figure 21*) act as main-line and booster pumps on most pipelines. With this kind of pump, fluid enters the case near the axis of a rotating impeller. The impeller accelerates the fluid and flings it outwards at high velocity. The surrounding casing captures the fluid, which then exits the pump at high pressure. Some pipelines also use positive displacement (PD) pumps for auxiliary or special purpose applications.

All pumps use a driver of some kind to supply power. Electric motors, internal combustion engines, and turbines are examples of typical drivers. Power is expressed in horsepower (hp). Moving a heavy product at a high flow rate requires more power than moving a lighter one at a low flow rate. Control center personnel plan and calculate shipping procedures to use the lowest horsepower possible to accomplish the job.

NOTE

In countries that use the metric system, power is expressed in watts (W) or kilowatts (kW), rather than horsepower (1 hp = 746 W, or 0.746 kW).

Pipeline pumps fall into three main categories:

- Main-line
- Booster
- Sump

1.3.1 Main-Line Pumps

Main-line pumps move the product through the pipeline, generating the pressure and flow required for meeting shipment schedules. Most main-line units are large centrifugal pumps (*Figure 22*) with high horsepower drivers. Each location will likely have a number of these. Main-line units operate individually, in parallel, or in series. In many installations, main-line pumps receive suction from a common line supplied by a tank manifold and supply flow to the pipeline through a common manifold. The manifold allows several tanks to supply the pumps during an operation.

The control center manages pumps, starting and stopping them as required. It also sets and adjusts suction and discharge setpoints, depending on how the pumps are working. When multiple units work together, control center personnel must ensure compatible flow rates. Check valves prevent the pumps from recirculating.

Pumps do not have to be of the same size or capacity. When operating in parallel, they must be adjusted so their head settings match, because unequal pressure will cause a shutdown. Pumps operating in series complement each other, so it's not necessary to match head settings. Total discharge is the sum of each pump's head.

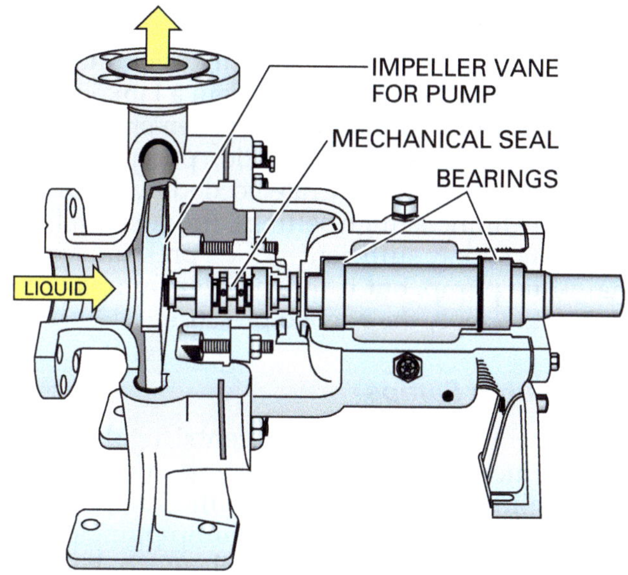

Figure 21 Centrifugal pump.

Figure 22 Multi-impeller centrifugal pump.

1.3.2 Booster Pumps

Booster pumps (*Figure 23*) are high-volume, low-head units that pump product from a tank into the suction side of a main-line unit. When operating in series, they can also act as injection pumps. Booster pumps provide the flow rate needed by main-line pumps running at high settings. Product level in a tank may supply sufficient suction pressure for the main-line pump, but if not, booster pumps come on line to supply the needed pressure.

1.3.3 Sump Pumps

Sump pumps are low-head centrifugal pumps used to move product from a sump into a tank or the pipeline. Product collects in the sump during normal operations. Automatic sump pumps come on and off in response to a level switch, keeping the sump product level below a certain point.

1.3.4 Managing Pressure and Flow

Safe and efficient pipeline operations depend on control center personnel monitoring and managing pressure. Control valves, installed in the pumping station's incoming and outgoing lines, adjust the pressure and flow. Transducers monitor and report pressure at the pump's suction and discharge sides to the SCADA system. Similarly, SCADA also collects data on line pressure and case pressure. Control center personnel use this information to adjust control valves to sustain steady state operating conditions on the outgoing line.

Control valves provide crucial pressure and flow regulation in systems with constant-speed main-line pumps. Valve throttling action at the suction and discharge sides keeps pressure and flow within appropriate ranges. Control center personnel determine the appropriate setpoints and adjust the valves from the control center.

Figure 23 Booster pumps on a manifold.

Station design may allow personnel to adjust pressure and flow at several different locations (*Figure 24*). Generally, the control center can manage valves at origin and delivery points, but not at booster stations.

Backpressure valves maintain line pressure above the product vapor pressure point in case of problems or operational changes. They also protect terminal and station low-pressure piping. These devices are preset and will operate as required.

1.3.5 Pressure Points

Suction pressure is generated on the pump's upstream side, while discharge pressure is generated on its downstream (pipeline) side. The pump controller, usually a PLC, manages both of these pressures by monitoring them and comparing their current values to programmed setpoints. Control valves throttle the flow to maintain safe and appropriate operating conditions.

Increasing discharge pressure increases flow, but it decreases suction pressure because of the greater demand on the supply. Suction must also be increased, or the pump may shut down. Decreasing discharge pressure increases suction pressure, but decreases flow because the pump is ejecting less product.

Product density can significantly affect pressure and flow as well. A pump will experience a density change when an interface passes through it. A density increase will cause discharge pressure to also increase. The pump control PLC may make adjustments to maintain appropriate suction and discharge. The control room may also modify settings through the SCADA system.

In a large pipeline, different segments may operate at different flow rates and pressures. Some sections may supply low-pressure equipment, such as metering and manifolds. On older pipelines, segments may be de-rated to a lower **maximum operating pressure (MOP)**. These must be handled separately. When starting the pipeline, control operators will probably open segments in a specific sequence rather than all at once. They will allow the initially opened segments to build up pressure before opening others.

Control center personnel must note pressure differences from one segment to the next, and be careful not to confuse them. Errors could cause an overpressure situation or a slack line condition. If there is a great difference in segment sizes, pressure management may be difficult. One section may require throttling while another may not. Pressure transducers provide

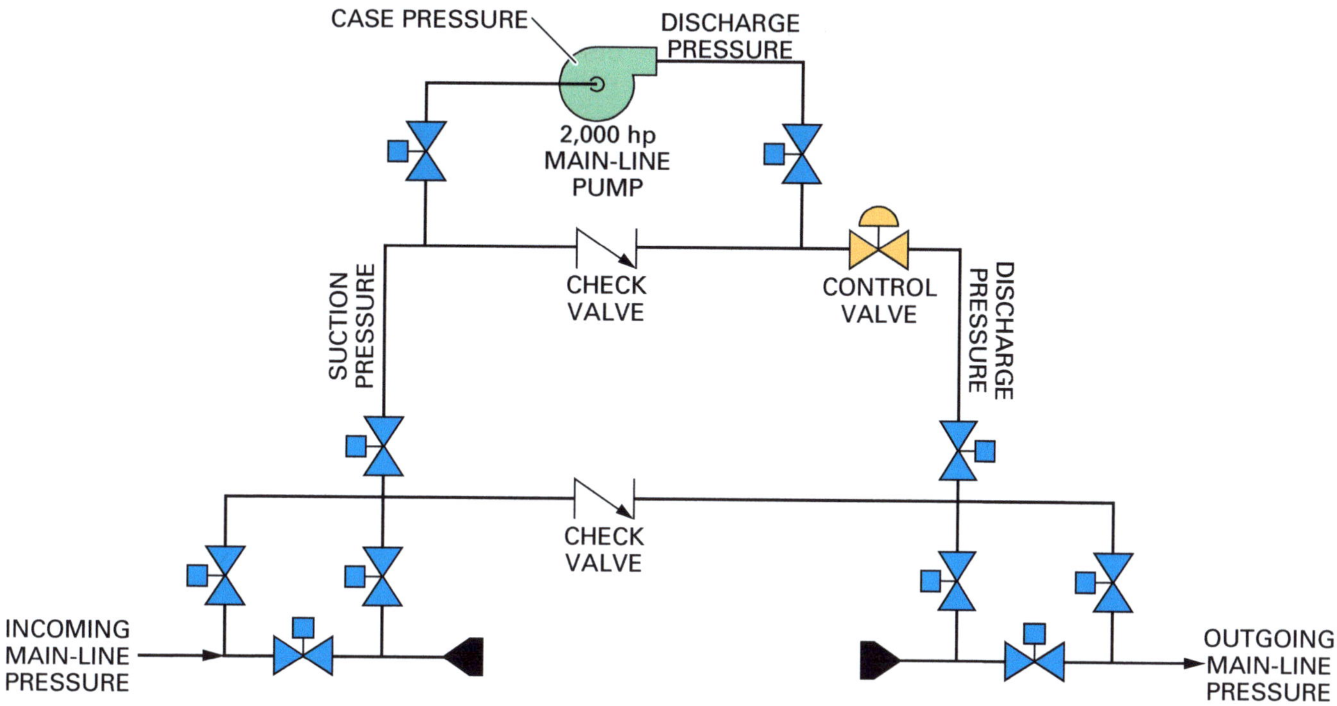

Figure 24 Simplified diagram of a pump station.

full information through the SCADA system so operators can make decisions.

A pump's discharge side setting determines the pressure within the pump case. It will always be higher than the pressure past the discharge side. Pump-case pressure can cause damage over time, but it might not be noticed immediately. As the condition worsens, high pump-case pressure may cause unstable pump controls, piping vibration, and high noise levels. The control center can deal with this problem by making pressure adjustments upstream and/or downstream. Pumps should be inspected periodically for signs of pressure-related damage.

1.3.6 Pumping Applications

Pipeline engineers select pumps based on product type and throughput requirements. Increasing or decreasing horsepower increases or decreases flow rate. Similarly, adding or removing pumps has the same effect. Control center personnel determine the best way to utilize the pumps under their control to make the shipment move on schedule.

If the pipeline climbs to a higher elevation, extra horsepower will be needed beyond that required to handle a level grade. Calculations determine the extra horsepower necessary to do the work. A station may need to start several additional main-line units to supply enough horsepower to go uphill (*Figure 25*).

Valves and Pigging

Pump stations may include technology designed to handle pipeline pigs. Pigs are cylindrical devices that fit snugly in the pipeline and are forced through it by the flowing product. They are used for one of several reasons:

- Cleaning or decontaminating the pipeline
- Inspecting the pipeline
- Pushing out gases or water
- Separating batches of product

Pigs come in a number of types and styles, depending on their purpose. Some are very simple, while others, such as "smart" pigs, contain sophisticated electronics that monitor and record conditions within the pipeline.

When pigs are used, appropriate technology is required to handle them. This includes valves for introducing and removing them, as well as detectors to sense their presence. The station shown in *Figure 24* includes valves to accommodate pigging.

Figure 25 Increased horsepower required for elevation.

If pumps come on or go off line unexpectedly, pressure and flow will probably change. The possibility of such occurrences increases with the shipment distance. Pressure and flow rates must be watched closely and adjusted to compensate for these changes.

1.3.7 Pump Timing

Systems made up of multiple stations, terminals, and tank farms require a specific pump start-up sequence. For example, a booster pump may have to be started before a main-line pump in order to provide sufficient suction. The booster pump delivers product to the suction side of the main-line unit. Failing to start the booster pump first could cause damage and force the main-line unit to shut down due to low suction (*Figure 26*).

> **NOTE**
> Raising suction pressure and discharge pressure also raises pump-case pressure.

On the other hand, if discharge or main-line pressure rises too high, the control center may shut down a booster pump or a main-line pump. These actions would reduce suction pressure, causing the pump-case and discharge pressures to drop in response.

The control center brings up an idle pipeline by starting the pumps at the originating station first and gradually adding downstream stations as the product stream and pressure allow. Prior to start-up, main-line block, terminal, and tank valves must be opened for flow to increase. Booster pumps come on line first. Main-line units then follow. After the initial start-up, each successive pump downstream comes on line. Pressure builds until the system reaches a steady state condition.

1.3.8 Pump Management and Power Optimization

No matter the type, power costs money. Almost every pipeline activity consumes power, so control center personnel work hard to manage power carefully and optimize its use. By using training, experience, and good judgment, personnel set up shipments to make the most economical use of the equipment. Some facilities compare utility rates when planning a delivery. They then schedule pumping routes to take advantage of the best rates and times of day.

By being thoroughly familiar with pump condition indicators, such as flow rate, differential pressure, and consumption of current (measured in amperes, or amps), operators can estimate efficiency. For example, a constant-speed pump's efficiency drops when throttle pressure increases. With a variable-speed pump, some speeds are more efficient than others for a given set of conditions. Any unexplained change requires investigation immediately. By staying on top of problems, the control center can reduce costs and keep operational efficiency high.

Control center personnel can use natural power sources, such as gravity, to accomplish some parts of their work. For example, a tank with a high level can deliver high flow rates to pumps without booster pump assistance. Gravity can also transfer product between tanks if they are at different heights. A booster pump may have to be used for part of the transfer, however.

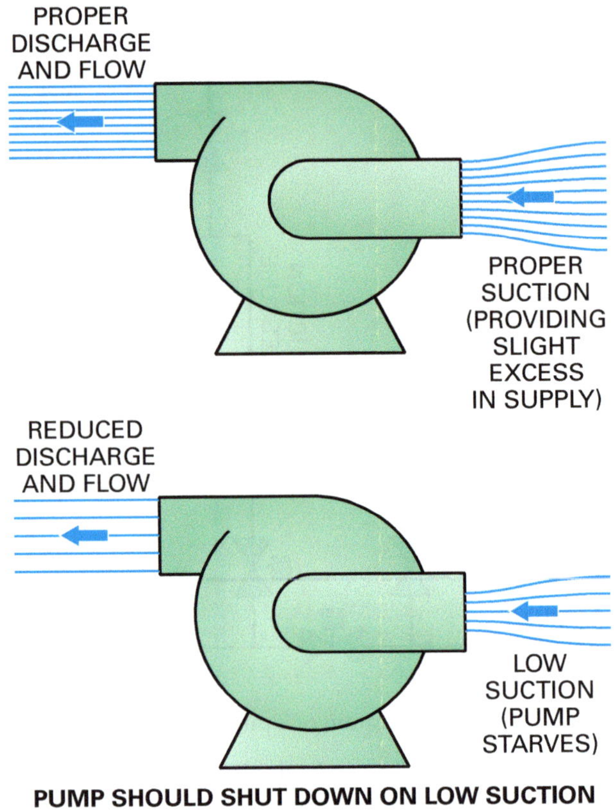

Figure 26 Effects of normal and low suction.

For variable-speed pumps, a **variable frequency drive (VFD)** can control the pump for maximum efficiency. VFDs manage their electric motor's speed by adjusting the voltage and frequency supplied to it. This technique controls pressure and flow without resorting to throttling, which wastes more power. It also offers an efficient way to reduce power when full power is not needed.

Power consumption can be calculated if a few key factors are known. The following formula is one such example based on the pipeline units used in the United States:

$$\text{power} = (\text{flow} \times \text{head}) \div (2{,}450 \times \text{efficiency})$$

Where

power = the power consumed in hp
flow = the pump's flow rate in bph
head = the pump's head in psi
2,450 = a conversion constant
efficiency = the pump's efficiency as a percentage

Another way to manage power is by switching pumps, an option available in installations that have multiple units. Pumps can be alternated based on power requirements in order to meet rate conditions or to perform maintenance, depending on company policy and procedure. One pumping unit can be brought on line and another unit taken off in order to improve pipeline hydraulics or to achieve line balance. Control center personnel manage suction and discharge at the origin station. They must also sequence valves and handle switching. Booster stations have their own control systems.

1.3.9 Drag Reducing Agents

Drag reducing agents (DRAs) are long-chain polymers that enhance flow by reducing pipeline friction. Personnel introduce DRAs into the pipeline through annular ring injectors in order to coat the inner surface (*Figure 27*). Pipelines treated with DRAs experience about 20 percent greater throughput and a corresponding 20 percent reduction in pressure. Consequently, DRA injections greatly reduce the amount of power required to pump certain products, especially crude oil. However, pipeline personnel must calculate the DRA cost and the injection expense to be sure that the energy savings are large enough to justify using a DRA.

NOTE

DRAs are also used during maintenance when a station must be bypassed.

DRA injections are planned by the scheduling department. By reducing friction, a DRA affects the pumping rate in positive ways. Adding a DRA results in a shorter pumping schedule and creates a negligible change in volume. Since DRAs are expensive and have a strong effect on operations, use care in following schedule orders.

Cautions and guidelines associated with using DRAs include the following:

- DRAs may only be used with compatible products, such as crude oil.
- DRAs may not be used with refined highly volatile liquid (HVL) products.
- DRAs injection must be stopped at least two hours before pumping specialty products, such as jet fuel.

Control center personnel can manage some injections from the control center. Others may require field personnel to be at the site. Control center personnel should consult their company's DRA product compatibility chart and the product schedule to verify that a planned DRA injection is compatible with the product. DRAs should not be injected if any of the following situations exist:

- An incompatible shipment will occur within a specific time period after the injection.
- An incompatible product interface has passed through the injection site less than a specific time before the injection is scheduled to start.
- A residue of product or DRA could cause some type of contamination.

Figure 27 DRA injection control valve.

Additional Resources

It is assumed that you will learn to perform this procedure under the direct supervision of your crew leader, foreman, or instructor. For independent study, refer to the recommended Additional Resources at the back of this module.

1.0.0 Section Review

1. A tank mixer helps prevent _____.
   a. evaporation
   b. stratification
   c. explosions
   d. contamination

2. Tank volume based on measurements that have been adjusted for factors like temperature and density is known as _____.
   a. net volume
   b. API volume
   c. SCADA volume
   d. gross volume

3. Which of the following valves can operate in the open, closed, and throttle positions?
   a. Block valves
   b. Safety valves
   c. Gate valves
   d. Control valves

4. When main-line pumps run in parallel, their heads _____.
   a. add together
   b. must match
   c. can be different
   d. subtract from each other

Section Two

2.0.0 Shutting Down a Liquid Pipeline

Objective

Describe the requirements and procedures for shutting down a liquid pipeline (CT64_2-17).

a. Identify the prerequisites, competencies, and abnormal operating conditions associated with shutting down a liquid pipeline.
b. Describe the procedures used to shut down a liquid pipeline.

Performance Task

Shut down a liquid pipeline (control center) (CT64_2-17).

- Identify potential abnormal operating conditions that may occur during performance of this CT, and know the appropriate actions to take in response to them.
- Utilize the appropriate personal protective equipment according to relevant company procedures.
- In accordance with company policies, be familiar with the desired pressure to keep on a non-running line and the desired pressure to keep on a static line.
- Shut down pump units or compressor units beginning at the origination point and proceeding downstream to the destination. Depending on the line profile, sufficient units at the source station and at the other stations may be kept operational. Then shut down in conjunction with valve closure at the delivery terminals or stations to satisfy the policies on maintaining minimum pressures on shutdown of the particular line.
- The closure of isolation valves at intermediate delivery points should occur in conjunction with the shutdown of that station's units unless otherwise specified in company policy or procedure.
- Perform proper operation according to company policy and procedures.
- Verify that the desired result was achieved.
- Complete the appropriate documentation as required by operator's procedures.

Trade Terms

Abnormal operating conditions (AOCs): Conditions within a pipeline system, identified by the operator, that may indicate (1) a malfunction of a component or (2) a deviation from normal operations that may indicate a condition exceeding design limits or result in a hazard to persons, property, or the environment.

Computational pipeline monitoring (CPM): A computer-based pipeline monitoring system that creates a real-time hydraulic model of the entire pipeline and also detects leaks and breaks.

Packed: The condition in which the pipeline is kept under pressure so its contents do not vaporize.

Static: The condition in which the pipeline is in shutdown and the product is not moving.

Surge: A wave of high pressure that occurs when fluid starts or stops moving suddenly. Surges occur both at start-up and shutdown, although from different causes.

Shutting down a liquid pipeline requires informed, thoughtful action. Some shutdowns are routine and allow plenty of time for planning, coordinating, and carrying out the required steps. However, shutdowns resulting from a sudden emergency may leave little or no time for planning; the pipeline must be stopped immediately and the emergency brought under control. The end result of any shutdown is the same: a static pipeline.

All shutdowns require careful timing, sequencing, and monitoring. Maintaining effective communication is vital. Many different pieces of equipment make up the pipeline, including pumps, valves, tanks, and monitoring technology. Personnel must know how to handle each one as the pipeline flow comes to a halt.

The fact that most stations are unstaffed makes this task even more demanding. Consequently, the control room relies heavily on remote monitoring and control through the SCADA system. Personnel must be proficient with this system and be able to use it quickly and efficiently.

2.1.0 Prerequisites, Competencies, and Abnormal Operating Conditions

Control room personnel must possess a variety of skills in order to safely shut down a liquid pipeline. They must also be aware of the particular abnormal operating conditions (AOCs) that may occur as they carry out these tasks.

2.1.1 Prerequisites/Competencies

Control room personnel must possess the following basic knowledge and skills prior to shutting down a liquid pipeline:

- Be familiar with SCADA displays and know how to interpret their informational pages as well as their alarm and alert pages.
- Know how to use the SCADA system to change or adjust pipeline settings.
- Understand how to interpret leak detection system displays—computational pipeline monitoring (CPM) and non-CPM—and use them to localize a leak.
- Know the maximum discharge pressure for each station as well as the pressure and flow values typical for a steady state condition. Know the MOP values for key points in the pipeline if this information is available.
- Have a basic understanding of pipeline hydraulics and the factors that influence it.
- Be thoroughly familiar with abnormal operating conditions and emergencies, understanding both their entry conditions and the proper responses to each.
- Understand and be skilled at communicating both within the control room and with field personnel.
- Understand and be skilled at documenting all activities using the proper methods and formats.

2.1.2 Abnormal Operating Conditions

Control room personnel shutting down a liquid pipeline must be able to identify the following potential abnormal operating conditions and know the proper response to each, should they occur:

- *A safety or relief device activates in response to an AOC or problematic condition* – The controller should identify the specific device involved as well as the conditions that caused it to activate. The controller must then take whatever actions are appropriate to deal with the situation and return the pipeline to a safe state. Relevant personnel must also be notified.
- *An alarm associated with a safety device activates* – The controller should identify the specific device involved as well as the conditions that caused it to activate. The controller must then take whatever actions are appropriate to deal with the situation and return the pipeline to a safe state. Relevant personnel must also be notified.
- *SCADA communications, control, or power fails* – The controller should activate backup systems and then determine what caused the original failure. The controller must take whatever actions are appropriate to deal with the situation and return the regular system to operation. Relevant personnel must also be notified.
- *Exposed pipe, missing or damaged supports* – If the line is operated from a central control center, contact the control center and any additional persons as required by company policy; at a minimum, notification should be made in accordance with company policy.
- *Fire or explosion, or ignition of flammable gases* – If the situation cannot be safely controlled immediately by hand, immediately evacuate the area, limit access, and make proper notifications. Notify local 911 as appropriate.
- *Damaged pipeline components (broken, bent, crushed, gouged, or burned)* – Minimize or eliminate pressures and/or flows as appropriate for the severity of the situation and make necessary notifications according to company policy.
- *Unexpected hazardous liquid or carbon dioxide encountered* – If the leak cannot be isolated safely at the site, the control center should be contacted to isolate or shut down the line. Evacuate the area.
- *Unexplained flow rate deviation* – Verify that the proper valves are aligned and opened. Always work to minimize any potential hazardous situation. Verify equipment readings. Verify that any units and/or valves operated prior to this deviation are still in the desired position.
- *Unexplained pressure deviation* – Assume the reading is correct and work to identify the cause. Always minimize the severity of the situation. If you are not immediately aware of the cause, you may need to shut down the line until the situation can be identified and corrected. Verify that any units and/or valves operated prior to this deviation are still in the desired position.
- *Unexplained status change* – Determine the cause and minimize the situation.

2.2.0 Pipeline Shutdown

Shutting down the liquid pipeline involves numerous tasks, many of which must happen in a specific and well-timed sequence. Personnel must learn to recognize the type of shutdown required, as well as its urgency. They should be able to perform all required shutdown tasks confidently and without hesitation. Above all, they must handle emergencies rapidly and safely.

2.2.1 Shutdown Overview

Pipeline shutdowns (and relevant company guidelines) are governed in 49 *CFR* 195.402. These procedures must be followed meticulously since deviations could result in unsafe conditions, personnel injuries, equipment damage, or environmental contamination. All operating conditions must remain within the system's design limits, and all hazardous liquids must stay safe. Control center personnel must be highly trained so their actions ensure maximum safety.

In order to select the correct shutdown procedure, personnel must determine the reason for the shutdown. A scheduled shutdown is considered *normal*. No urgency surrounds this procedure, so personnel can plan it carefully and coordinate with all staffed facilities to ensure that it happens smoothly. Pumps shut down in the correct sequence, block valves close, and backpressure regulators maintain appropriate pressure levels in each part of the line.

On the other hand, shutdowns caused by abnormal operating conditions (called *abnormal shutdowns*) carry far greater urgency. While control center personnel still follow well-defined procedures, they must act rapidly. As with many emergency procedures, personnel must memorize the proper responses so they can act quickly, without having to consult procedure manuals.

> **NOTE**
> According to 49 *CFR* 195.446, anyone can shut down the pipeline in an abnormal operating condition. This policy maximizes pipeline safety.

Basic procedures are essentially the same for all shutdown types. However, shutdowns that are not scheduled have additional guidelines. These will be discussed after the standard shutdown preliminaries and procedures are explained.

2.2.2 Preliminaries

Before starting a scheduled shutdown, control center personnel must receive authorization. This will usually come from the scheduling department in the form of orders. A completed delivery operation is the usual reason for the shutdown. Scheduled (non-emergency) maintenance is another possibility. In these situations, personnel can shut down the pipeline in the safest and most optimal manner for the equipment and the product in the line.

Control center personnel plan the shutdown guided by company policy and procedures. They work out and then execute the shutdown sequence (shutting down pumps and closing valves) to bring the pipeline to a static condition. The control center must be sure to notify all appropriate personnel at staffed stations. Unstaffed stations must be monitored and controlled through the SCADA system.

Proper timing is essential for a safe shutdown. Improperly timing device shutdowns or valve closures can result in a **surge**. These, in turn, can lead to overpressure conditions or relief events.

For refined products and crude oil, the control center will generally try to leave the pipeline in a **packed** condition. This avoids complications and offers two advantages: the leak detection system continues to function and product contamination is less likely. However, pipeline conditions must still be watched closely, so that nothing goes outside of design limits. Highly volatile liquids (HVLs) must stay under sufficient pressure to keep them liquid even at the highest elevations. Some companies specify that the pressure should be 200 psig (1,380 kPa [gauge]) above the product's vapor pressure. If the line cannot be left in a packed condition, consult your company's guidelines for properly handling the issues that may result from this situation.

2.2.3 Normal Shutdown Procedures

Although every pipeline system is unique, scheduled shutdown procedures tend to have a common pattern. Always follow company policy and procedures. Control center personnel must protect life, the environment, and the pipeline's safety. The following steps summarize the procedures involved in shutting down the pipeline:

> **CAUTION**
> Whenever you are working in the control room, you must be alert to the possibility of abnormal operating conditions and be prepared to initiate the proper responses from memory.

Step 1 At a predetermined time before the actual shutdown, notify the origin and delivery facilities to prepare for shutdown.

Step 2 Adjust pressure and flow regulators to keep pipeline conditions within appropriate operating pressure limits.

> **NOTE**
> Timing is the key to a successful shutdown. It's crucial that the pressure stay at optimal levels throughout the pipeline as pumps go off line. Timing the shutdown improperly could result in overpressure conditions and relief events.

Step 3 Have the origin adjust the product flow to appropriate levels that will keep the pipeline within the correct limits.

Step 4 Have the receiving terminal adjust backpressure as required to keep the pipeline packed during the shutdown.

Step 5 Confirm that all staffed locations are ready for the shutdown. Unstaffed locations should be checked by the SCADA system to confirm that their systems are ready.

Step 6 Have the origin shut down all pumps.

Step 7 Shut down all injection pumps, including portable units, either through SCADA or by notifying station personnel in staffed facilities.

Step 8 Shut down all pumps in sequence from the origin to the receiving terminal.

Step 9 Using SCADA, close pipeline block valves associated with the shutdown procedure.

Step 10 Notify the receiving terminal that the pipeline is in shutdown and that the upstream station is blocked.

Step 11 Notify all staffed locations that the pipeline is in shutdown. Notify the receiving terminal to complete their shutdown procedures, if applicable.

Step 12 If known, notify staffed stations when the pipeline will be re-started.

Step 13 Complete any records, logs, or forms required to document the pipeline shutdown.

2.2.4 Monitoring

Pipeline conditions can change after shutdown, so the control center must continue monitoring even with a static line. Pressure, in particular, must stay within limits. For example, the product may expand if it heats up; with the line already packed, pressure could exceed the MOP.

The pipeline should also be monitored to confirm that it stays packed. The leak detection system requires a pressurized line. If the pressure drops, the system may not detect leaks, which could lead to eventual product loss.

2.2.5 Maintenance Shutdowns

At times, pipelines must be shut down for maintenance. Components and systems needing maintenance include the main-line pump, the main-line valve, and the relief system. Not all maintenance issues require immediate shutdown. In some cases, the pipeline can continue operating until a maintenance crew is available. In other cases, however, a critical failure requires immediate shutdown. Pipeline personnel handle this situation much like an abnormal shutdown.

Once a work order comes through, control center personnel will coordinate with the scheduling department to select a time for pipeline shutdown. With the crew on site and all work permits in order, the control center will shut down the line. The control center must stay in contact with the maintenance crew so the pipeline start-up can be scheduled.

2.2.6 Abnormal Shutdowns

Certain events create immediate and serious hazards that require rapid pipeline shutdown. Hazards to people, property, and the environment all fall into this category. Examples include mechanical failure, electrical failure, fire, explosion, pipeline accidents, and product release. Weather events and social emergencies, such as riots, may also necessitate immediate shutdown.

NOTE: The control center must give first priority to the safety of people and the environment.

In emergencies such as product releases, explosions, and fires, all control center personnel have the authority to shut down the pipeline. Regardless of the reason, shutdown must happen quickly, efficiently, and safely. Personnel must act by following the appropriate company emergency procedure for such a situation. There will be no time for consultation, discussion, or searching manuals. The control center must take the correct course of action immediately.

Emergency procedures are site-specific and probably include a certain amount of flexibility so the control center can adjust their responses to the specific situation. Company policy must always be followed, and carried out promptly and confidently. Personnel must know all emergency procedures by heart and be ready to carry them out.

NOTE: Many facilities emphasize training and drilling as a means of being ready for abnormal operating conditions. Take these activities seriously and use them to improve your own performance.

In cases involving fire, explosion, or product release, the control center must shut in the involved pipeline section by shutting down the originating station, and continuing downstream until all pumps are off line. All automated block valves should then be closed to complete the shutdown. If necessary, field personnel must be mobilized and directed to critical locations.

Certain product releases may be handled differently. If they have visual confirmation, control center personnel have the option of trying to drain down the pipeline segment containing the release by shutting down all pumps and closing all valves upstream. The line should be blocked in as well, in order to prevent problems.

Depending on the situation, the control center may have to notify local safety services, fire, police, or governmental agencies. The operator should follow company policy concerning such notifications. The role of the local authority should be well-defined by policies. If the emergency becomes uncontrollable, local regulations may require the control center to transfer command to a local incident command center, or a local incident commander.

Additional Resources

It is assumed that you will learn to perform this procedure under the direct supervision of your crew leader, foreman, or instructor. For independent study, refer to the recommended Additional Resources at the back of this module.

2.0.0 Section Review

1. Control room personnel must know how to interpret SCADA pages as well as how to use the SCADA system to _____.
   a. reset the CPM system
   b. adjust pipeline settings
   c. complete required documentation
   d. monitor system security

2. If an unexpected change in flow occurs, the pipeline controller must first _____.
   a. record the alarm in the control center logbook
   b. check SCADA for related problems
   c. alert higher management
   d. determine the cause of the change

3. In most cases, after shutdown the pipeline will be static and _____.
   a. drained
   b. vented
   c. packed
   d. relieved

SUMMARY

A liquid pipeline is comparable to a huge machine, made up of many different parts. Pumps, valves, tanks, and pipes are its main components. However, this machine spans hundreds of miles and must perform reliably with minimal human presence; operating it requires more than just equipment—it requires control. Pipelines, therefore, also include sophisticated systems that monitor the equipment and allow personnel to manage almost everything from a central location. It is through this control system that personnel shut down the pipeline.

In order to halt the pipeline, operators must understand its key components. Tanks store product, keeping it in good condition until it's ready to move through the pipeline. Tanks also receive deliveries and provide a place for relief devices to vent. Managing tanks and keeping them from overflowing are major control center responsibilities.

Valves control product as it moves through tanks, manifolds, and the pipeline itself. The pipeline contains numerous types and sizes of valves, many of which can be controlled remotely. Personnel must know how to operate valves and check their status through the SCADA system. They must also know how to manage them to maintain product quality and prevent contamination.

Pumps move product from one part of the pipeline to another. Most pipeline pumps are centrifugal. Properly controlling them leads to efficient energy use and lower operating costs. Similarly, using drag reducing agents can save money by lowering pipeline friction.

Successful shutdown requires assessing the situation so the proper procedures can be followed. Normal shutdowns are different from abnormal shutdowns, in which rapid response is the highest goal. Regardless of the reason for the shutdown, pipeline control center personnel must follow company procedures smoothly and efficiently to ensure that their actions protect people, equipment, and the environment.

Review Questions

1. Filling an empty tank can create a large amount of _____.
   a. sediment and water
   b. transmix
   c. static electricity
   d. stratification

2. The process of allowing product to flow into and out of a tank at the same time is known as _____.
   a. floating the tank
   b. mixing the tank
   c. gauging the tank
   d. dewatering the tank

3. Valves that help prevent a slack line condition are called _____.
   a. hand valves
   b. safety valves
   c. block valves
   d. backpressure valves

4. What control system feature prevents inappropriate valve alignments and ensures that valve sequences happen in the correct order?
   a. SCADA
   b. Valve interlocks
   c. Safety valves
   d. Fieldbus networks

5. What kind of pressure is generated on the downstream side of a pump?
   a. Discharge pressure
   b. Mainstream pressure
   c. Pump-case pressure
   d. Suction pressure

6. How long must a DRA injection be stopped before special products can be run through the pipeline?
   a. 1 hour
   b. 2 hours
   c. 5 hours
   d. 10 hours

7. One type of leak detection system that control room personnel must be able to interpret is _____.
   a. PLC
   b. AOC
   c. CPM
   d. AOPS

8. If a relief device activates in response to an abnormal operating condition, what is the first thing that a pipeline controller must do?
   a. Halt the current operation
   b. Determine the cause of the problem
   c. Notify relevant personnel
   d. Identify the device involved

9. A non-urgent, scheduled shutdown is considered _____.
   a. everyday
   b. normal
   c. typical
   d. common

10. During a shutdown, the first pumps to go off line are those at the _____.
    a. booster stations
    b. injector stations
    c. origin station
    d. receiving terminal

Trade Terms Quiz

Fill in the blank with the correct term that you learned from your study of this module.

1. A pipeline instrument that determines the product's specific gravity is called a(n) ________.
2. The tendency for liquid mixtures to separate over time, forming distinct layers according to density is known as ________.
3. A temporary storage tank used for holding specific grades of product, product mixtures, or the output of a pressure relief system is a(n) ________.
4. A control system used in the pipeline industry that collects and assembles data and also allows operators to control the pipeline remotely is called a(n) ________.
5. A control system feature that prevents unsuitable valve alignments and ensures that valve sequencing occurs in the correct order is referred to as ________.
6. The highest pressure at which a liquid pipeline may be operated is the ________.
7. The process of moving product directly from one pipeline to another without transferring it into a tank first is known as ________.
8. A tank level monitoring system that acts automatically and independently of other systems to prevent tank overfilling is called a(n) ________.
9. The region between batches of dissimilar product in which both products combine to form a transmix is called the ________.
10. A system that measures and reports the liquid level within a tank is called a(n) ________.
11. The condition in which pipeline variables are relatively constant over a period of time is called ________.
12. The API-defined tank level at which an alarm will activate and tank filling must halt within a specified time interval is known as the ________.
13. A pipeline instrument that uses light absorption to detect the interface between batches of dissimilar product is called a(n) ________.
14. A wave of high pressure that occurs when fluid starts or stops moving suddenly is known as a(n) ________.
15. The condition in which the pipeline is kept under pressure so its contents do not vaporize is known as ________.
16. A mechanical sensing system that uses a float to detect the product level within a tank and control an outside display is called a(n) ________.
17. The practice of using a tank as a buffer between pipelines by allowing product to enter and exit the tank at the same time is known as ________.
18. The fluid volume required to fill a pipeline section is called the ________.
19. The API-defined tank level at which a tank begins to overflow is the ________.
20. An electric motor paired with an electronic controller that adjusts motor speed by varying the power supply voltage and frequency is called a(n) ________.
21. The condition in which the pipeline is in shut down and the product is not moving is called ________.
22. An unusable mixture of two different products, usually created by switching from one incompatible product to another, is referred to as ________.

23. The API-defined tank level considered the highest point to which liquid should rise during normal operations is called the _________.

24. Conditions within a pipeline system that may indicate (1) a malfunction of a component or (2) a condition exceeding design limits or resulting in a hazard to persons, property, or the environment are called _________.

25. A computer-based pipeline monitoring system that creates a real-time hydraulic model of the entire pipeline and also detects leaks and breaks is called _________.

Trade Terms

Abnormal Operating Conditions (AOCs)
Automatic overfill protection system (AOPS)
Automatic tank gauge (ATG)
Breakout tank
Computational pipeline monitoring (CPM)
Critical high (CH)
Floating
Gravitometer
High-high (HH)
Interface
Line fill
Maximum operating pressure (MOP)
Maximum working (MW)
Optical interface detector (OID)
Packed
Supervisory Control and Data Acquisition (SCADA)
Static
Steady state
Stratification
Surge
Tank gauging system
Tightlining
Transmix
Valve interlock
Variable frequency drive (VFD)

Trade Terms Introduced in This Module

Abnormal operating conditions (AOCs): Conditions within a pipeline system, identified by the operator, that may indicate (1) a malfunction of a component or (2) a deviation from normal operations that may indicate a condition exceeding design limits or result in a hazard to persons, property, or the environment.

Automatic overfill protection system (AOPS): A tank level monitoring system that acts automatically and independently of other systems to prevent tank overfilling. Also, an API-defined tank level at which the AOPS will activate.

Automatic tank gauge (ATG): A mechanical sensing system that uses a float to detect the product level within a tank and control an outside display.

Breakout tank: A temporary storage tank used for holding specific grades of product, product mixtures, or the output of a pressure relief system.

Computational pipeline monitoring (CPM): A computer-based pipeline monitoring system that creates a real-time hydraulic model of the entire pipeline and also detects leaks and breaks.

Critical high (CH): The API-defined tank level at which a tank begins to overflow.

Floating: The practice of using a tank as a buffer between pipelines by allowing product to enter and exit the tank at the same time, although not necessarily at the same rate.

Gravitometer: A pipeline instrument that determines the product's specific gravity.

High-high (HH): The API-defined tank level at which an alarm will activate and tank filling must halt within a specified time interval.

Interface: The region between batches of dissimilar products where both products combine to form a transmix.

Line fill: The fluid volume required to fill a pipeline section.

Maximum operating pressure (MOP): The highest pressure at which a liquid pipeline may be operated.

Maximum working (MW): The API-defined tank level considered the highest point to which liquid should rise during normal operations.

Optical interface detector (OID): A pipeline instrument that uses light absorption to detect the interface between batches of dissimilar product.

Packed: The condition in which the pipeline is kept under pressure so its contents do not vaporize.

Supervisory Control and Data Acquisition (SCADA): A control system used in the pipeline industry that collects and assembles data, and also allows operators to control the pipeline remotely

Static: The condition in which the pipeline is in shutdown and the product is not moving.

Steady state: The condition in which pipeline variables, such as pressure and flow, are relatively constant over a period of time.

Stratification: The tendency for liquid mixtures to separate over time, forming distinct layers according to density.

Surge: A wave of high pressure that occurs when fluid starts or stops moving suddenly. Surges occur both at start-up and shutdown, although from different causes.

Tank gauging system: A system that measures and reports the liquid level within a tank.

Tightlining: The process of moving product directly from one pipeline to another without transferring it into a tank first.

Transmix: An unusable mixture of two different products, usually created by switching from one incompatible product to another (stands for "transportation mixture").

Valve interlock: A control system feature that prevents unsuitable valve alignments and ensures that valve sequencing occurs in the correct order.

Variable frequency drive (VFD): An electric motor paired with an electronic controller that adjusts motor speed by varying the power supply voltage and frequency.

Additional Resources

The following recommended resources can provide additional helpful information related to the requirements for this covered task:

NCCER Module AOCCC-17, *Abnormal Operating Conditions Control Center*

NCCER Module AOCFG-17, *Abnormal Operating Conditions Field and Gas*

NCCER Module 12406-16, *Programmable Logic Controllers*, from *Instrumentation Level Four*

NCCER Module 64205-02, *Supervisory Control Systems*

NCCER Module CT23_1-17, *Maintain/Repair Relief Valves*

NCCER Module CT24_1-17, *Inspect, Test, and Calibrate Pressure Limiting Devices*

NCCER Module CT44_3-17, *Inspect, Test, and Maintain Liquid Pipeline Leak Detection Flow Computer*

NCCER Module CT44_4-17, *Inspect, Test, and Perform Corrective and Preventative Maintenance of Tank Gauging for Leak Detection*

NCCER Module CT44_7-17, *Inspect, Test, and Maintain Gravitometers / Densitometers for Hazardous Liquid Leak Detection*

The following websites offer resources for products and training:

American Petroleum Institute, **www.api.org**

Pipeline and Hazardous Materials Safety Administration, **www.phmsa.dot.gov**

Figure Credits

ABB Module Opener, Figure 5

© Ginton/Dreamstime.com, Figure 1

Chemineer HS agitator. Photo provided by National Oilwell Varco., Figure 2

© iStockphoto.com/ Olivier Lantzendorffer, Figure 4

© Burnstuff2003/Dreamstime.com, Figure 6

U.S. Tape Company, Inc., Figure 7

Emerson Process Management, Figure 9

Apollo Valves and Elkhart Products Corporation, Figure 11

Rockwell Automation, Inc., Figure 14

Photos Courtesy of Entergy New Orleans, Inc. © 2016 All Rights Reserved, Figure 15

OMNI Flow Computers, Inc., Figure 19

© Stepan Popov/Dreamstime.com, Figure 22

Satam, Figure 23

Tim Mills, Figure 27

Section Review Answer Key

Answer	Section Reference	Objective
Section One		
1. b	1.1.1	1a
2. a	1.1.4	1a
3. d	1.2.1	1b
4. b	1.3.1	1c
Section Two		
1. b	2.1.1	2a
2. d	2.1.2	2a
3. c	2.2.2	2b

NCCER CURRICULA — USER UPDATE

NCCER makes every effort to keep its textbooks up-to-date and free of technical errors. We appreciate your help in this process. If you find an error, a typographical mistake, or an inaccuracy in NCCER's curricula, please fill out this form (or a photocopy), or complete the online form at **www.nccer.org/olf**. Be sure to include the exact module ID number, page number, a detailed description, and your recommended correction. Your input will be brought to the attention of the Authoring Team. Thank you for your assistance.

Instructors – If you have an idea for improving this textbook, or have found that additional materials were necessary to teach this module effectively, please let us know so that we may present your suggestions to the Authoring Team.

NCCER Product Development and Revision

13614 Progress Blvd., Alachua, FL 32615

Email: curriculum@nccer.org
Online: www.nccer.org/olf

❑ Trainee Guide ❑ Lesson Plans ❑ Exam ❑ PowerPoints Other ______________________

Craft / Level: Copyright Date:

Module ID Number / Title:

Section Number(s):

Description:

Recommended Correction:

Your Name:

Address:

Email: Phone:

CT64_3-17

Monitor Pressures, Flows, Communication, and Line Integrity and Maintain Them Within Allowable Limits on a Liquid Pipeline System (Control Center)

Overview

The control center is essentially the heart of the pipeline. Control center operators are responsible for monitoring pipeline operations—a critical task for keeping the pipeline stable. The sooner operations personnel become aware of developing problems, the more likely it is that they can handle them before they become serious.

CT64_3-17
Monitor Pressures, Flows, Communication, and Line Integrity and Maintain Them Within Allowable Limits on a Liquid Pipeline System (Control Center)

Objectives

When you have completed this module, you will be able to do the following:

1. Describe key liquid pipeline monitoring components and their operations.
   a. Describe the SCADA system and explain its role in monitoring and controlling the liquid pipeline.
   b. Summarize various monitoring tools and the computational pipeline monitoring (CPM) system.
   c. List and describe common liquid pipeline safety devices and alarms.
   d. Describe the pipeline purging process and the purposes that it serves.
   e. Explain liquid pipeline surveillance and weather monitoring techniques.
2. Describe the requirements and procedures for monitoring and maintaining the liquid pipeline (CT64_3-17).
   a. Identify the prerequisites, competencies, and abnormal operating conditions associated with monitoring and maintaining the liquid pipeline.
   b. Describe the procedures used to monitor and maintain the liquid pipeline.

Performance Task

Under the supervision of your instructor, you should be able to do the following:

1. Monitor pressures, flows, communications, and line integrity and maintain them within allowable limits on a liquid pipeline system (CT64_3-17).
   - Identify potential abnormal operating conditions that may occur during performance of this CT, and know the appropriate actions to take in response to them.
   - Utilize the appropriate personal protective equipment according to relevant company procedures.
   - Be familiar with the normal operating pressures, flows, and linefill for the pipeline under similar conditions. Verify that the flows, pressures, and product characteristics are within normal range.
   - Continuously monitor, by whatever means available (SCADA, pressure gauge, meter), the pressures and flow rates of the pipeline to look for any changes. Monitor the line balance for changes by monitoring volume in and volume out for liquid pipelines and line pack in gas pipelines. Monitor storage facilities. Verify continuous communications and that data is accurate and updated.
   - If pressures, flows, and/or line balance are not within normal range, determine the cause of variation from normal and correct the situation if possible. For communications variations, maintain monitored operations by using alternative communication methods. If necessary, seek assistance from and make notifications to proper personnel.
   - Perform proper operation according to company policy and procedures.
   - Verify that the desired result was achieved.
   - Complete the appropriate documentation as required by operator's procedures.

Trade Terms

Abnormal operating conditions (AOCs)
Actuator
Fieldbus
Gravitometer
Human-machine interface (HMI)
Interface
Lower explosive limit (LEL)
Maximum operating pressure (MOP)
Monitors
Optical interface detector (OID)
Packed
Programmable logic controller (PLC)
Purging
Remote terminal unit (RTU)
Supervisory Control and Data Acquisition (SCADA)
Slug
Steady state
Supervisory computer
Transmix
Voting circuit

Industry Recognized Credentials

If you are training through an NCCER-accredited sponsor, you may be eligible for credentials from NCCER's Registry. The ID number for this module is CT64_3-17. Note that this module may have been used in other NCCER curricula and may apply to other level completions. Contact NCCER's Registry at 888.622.3720 or go to **www.nccer.org** for more information.

Contents

Figures

Section One

1.0.0 Liquid Pipeline Monitoring Technology

Objective

Describe key liquid pipeline monitoring components and their operations.

a. Describe the SCADA system and explain its role in monitoring and controlling the liquid pipeline.
b. Summarize various monitoring tools and the computational pipeline monitoring (CPM) system.
c. List and describe common liquid pipeline safety devices and alarms.
d. Describe the pipeline purging process and the purposes that it serves.
e. Explain liquid pipeline surveillance and weather monitoring techniques.

Trade Terms

Abnormal operating conditions (AOCs): Conditions within a pipeline system, identified by the operator, that may indicate (1) a malfunction of a component or (2) a deviation from normal operations that may indicate a condition exceeding design limits or result in a hazard to persons, property, or the environment.

Actuator: A powered mechanism that operates a valve.

Fieldbus: A simple digital communications network with a set of communications standards designed to connect instruments, controllers, and other intelligent devices together.

Gravitometer: A pipeline instrument that determines the product's specific gravity.

Human-machine interface (HMI): A display station attached to the SCADA system that allows users to monitor and/or interact with the pipeline. Also known as a man-machine interface (MMI).

Interface: The region in between batches of dissimilar products where both products combine to form a transmix.

Lower explosive limit (LEL): The lowest gas/vapor concentration in air that could explode if an ignition source were present.

Maximum operating pressure (MOP): The highest pressure at which a liquid pipeline may be operated.

Monitors: Devices that watch a process parameter, such as temperature, flash point, or sediment and water, and generate informational reports or alarms based on these parameters.

Optical interface detector (OID): A pipeline instrument that uses light absorption to detect the interface between batches of dissimilar product.

Programmable logic controller (PLC): A compact industrial computer largely used for machine control and other dedicated tasks requiring intelligence.

Purging: The process performed when changing between incompatible products in a pipeline that prevents their mixing. Also refers to a maintenance procedure in which flammable vapors are forced out of the pipeline by an inert gas.

Remote terminal unit (RTU): A computer-like device that connects field devices to the supervisory computer in a SCADA system; RTU can also stand for remote telemetry unit or remote telecontrol unit.

Supervisory Control and Data Acquisition (SCADA): A control system used in the pipeline industry that collects and assembles data, and also allows operators to control the pipeline remotely.

Slug: Refers to a quantity of product that contaminates another incompatible product. Can also refer to a volume of inert gas injected into a pipeline during a maintenance purge.

Steady state: The condition in which pipeline variables, such as pressure and flow, are relatively constant over a period of time.

Supervisory computer: A large computer in a SCADA system that collects data from and issues commands to field devices such as RTUs and PLCs.

Transmix: An unusable mixture of two different products usually created by switching from one incompatible product to another (stands for "transportation mixture").

Voting circuit: An alarm circuit connected to several sensors. It will not alarm unless a majority of the sensors activate. Also known as a majority circuit.

Keeping the pipeline running smoothly and dealing with problems are the dual roles of pipeline control operators. Equipment malfunctions can cause conditions to reach operating limits quickly. Although abnormal conditions do not occur frequently, control operators must react to them promptly.

To make good decisions, operators depend on a steady flow of information. Remote sensors collect data about pipeline conditions. They pass this information along to computers that store and process it. Displays present the information in easy-to-interpret forms. Safety devices and alarms alert operators of potentially hazardous situations. All of this technology makes it possible for operators to monitor pipeline conditions because they can view the entire pipeline from their seat in the control room (*Figure 1*). Using this information, they can make informed decisions concerning everyday situations as well as emergency events.

Every task performed by control personnel must comply with company and industry practices and policies. These policies are well-defined and cannot be set aside; personnel must not deviate from them. While a procedure may sometimes be suspended temporarily, such situations are limited. A key standard to know is 49 *CFR* 195.402, which specifies immediate response to and mitigation of pipeline hazards.

1.1.0 Pipeline Supervisory Control

The pipeline is a huge system with many different pieces spread over a vast geographic area. Creating a single picture of it all is the job of the **Supervisory Control and Data Acquisition (SCADA)** system. SCADA has two main purposes: to enable operators to manage the pipeline (supervisory control), and to collect information (data acquisition).

Unlike other industrial control systems, SCADA doesn't make numerous autonomous decisions. Instead, it relies on human operators to make decisions by using the information that it collects and assembles for them.

SCADA helps to reduce the operator's workload as well. Remote control eliminates the need for frequent travel between facilities, leading to more efficient time management. Similarly, SCADA makes it possible for many facilities to operate without a regular staff. It also provides increased safety during severe weather since operators can control the equipment from a safe location. As required by 49 *CFR* 195.402, pipeline systems must have well-defined procedures and safety protocols in place. Supervisory systems allow human operators to satisfy these requirements more easily and efficiently.

A typical SCADA system (*Figure 2*) consists of four component categories: field device, **supervisory computer**, **human-machine interface (HMI)**, and communications system. Together, these gather data from equipment in the field, send it to centralized locations for analysis, and provide a means for human operators to examine the data and interact with the pipeline.

1.1.1 Field Devices

In the field, sensors gather information reflecting the pipeline's current state. Typical variables include pressure, temperature, and flow. In many cases, this information goes to field control devices (controllers) that make local decisions to

Figure 1 A typical control room.

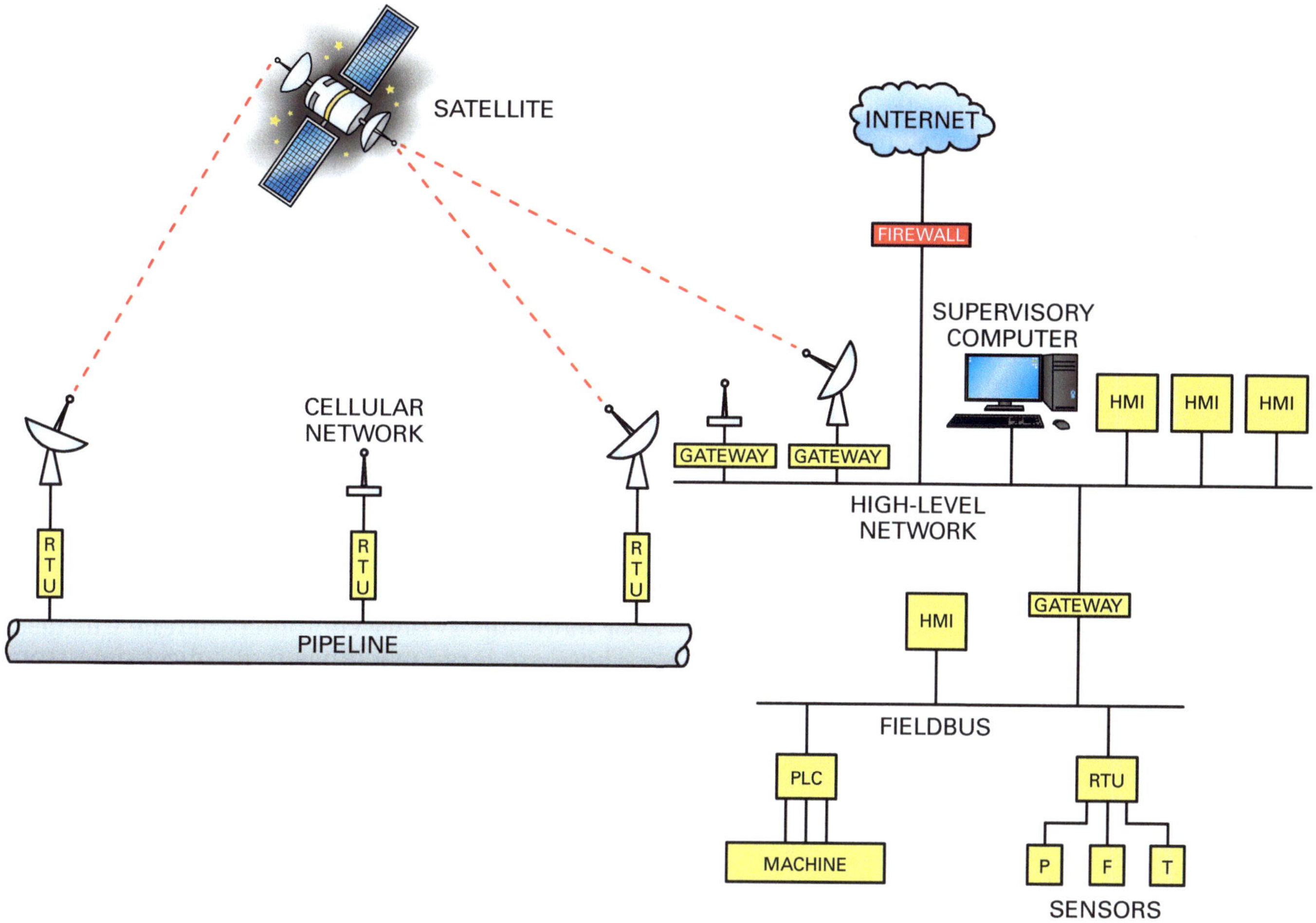

Figure 2 Components of a SCADA system.

modify these parameters. Using an **actuator**, the controller opens, closes, or adjusts the appropriate valve to alter pipeline conditions.

In many cases, however, a higher level of control may be required. Controller parameters must sometimes be changed or the actuator's current state overridden. To make this possible, many controllers are connected to compact, computer-like devices that are part of the SCADA system. These devices fall into one of two types: the **remote terminal unit (RTU)** or the **programmable logic controller (PLC)**.

1.1.2 RTUs

An RTU (*Figure 3*) is a small control computer that is mounted in a protective industrial enclosure. Like any computer, it contains a processor and memory, and also runs a control program. RTUs include interfaces to connect to sensors, transmitters, controllers, and actuators. They also contain a system for communications with a SCADA supervisory computer.

Functionally, the RTU is both a controller and an interface between the field devices and higher-level control systems. Its main purpose is to gather data and forward it to the supervisory computer. This data stream is known as *telemetry*. The supervisory computer can, in turn, send the RTU commands to modify the pipeline's behavior.

Depending on the application, the RTU could be a fairly simple device that has only limited control capabilities, or it may be highly sophisticated

Figure 3 Typical RTU.

and capable of autonomous control. In older systems, RTUs were rudimentary and mostly gathered data, rather than performing significant control functions. Due to the development of other kinds of industrial control systems, RTUs have changed in recent years. Today, many of the newer ones are quite powerful and are similar to other industrial control computers.

1.1.3 PLCs

PLCs are another type of industrial control computer. However, they were developed to address a different need than RTUs—machine control. Prior to the introduction of PLCs in the early 1970s, machine controllers contained complicated circuits made from relays and other electromechanical components. These systems worked well but required a lot of time and effort to alter if changes became necessary.

Early PLCs replaced relay-based controllers with a small computer that mimicked the relay logic through software. Changes now involved reprogramming the PLC rather than rebuilding a circuit. Over time, PLCs developed into sophisticated devices that began to be used in more and more applications. Modern PLCs are compact but powerful computers designed to operate in harsh industrial environments. Like RTUs, they too contain a processor with memory and run a control program. PLCs have flexible input/output (I/O) capability and can work with a wide variety of signal types. Large PLCs can process thousands of inputs and outputs (*Figure 4*).

Even more significantly, modern PLCs have extensive communications capabilities. They can talk to each other and to other devices, including large computer systems. While early PLCs were extremely limited, modern PLCs can interact with virtually any device.

PLCs execute their programs through what is known as a *scan*. At the beginning of the scan, the PLC processor reads all of its inputs and stores their values in memory. It then runs through its program logic, which uses the input values to make decisions and generate output values in another section of memory. As the final step, the PLC sends the output data from memory to its outputs. The scan process then repeats. A complete scan takes anywhere from a fraction of a millisecond to a few milliseconds, depending on the processor's power and the program's complexity.

Figure 4 A PLC with various I/O modules.

In a pipeline setting, PLCs often perform sophisticated local control tasks in which they have to make autonomous decisions. Unlike RTUs, they tend to take a more active role in controlling the pipeline. Since PLCs have powerful communications capabilities, they can talk to SCADA and function as a piece of the much larger system.

1.1.4 Supervisory Computers

Individual PLCs and RTUs see only a small part of the pipeline, but pulling their data together in a central location makes it possible to create a picture of the entire pipeline and its activities. One or more supervisory computers handle this task. These are large machines that collect data from many different RTUs and PLCs, format it, and supply it in various forms to those who manage the pipeline.

Supervisory computers communicate with RTUs and PLCs over a network of some type. They run sophisticated software that monitors field devices and collects their data. This data becomes the basis for reports. These can be simple screens presented on an HMI, or they can be downloaded in the form of spreadsheets by a standard PC. Pipeline operators use HMI-presented reports to make control decisions. Managers and executives use downloadable reports to make higher-level decisions about the pipeline and to make projections that will help them plan future decisions.

Besides collecting and assembling data to help people make decisions, supervisory computers perform a number of other tasks. A significant one is the historian function, which archives data over long periods of time. Historical data can reveal subtle pipeline trends, facilitating more sophisticated management decisions.

Finally, the supervisory computer centralizes pipeline control. Operators interact with HMIs, which, in turn, interact with the supervisory computer. Based on these interactions, the supervisory computer issues appropriate commands to the correct RTU or PLC, which modifies the pipeline's behavior.

1.1.5 HMIs

Every control system contains one or more HMIs, which allow operators to monitor the process and make control decisions. In some systems, HMIs are called *man-machine interfaces* (MMIs). An HMI can be relatively simple, such as a small, text-only LCD display with a few buttons or perhaps a simple keypad. More sophisticated HMIs have full-color graphical displays with touchscreens for operator input, such as the one shown in *Figure 5*. These are often used by field personnel to interact with more sophisticated field devices, such as PLCs.

Control rooms usually contain many HMIs, including large wall-mounted ones and smaller panel-mounted units (*Figure 6*). These graphically display key process-control components along with important data, status values, and alarms. HMIs communicate with the supervisory computer through the SCADA network.

When interacting with field personnel, remember that field HMIs are usually smaller and cannot show as much information on a single page as the larger control room units. Field personnel may have to navigate through multiple screens to locate information that appears on a single page in the control room. Be aware of these differences so you can communicate effectively with field personnel.

Figure 5 A small graphical HMI.

Figure 6 Control room with large and small SCADA displays.

1.1.6 Communications Technologies

For SCADA components to be effective, they must communicate with each other. For this reason, all SCADA systems have several communications networks that link their components into a unified whole, allowing data to circulate freely.

At the field level, a SCADA system consists of many devices that interact directly with the pipeline. Sensors, actuators, controllers, RTUs, and PLCs are all examples of field devices. Field device communication is fairly basic and involves only small amounts of data. There are usually less than a few hundred devices in a given area. For this reason, communications systems linking field devices tend to be small and simple.

A **fieldbus** is a low-level network (*Figure 7*) designed to function over a relatively limited physical area. It uses a simple cabling system to tie field devices together. A typical fieldbus might link a PLC or RTU and a number of other devices. A local operator HMI might also be connected to the fieldbus. Because fieldbuses cannot operate over large areas, a second class of network, called the high-level network, handles this task.

A high-level network must be fast, robust, and capable of handling large volumes of data efficiently and reliably. There are many different technologies available to implement high-level networks. Some of these are used in plants, offices, and on corporate campuses. This kind of technology usually relies on cabled connections (copper or fiber optic).

In some situations, a cabled network just isn't practical. It may not be possible to run cables in certain places, or the distance might be prohibitive. Wireless networking can solve some of these problems by using radio-frequency (RF) technology to carry digital data. Short-range wireless networking, commonly known as Wi-Fi™, is often used in offices and factories as a means of supporting mobile devices. But it does not have the range needed for the pipeline industry. In these situations, satellite radio, cellular phone communications, microwaves, and other types of long-range RF communication are used.

1.1.7 Alarms versus Alerts

One of SCADA's significant features is its alarm system. In a typical control center, one or more panels will probably be dedicated to displaying alarm information of various kinds. It's important to dis-

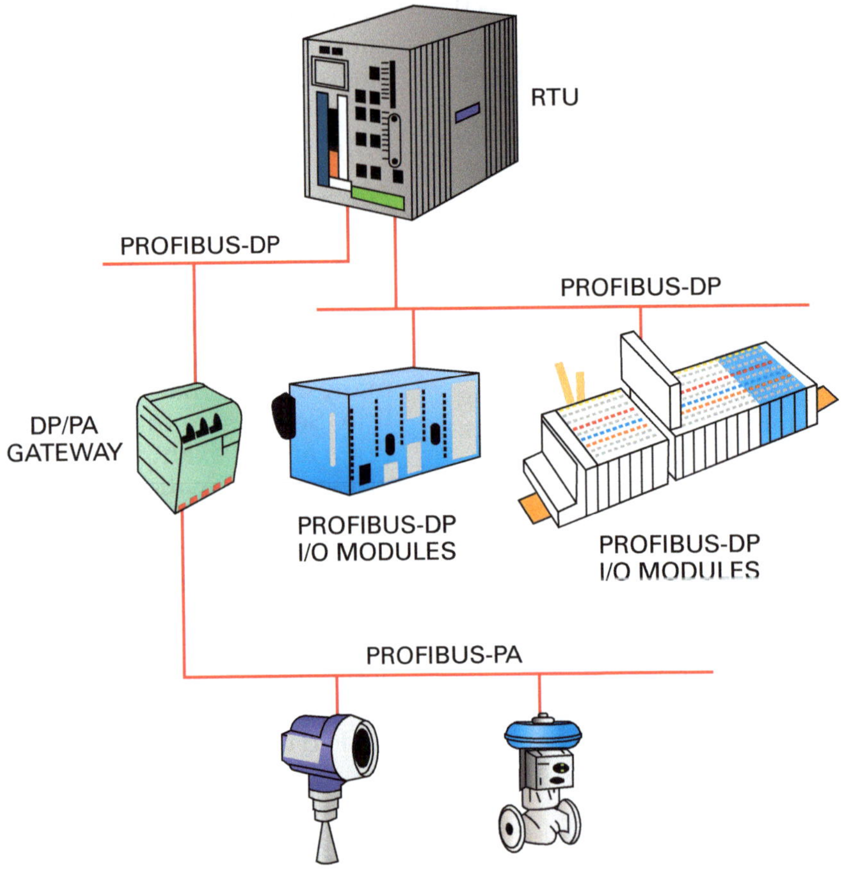

Figure 7 A fieldbus ties local devices together.

tinguish between the terms *alarm* and *alert*. While they are related, there are some crucial differences.

An alarm is a status event that results from a problem, abnormality, or emergency that occurs in the pipeline system. All alarms have a mandatory response, known as an *action*, associated with them. Control center personnel must respond to all alarms according to company procedure.

An alert is also a status event, but it is more informational and usually does not have an associated action. For example, when a valve travels between the open and closed positions, an alert may appear on an HMI. Control center personnel monitor alerts to be aware of conditions along the pipeline, but alerts do not have the same urgency as alarms. Some systems use the term *notification* for an alert.

SCADA gathers information from the pipeline and uses it to generate alarms and alerts, which it displays on various HMIs (*Figure 8*). Some systems have separate panels for alarms and alerts to reduce operator confusion.

Some systems flash alarm codes to get the operator's attention. Color codes may classify different kinds of alarms and alerts. Alarm priority may also be color-coded so operators can quickly decide which alarms to handle first. Typical priority levels are high, medium, and low. Some control centers have audible alarms as well as visual ones. These may use different sounds to distinguish alarm priorities.

NOTE

A number of standards govern alarms, alerts, and mandatory operator responses. Examples include 49 *CFR* 195.403, 49 *CFR* 195.446, and API *Recommended Practice 1168*.

One goal of current standards is to reduce excessive alarms and alerts. These so-called "floods of alarms" can overwhelm control center personnel, leading to poorer handling of abnormal conditions.

1.2.0 Monitoring the Pipeline

Control center personnel spend much of their time monitoring the pipeline in various ways. They watch process parameters such as temperature, pressure, and flow. They must assess alerts and handle alarms. Quality control monitoring requires checking other parameters such as sediment and water (S&W), flash point, and API gravity. Finally, personnel must coordinate the events that make up the work schedule. All of these tasks rely on different kinds of monitoring tools present in the control center.

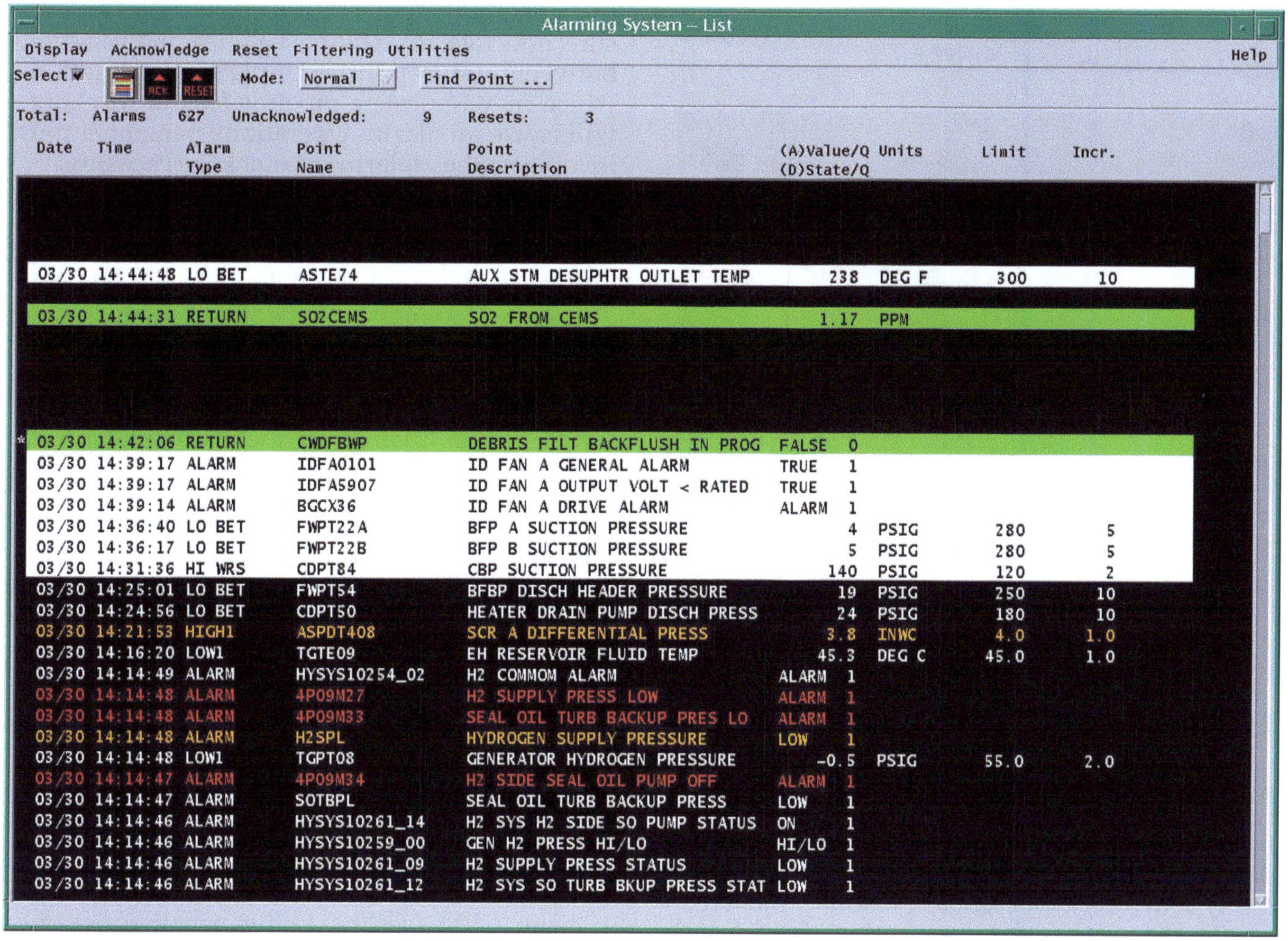

Figure 8 An alarm screen.

1.2.1 Monitoring Pressure and Flow

Pressure and flow are the two primary indicators of pipeline condition (AOC). Deviations in either can signal a developing problem that may escalate into **abnormal operating conditions (AOCs)**. While SCADA will alarm if either parameter goes out of range, personnel must still be familiar with your pipeline's normal pressure and flow values. Serious consequences may result if the **maximum operating pressure (MOP)** is exceeded.

Abnormal pressure and flow values could indicate a possible product release or an approaching equipment shutdown. For example, if pressure begins to build at one point in the line, the SCADA readings will show it immediately. The rise could indicate an unintended valve closure that might jeopardize pipeline operations. If the valve closure goes unnoticed, an emergency condition will develop, which could endanger public safety, the environment, and pipeline operations.

In this situation, operations personnel must attempt to reopen the valve immediately. Sending an Open command by the SCADA system or dispatching field personnel in a staffed location to locally control the valve (*Figure 9*) may be the solution. If the valve cannot be reopened quickly, pipeline shutdown must begin. Operations personnel rely on their best judgment as well as their experience and knowledge of the facilities.

1.2.2 Monitoring after Corrections

Operations personnel must monitor the line closely after making corrections. When abnormal conditions have ceased at critical locations, personnel must check for deviations from normal. Pump stations, origin and receiving locations, injections and deliveries, and meter stations must be watched closely to verify that the pipeline is safe and has returned to a **steady state**.

1.2.3 Monitoring Communications

Operations personnel also monitor the conditions within the communications system as required by 49 *CFR* 195.402 because it is the control line for the pipeline. Normally, SCADA manages this, but

Figure 9 Valve control panel.

if partial or total loss of communications occurs, personnel must take alternative action to control the system safely. Operations personnel must recognize the signs of communications failure and take corrective actions as quickly as possible.

Furthermore, 49 *CFR* 195.446 specifies that personnel must have a well-defined manual control plan in place to manage the pipeline should SCADA communications fail entirely. Control room personnel must review this plan periodically and be ready to switch to it if necessary.

1.2.4 Monitoring Schedules

Another aspect of monitoring that may not seem immediately obvious is monitoring schedules. While process parameters, alarms, and alerts are of high importance, control room personnel must also manage how product is batched, queued, and dispatched. They must be fully aware of field operations since these affect the information appearing on their monitors. In other words, control room personnel deal with more than alarms and alerts—management is a big aspect of the job.

1.2.5 Computational Pipeline Monitoring

Computational pipeline monitoring (CPM) is a computerized leak detection program required by 49 *CFR* 195.444. With this program, computers carry on a continuous monitoring process for pressure, flow rate, line volume balance, and variations between status points. Unplanned deviations, which may indicate a pipeline leak or discharge, will cause an alarm. Operations personnel must investigate these alarms as quickly as possible.

Using technology such as flow computers (*Figure 10*), the CPM system performs line balance calculations to compare the volume received into the system with the volume being delivered. The system adjusts for product retained in relief tanks, product removed for other uses, and any factors that would affect the final total. Averaging eliminates the occasional data blip. If a discrepancy appears after all of these calculations, the system alerts operations personnel to the possibility of a product release.

The CPM system also accesses SCADA data and uses it to build a simulated model of the pipeline's hydraulic state. This model provides operators with the up-to-date information that they need to work with the pipeline.

> **NOTE**
> In most systems, there are multiple ways to detect leaks. Pipeline controllers can often identify leaks through the SCADA system, as well as through CPM.

Leak detection and line break detection are a major CPM feature. Using negative pressure wave detectors, CPM detects leaks and pinpoints their exact location. When a line break occurs, pressure suddenly drops and then rebounds. These rapid changes create a wave that travels at the speed of sound through the pipeline in both directions. Sensors located every few miles detect the wave. CPM then calculates the break's location and raises an alarm within seconds.

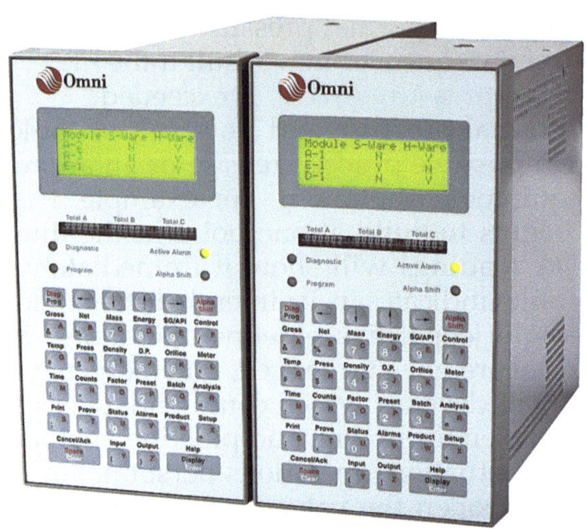

Figure 10 A pipeline flow computer.

CPM line break detection is illustrated in *Figure 11.* It is displayed in graphical form on a screen in the control center.

> **NOTE**
> As with all alarm systems, false alarms are a possibility with the negative pressure wave program. For example, field personnel calibrating a pressure transmitter may trigger a false CPM alarm. Nevertheless, all CPM alarms must be investigated as quickly as possible.

> **NOTE**
> A communications failure at any point in the CPM system will affect its accuracy. Faults should be corrected as quickly as possible to maintain CPM's accuracy and usefulness.

Another useful task that CPM performs, called *trending,* is showing the pipeline's behavior over a specific time period. Using data obtained from various sensors, CPM builds a graphical pipeline profile. The system also calculates and demonstrates the rate of change taking place so that operations personnel can spot a problem that might be developing.

1.2.6 Release Alarm Documentation

Operations personnel must log all release alarms, and their associated data, in the release alarm record. After the event has been successfully controlled, appropriate personnel must decide if investigation and analysis reports need to be filled out or if any further action is required. After a shutdown due to high pressure, engineers must perform a hydraulic analysis upstream to confirm that no problems occurred elsewhere. They must also check for damage caused by the event. It's essential that they examine all records of the incident to see if the MOP was exceeded at any point.

If so, management and government agencies must be notified and all required paperwork must be completed and confirmed. Computer logs, incident reports, event notifications, and control center logbooks must be updated. All other related documentation must be checked for

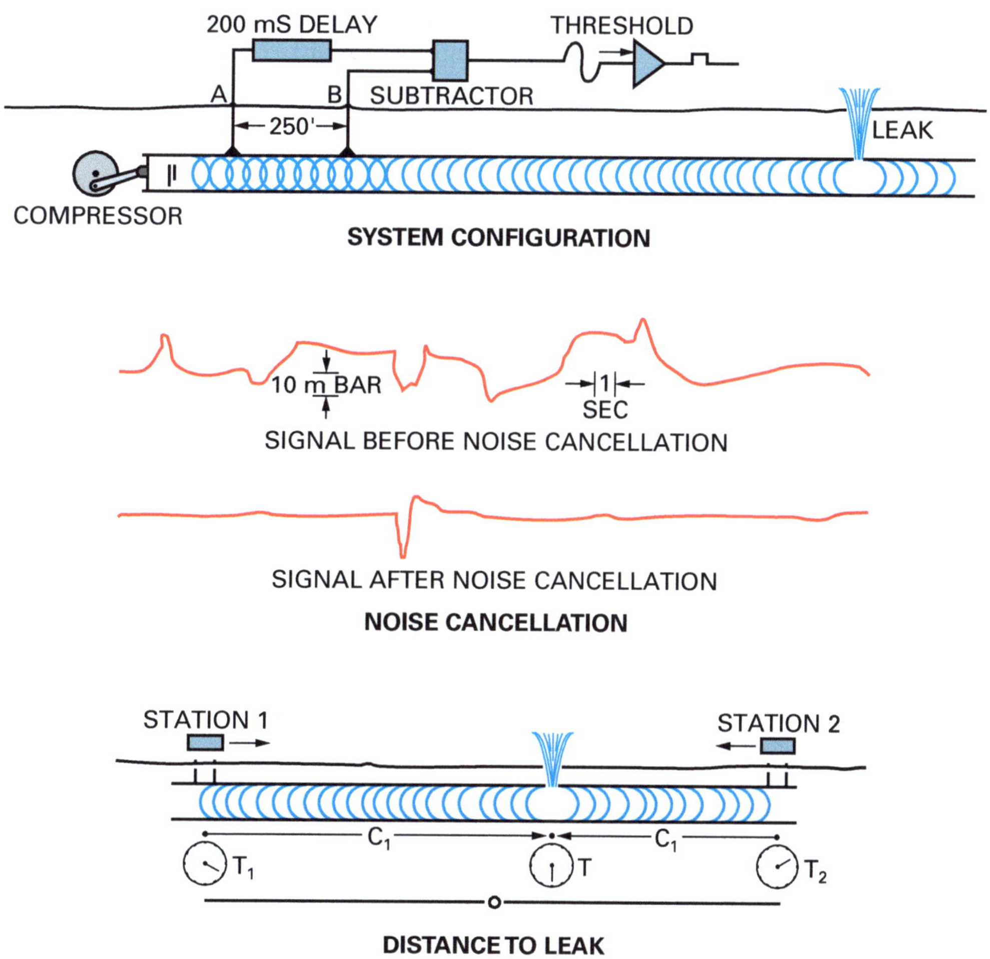

Figure 11 CPM line break detection.

accuracy and filed. Once they are satisfied, the engineers will approve the pipeline as ready for restart.

1.3.0 Safety and Alarm Devices

There are a variety of abnormal situations that could develop affecting pipeline operations. Fire, hazardous vapors, or leaks can easily occur anywhere along the pipeline or in pipeline facilities. Overpressure can develop. An intruder can break into an unstaffed facility. Often, personnel may not be physically present when these situations develop.

Instead, a series of safety and alarm devices monitor the area and either respond to the problem automatically or signal operations personnel that an abnormal situation has developed. Most detectors and alarm circuits report to SCADA, which then passes the information along to the control room.

Safety and alarm devices are found throughout the pipeline and its storage facilities. These devices alert control operators of abnormal conditions, and some even prevent or relieve them. Such devices include the following:

- Hazardous atmosphere detectors
- Fire and thermal detectors
- Intrusion detectors
- Thermal relief devices
- Pressure relief devices
- Tank and sump level detectors
- Leak detectors

Most safety devices are associated with an alarm of some kind. When the alarm first activates, a status indicator will flash. Operations personnel must respond to the alarm. Warning and danger alarms remain active (latched) until control room or field personnel reset them.

WARNING!

Never ignore any safety device alarm. If you ever receive an alarm with which you are unfamiliar, contact appropriate personnel according to your company's policy.

NOTE

All safety alarms must be properly documented. In many cases, a "lessons learned" debriefing will occur after an alarm, and its results will be documented along with the alarm itself. The goal of this activity is to prevent avoidable alarms in the future.

1.3.1 Hazardous Atmosphere Detectors

The pipeline environment contains many hazardous liquids, vapors, and combustible gases. Maintaining the safety of personnel and equipment is very important, so combustible vapor/gas detectors (*Figure 12*) are used to sense the presence of hazardous vapors or atmospheres around pipeline equipment and buildings. These devices detect a buildup of specific substances during normal operations or in the event of an accidental release. They can also determine the extent of an accidental release.

Combustible vapor/gas detectors can be wall- or panel-mounted. An example of a panel-mounted one is shown in *Figure 13*. Many of these continually monitor pump rooms and other stations for hazardous vapors. These devices communicate with the SCADA system to trigger an alarm at the control center if the vapor concentration exceeds the alarm setpoint.

Since the explosive ranges of flammable liquids and gases are very wide, manufacturers calibrate detectors to the **lower explosive limit (LEL)** of the substance. For safety, flammable gas and vapor concentrations must remain well below the LEL. Most calibrations are expressed as a percentage of the LEL, ranging from 0 to 100 percent. For example, a detector may be calibrated to alarm when it detects a concentration of 20 percent of the LEL.

Figure 12 Combustible vapor/gas detector.

NOTE

By law, combustible gas/vapor detectors must be set to alarm at no higher than 25 percent of the LEL. Many are set to lower levels as an added safety precaution.

In addition to combustible gases and vapors, the pipeline environment may also contain toxic or hazardous gases that could injure or kill personnel. For this reason, installations will include other types of gas detectors as well. These may detect just a single gas or they may detect several. Typical gases include hydrogen sulfide (H_2S), methane (CH_4), carbon monoxide (CO), and carbon dioxide (CO_2).

CAUTION

Toxic gases harm the body by poisoning or injuring a biological system. For example, hydrogen sulfide aggressively attacks several different body systems and can cause a great deal of damage. Carbon monoxide attaches to hemoglobin in red blood cells, preventing them from carrying oxygen.

Gases can also be hazardous without being toxic. Methane is not poisonous, but is highly flammable and potentially explosive. Lower concentrations of carbon dioxide are harmless, but high concentrations can be toxic.

Gases that create a dangerous environment by displacing oxygen are called asphyxiants. Any gas can asphyxiate. "Safe" gases, such as nitrogen and helium, are especially risky since they are not easily detected. For this reason, gas detectors and oxygen monitors are crucial safety devices in the pipeline industry.

Figure 13 Panel-mounted combustible vapor/gas detector.

Gas detectors use a number of different technologies to sense the target substance. Some, like the one shown in *Figure 14*, use a semiconductor that changes electrical resistance when in the presence of a specific gas or vapor. Others shine a beam of infrared light through a test sample (*Figure 15*). If the target substance is present, it absorbs specific wavelengths and allows others to pass—a kind of chemical fingerprint.

1.3.2 Fire and Thermal Detectors

Fire and thermal detectors sense the presence of smoke, flames, or excessive heat. In addition to activating an alarm, many of these devices perform automatic responses as well. The following are some typical examples:

- Activating an emergency ventilation system
- Isolating potential fuel sources
- Shutting down equipment
- Discharging a chemical fire suppressant

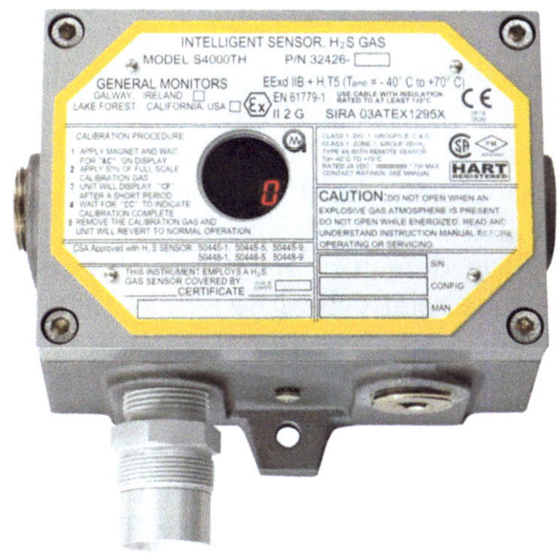

Figure 14 A semiconductor hydrogen sulfide gas detector.

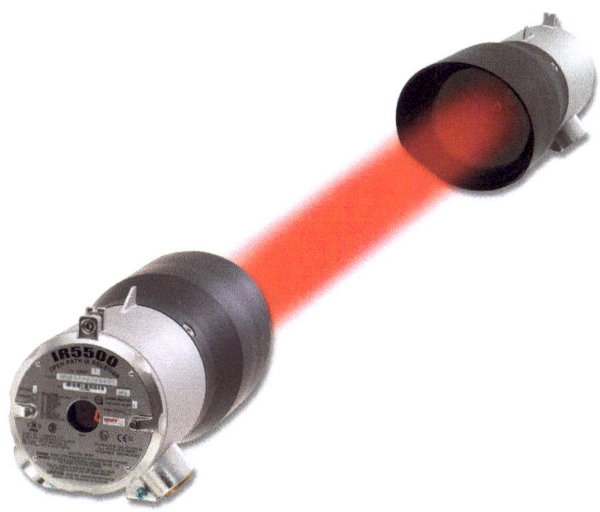

Figure 15 An infrared combustible gas detector.

NOTE

Since many facilities are unstaffed, control room personnel must make decisions about how to handle a fire at a remote location. Alarm systems will probably alert local firefighters automatically. But the control room will have to decide whether to shut down the facility, either with an emergency shutdown (ESD) in which all valves are closed, or a main breaker trip in which power to the facility is cut.

Smoke detectors sense the presence of smoke particles in the air rather than detecting the fire itself. Units range from the simple ones found in homes to industrial-grade ones found in pipeline settings. Smoke detectors are also used in offices or wiring closets, above suspended ceilings, and in equipment rooms.

Smoke-sensing technology divides into two main categories: ionization and photoelectric. An ionization detector uses a small radioactive source to make the air in the sensor chamber electrically conductive. When smoke particles enter the chamber, the air's conductivity changes, triggering the detector circuitry. Photoelectric detectors use a light beam/light sensor combination to detect smoke. When it enters the sensor chamber, it reduces the light beam's intensity, triggering the alarm.

Ionization detectors are better at sensing fires that produce relatively little smoke, while photoelectric detectors are better at sensing smoldering fires. For this reason, some smoke detectors contain both types of sensors.

Rather than sensing smoke, some detectors spot the flame itself. Ultraviolet (UV) sensors (*Figure 16*) fall into this category. They sense the invisible ultraviolet light generated by the flame rather than its visible light, reducing false alarms.

Figure 16 An ultraviolet flame detector.

UV sensors are very quick, detecting a flame in a fraction of a second.

UV flame detectors can be fooled by other energy sources that emit ultraviolet light, such as welding arcs, camera flashes, and electrical sparking. In cases where a false alarm would be detrimental, several detectors may be wired to a **voting circuit**. Only if a majority of detectors report a flame will the system alarm.

NOTE

UV flame detectors work on the line-of-sight principle. If they cannot see the flame, they won't detect it. For this reason they should be inspected regularly to confirm that nothing is blocking the sensor or preventing it from monitoring the area of interest.

A third way of detecting a fire is by sensing its heat. Thermal detectors are temperature sensors that close a pair of contacts if the temperature rises rapidly, usually a sign of a fire. Besides being used for general fire monitoring, they may also be installed inside pieces of equipment that are likely to overheat and catch fire. For example, variable frequency drive (VFD) motor controllers often contain thermal detectors since they can overheat easily.

Thermal detectors are available with a variety of temperature setpoints. The engineer who designs the safety system will select an appropriate setpoint for the particular environment. Never paint or coat a thermal detector as that will change its temperature setpoint, possibly making the detector useless.

1.3.3 Intrusion Detectors

Since many pipeline facilities are now unstaffed, security relies far more on technology than it did in the past. Along with alerts and alarms from pipeline sensors, control room personnel must monitor security sensors that watch over the pipeline and its various facilities.

Security sensors fall into a number of categories. The most common type are intrusion sensors that monitor doors, windows, and gates. These are often simple contact switches that open and close in response to the entry point that they are monitoring. Infrared motion detectors (*Figure 17*) sense intruders by the body's thermal signature.

Video surveillance with automatic motion detection is also a significant part of security, and will be discussed later in this module.

Figure 17 A passive infrared motion detector.

1.3.4 Thermal Relief Devices

Liquids expand when heated. If liquid contained within a line expands too much, it could rupture the pipe. Thermal relief devices release pressure caused by heat expansion, usually by allowing the product to escape into a breakout tank. A manifold thermal relief device is one such example. Other devices, such as bypass relief devices, provide relief if a controller can't operate a valve.

Closely related to these devices are those that protect against damage due to high temperature. Mainline and booster pumps can overheat, resulting in damage. High-temperature protection devices either shut down the pump or open a bypass valve to deal with the situation.

1.3.5 Pressure Relief Devices

An overpressure situation can rapidly lead to a catastrophic failure. Pipelines and tanks can rupture from excessive internal pressure. It's also possible for a tank to be crushed by atmospheric pressure if a vacuum (internal pressure below atmospheric pressure) develops inside. For this reason, the pipeline and its accompanying equipment will include a variety of devices designed to deal with pressure problems.

Pressure relief valves are the simplest and most straightforward way to deal with excessive pressure. These are valves that have a specific pressure value at which they will open. They relieve pressure by releasing product from the pipeline, usually into a breakout tank. Excessive air pressure in tanks is usually vented to the atmosphere. Vacuum-related pressure problems are relieved in the opposite way, by admitting air to eliminate the vacuum.

Key pressure relief devices usually have electronic circuits that report to SCADA so the control room can monitor pressure relief events.

Pressure switches are closely are related to pressure relief valves, and are sometimes known as *HULP switches* (high upstream line pressure switches). Instead of releasing product to relieve pressure, they operate block valves that divert excessive pressure downstream. In many cases, a HULP switch will be paired with a pressure relief valve. The switch will trigger first. Only if its actions fail to reduce pressure sufficiently will the relief valve open and take more drastic action.

1.3.6 Tank and Sump Level Detectors

Tanks and sumps hold liquid, either for normal storage or as a temporary measure. Regardless of the application, tank and sump levels must be monitored to be sure that they remain within the desired range. An overfilled or underfilled tank can create a variety of problems. Level sensors report the liquid height within the tank to SCADA, which then makes decisions about alarms or alerts.

1.3.7 Leak Detectors

Given that pipelines can be hundreds or even thousands of miles long, leak detection and management is a high priority. Most pipelines use a number of different strategies to identify and locate leaks. Many of these rely on calculations rather than direct detection.

Recall that CPM includes a leak/release detection feature. Using negative pressure wave detectors, it can identify and locate pipeline breaks. Using its line balancing calculator, it compares volumes entering and leaving a section of the pipeline. Discrepancies in these values suggest the presence of a leak.

Since the SCADA system has access to large amounts of data about metering operations, it performs similar calculations to those performed by the CPM system. Using meter in / meter out data, it can detect discrepancies that may indicate leaks. Other data points, such as pressure, flow, and various alarms, can enhance these calculations for even better accuracy.

Some pipeline leak detectors directly sense product seeping into the ground from a damaged pipe. One such detector is made from a perforated PVC pipe that's buried below the pipeline. Product leaking into the soil enters the detector pipe, where circulating air moves its vapors past sensors installed at regular intervals along the pipe. By noting which sensors trigger as well as the order in which they trigger, the detection system can deduce the location of the leak as well as its extent. The downside of this direct detection method is cost and the need to maintain the detection system.

Seal leak detectors are used on large mainline pumps and other high-speed devices. While seals are very durable, they can fail due to age, wear, and high pressures. If a seal fails and it's not discovered in time, the equipment can be seriously damaged. A seal leak detector has a collector located under the shaft near the seal. A tube runs to a sump and contains a flow detector. If the seal fails, liquid will run into the collector, flow through the tube, and drain into the sump. The sensor detects the flow and raises an alarm.

In many newer facilities, PLCs handle seal leak detection rather than dedicated detectors. A PLC has the advantage that it can monitor many different devices at the same time and funnel all of the information to the SCADA system through a single communications channel.

1.4.0 Pipeline Purging

There are many products that cannot be mixed because the combination might be dangerous or simply fail to meet standard requirements. For example, mixing gasoline with a shipment of fuel oil is dangerous because it has a lower flash point than fuel oil should have. Mixing regular gasoline with a shipment of premium gasoline is not dangerous, but the mixture doesn't have the required quality level. If ultra-low-sulfur diesel becomes contaminated with other grades of diesel, it violates the strict quality standards for diesel fuels.

This situation is more common than it might appear to be. Pipeline facilities process and store numerous products, many of which must pass through certain common systems. For this reason, pipeline personnel must use a technique called **purging** when switching between incompatible products.

A different issue comes up when pipelines have to be drained for maintenance. Since pipeline products are hazardous and flammable, the emptied pipeline will contain a mixture of air and product vapor. This dangerous combination must be removed before the pipeline can be worked on. The procedure used to do this is also called *purging*, though it does not follow the same process as the purging done when switching between incompatible products.

1.4.1 Purging between Incompatible Products

The first and more common purging operation helps to maintain product quality. As previously discussed, certain pipeline products cannot be mixed without creating problems. In some cases the mixture would dangerous, and in other cases, the mixture wouldn't have the required quality level. For this reason, when the pipeline schedule calls for two incompatible products to be run through the system in sequence, control center personnel must include a purging procedure between products, preventing quality or safety issues.

The procedure itself is fairly straightforward. Once the first product has run through the pipeline and delivery has finished, the control center switches valves so that the pipeline empties into a storage tank known as a **transmix** tank. A certain amount of the new product, called a **slug**, then runs through the system, displacing all of the remaining original product. The resulting mixture empties into the transmix tank. This unusable product is either stored for eventual removal or separated back into the original two products and recycled. Once the purge has completed, the second product delivery can begin.

1.4.2 Purging for Maintenance

Maintenance procedures call for particular care if product has been in the system. When a pipeline or tank is drained for maintenance, product residue combines with air to form a flammable, and possibly explosive, mixture. Maintenance procedures cannot be performed in this kind of environment, or the mixture could ignite.

Before beginning maintenance, the hazardous mixture must be forced out of the pipeline. Typically, an inert (non-reactive) gas performs the purging process. This volume of gas is also called a slug, but should not be confused with the kind used for purging between incompatible products. Nitrogen gas is the most common purge gas since it is both inert and inexpensive. Because it doesn't support combustion, pipeline mechanical personnel often use it when the pipeline must be cut.

In some cases, a second purge operation must be carried out before refilling the pipeline with product. This would be required only if it would create a problem for the product to come in contact with air.

1.4.3 Monitoring Purging Operations

When purging between incompatible products, a number of pipeline monitoring tools help make the task easier. The key to keeping product quality high and wasted product low is detecting the **interface** between each product batch.

A **gravitometer** measures the product's specific gravity as it passes through the pipeline. Calculations convert this value to API gravity. In some cases, adjacent batches will have significantly different gravities, making detecting the interface relatively simple. The sudden change in gravity signals the transition.

Not all products have different gravities. Some have gravities that overlap or are very similar. Dyed batches of fuel may have identical gravities and differ only by color, making a gravitometer useless. An **optical interface detector (OID)** solves this problem. It passes a beam of light through the product and analyzes the way in which the liquid absorbs certain wavelengths. The result is a kind of product fingerprint. Many pipelines use a combined gravitometer and OID since the pair is much more versatile than either device alone. Consequently, interface identification and detection becomes much easier and more reliable.

All pipeline purges, whether done to maintain product quality or to create a safe maintenance environment, must follow company procedures. The control center typically coordinates and monitors the operation and notifies all involved personnel.

In the case of maintenance, all appropriate records must be completed and all permits, drawings, work orders, and records of the operation must be stored. If any system changes resulted from the maintenance activity, SCADA screens should be updated to reflect them.

1.5.0 Pipeline Surveillance

Thanks to automation, most modern pipeline facilities are unstaffed. Aside from maintenance visits or emergencies, they may operate for weeks without human attention. Similarly, the pipeline itself may run for many miles through empty countryside. For these reasons, control center personnel need the means to keep an eye on remote facilities. A number of different monitoring and surveillance techniques have become popular for this purpose.

1.5.1 Video Monitoring

Unstaffed facilities have the advantage of costing less to operate since they don't require regular human attendance. But the ability to visually inspect equipment can be important. A status indicator on a SCADA display can tell the operator that a valve has failed to open, but a visual inspection may be required to determine the cause.

Similarly, an unstaffed facility can be a magnet for intruders, thieves, and vandals. While security systems can alert the control center that a sensor has triggered, they can't tell the difference between an animal, a person, and a simple malfunction; only visual surveillance can accomplish this.

On-site video cameras provide this needed service. Mounted at key locations, they deliver a steady stream of visual information to the control center. Modern digital video systems (*Figure 18*) are greatly superior to older analog closed circuit televisions (CCTVs). They send their data over a network in a compressed digital format. The quality is much higher, and the cameras can be remotely commanded to pan and zoom. These capabilities make it much more feasible to examine problem equipment from the comfort of the control room.

Digital video cameras greatly improve security as well. Using infrared illumination, they can monitor dark locations. Even more importantly, they store their data digitally on a server rather than on tape, as was the case with older systems. The result is more data and higher-quality information. Modern video systems are often integrated into security packages that automatically monitor a scene for sudden changes, which could indicate an intruder moving around.

Figure 18 A digital video surveillance camera.

1.5.2 Aerial Surveillance

Aerial surveillance is another technique often used to examine the many miles of pipeline. Small aircraft can fly its length, with the pilot or an observer watching for abnormal conditions. Control center personnel should make it a priority to act on the results of aerial surveys. In some cases, immediate action may be necessary. In other cases, scheduled repair may be acceptable.

1.5.3 Monitoring Weather Conditions

Because inclement weather can impact the pipeline, operations personnel must monitor weather conditions regularly. Electrical storms, tornados, violent thunderstorms, and blizzards create unique hazards. Facilities located in hurricane-prone areas, such as the Gulf Coast, must be especially weather-conscious.

Depending on the particular weather event, facilities may have to be shut down or have their operations altered. Personnel may not be able to work outside safely, and some facilities may not be appropriate shelters in severe weather situations. Many pipeline functions depend on moderate conditions. High winds, rain, snow, ice, and very cold temperatures could seriously hinder some operations.

WARNING:

Always seek shelter during severe weather.

During a weather event that spans several days, such as a hurricane, personnel need to monitor conditions constantly in order to make appropriate decisions. Hurricanes can change their course, strengthen, or weaken. Each change could influence a decision to continue pipeline operations or to shut them down.

To provide the information needed to make these decisions, facilities use a variety of weather-monitoring tools. Weather websites, such as those operated by the National Oceanic and Atmospheric Administration (NOAA), provide information about nearly any location. Many control centers display the local weather channel for each area that the pipeline passes through. In this way, control personnel are made aware not only of weather conditions in their location, but are also informed of conditions that will affect remote facilities and equipment.

Some installations have small on-site weather stations (*Figure 19*) so local weather conditions may be monitored directly. These stations typically monitor the following weather parameters:

- Outside temperature
- Inside temperature
- Wind direction and speed (with high wind alarm)
- Wind chill factor (with low wind chill alarm)
- Daily rainfall
- Barometric pressure
- Humidity (inside, with alarm)

Drone Surveillance

In the near future, it's possible that aerial surveillance may be taken over by unmanned drones rather than traditional aircraft. Piloted from the ground, drones can fly over the pipeline and deliver a steady stream of images from their onboard camera. In many ways, they are well suited for this task since they are small, light, and inexpensive.

However, drone technology will have to improve before this method can become feasible. Licensing of drone pilots and relevant federal regulations are still open questions. But don't be surprised if one day you find your aerial pipeline surveillance being done by a drone.

Figure 19 An industrial weather station.

1.5.4 Operation During Weather Events

Given the importance of reliable pipeline operation during weather events, some companies maintain backup control facilities with a backup SCADA system. If weather forces personnel to evacuate the regular control center, they can relocate to the backup site. In locations where this scenario is common, some companies have mobile control centers built into large vans. Personnel continue to manage the pipeline as the van takes them to safety or to an alternative control center.

NOTE: If a company maintains backup or mobile control centers, they must be tested regularly to ensure that they will be reliable during an emergency. Personnel must practice switching over to the alternative control center to ensure that they can do it smoothly and reliably during a real weather event.

Additional Resources

It is assumed that you will learn to perform this procedure under the direct supervision of your crew leader, foreman, or instructor. For independent study, refer to the recommended Additional Resources at the back of this module.

1.0.0 Section Review

1. Which SCADA control room component allows personnel to interact with the pipeline?
   a. A PLC
   b. An HMI
   c. An RTU
   d. An IPX

2. Field devices might connect to RTUs and PLCs through a(n) _____.
   a. fieldbus
   b. APU
   c. high-level network
   d. API

3. The CPM system identifies line breaks by detecting _____.
   a. product vapors
   b. product type
   c. changes in product API gravity
   d. negative pressure waves

4. By law, combustible gas/vapor detectors must be set to alarm at no higher than what percent of the lower explosive limit (LEL)?
   a. 20%
   b. 25%
   c. 30%
   d. 35%

5. There are two types of purging operations; one may be done as a safety prelude to maintenance, and the other may be done during normal operations to _____.
   a. maintain product quality between batches of incompatible product
   b. prepare the pipe for pigging
   c. prepare a valve for inspection
   d. adjust product API gravity

6. Video surveillance enables the control center to examine parts of the system visually and to _____.
   a. monitor remote pressure gauges
   b. detect intruders and vandals
   c. observe local SCADA panels
   d. detect pipeline motion

SECTION TWO

2.0.0 MONITORING AND MAINTAINING THE LIQUID PIPELINE

Objective

Describe the requirements and procedures for monitoring and maintaining the liquid pipeline (CT64_3-17).

a. Identify the prerequisites, competencies, and abnormal operating conditions associated with monitoring and maintaining the liquid pipeline.
b. Describe the procedures used to monitor and maintain the liquid pipeline.

Performance Task

1. Monitor pressures, flows, communications, and line integrity and maintain them within allowable limits on a liquid pipeline system (CT64_3-17).
   - Identify potential abnormal operating conditions that may occur during performance of this CT, and know the appropriate actions to take in response to them.
   - Utilize the appropriate personal protective equipment according to relevant company procedures.
   - Be familiar with the normal operating pressures, flows, and linefill for the pipeline under similar conditions. Verify that the flows, pressures, and product characteristics are within normal range.
   - Continuously monitor, by whatever means available (SCADA, pressure gauge, meter), the pressures and flow rates of the pipeline to look for any changes. Monitor the line balance for changes by monitoring volume in and volume out for liquid pipelines and line pack in gas pipelines. Monitor storage facilities. Verify continuous communications and that data is accurate and updated.
   - If pressures, flows, and/or line balance are not within normal range, determine the cause of variation from normal and correct the situation if possible. For communications variations, maintain monitored operations by using alternative communication methods. If necessary, seek assistance from and make notifications to proper personnel.
   - Perform proper operation according to company policy and procedures.
   - Verify that the desired result was achieved.
   - Complete the appropriate documentation as required by operator's procedures.

Trade Terms

Packed: A condition in which the pipeline is kept under pressure so its contents do not vaporize.

While control room personnel are responsible for a variety of tasks, much of the day-to-day activities revolve around monitoring operations. Monitoring is essential in order to maintain the pipeline in a stable condition. Pipeline control personnel perform many different tasks to achieve steady state. Ordinary operations are intermixed with alerts, alarms, and abnormal operating conditions (AOCs). Since each day is unique, there is no particular sequence to the individual tasks involved in monitoring and maintaining a pipeline.
However, each of these tasks has very specific procedures and a definite order. AOCs, in particular, have very specific guidelines and procedures associated with them. Pipeline control personnel must be both highly flexible and highly trained to achieve the goal of a reliable and stable pipeline.

2.1.0 Prerequisites, Competencies, and Abnormal Operating Conditions

Control room personnel must possess a variety of skills in order to successfully monitor and maintain a liquid pipeline. They must also be aware of the particular AOCs that may occur as they carry out these tasks.

2.1.1 Prerequisites/Competencies

Control room personnel must possess the following basic knowledge and skills prior to monitoring and maintaining a liquid pipeline:

- Be familiar with SCADA displays and know how to interpret their informational pages as well as their alarm and alert pages.
- Know how to use the SCADA system to change or adjust pipeline settings.

- Understand how to interpret the leak detection system displays (both CPM and non-CPM) and use them to localize a leak.
- Know the maximum discharge pressure for each station as well as the normal range of pressure and flow values for the pipeline (associated with a steady state condition). Know the MOP values for key points in the pipeline if this information is available.
- Have a basic understanding of pipeline hydraulics and the factors that influence it.
- Be thoroughly familiar with AOCs and emergencies, understand their entry conditions, and know the proper responses to each.
- Understand and be skilled at communicating both within the control room and with field personnel.
- Understand and be skilled at documenting all activities using the proper methods and formats.

2.1.2 Abnormal Operating Conditions (AOCs)

Control room personnel monitoring and maintaining a liquid pipeline must be able to identify the following potential abnormal operating conditions and know the proper response to each, should they occur. The following are potential AOCs, and the appropriate action the controller should take in response to them:

- *A safety or relief device activates in response to an abnormal or problematic condition* – The controller should identify the specific device involved as well as the conditions that caused it to activate. The controller must then take whatever actions are appropriate to deal with the situation and return the pipeline to a safe state. Relevant personnel must also be notified.
- *An alarm associated with a safety device activates* – The controller should identify the specific device involved as well as the conditions that caused it to activate. The controller must then take whatever actions are appropriate to deal with the situation and return the pipeline to a safe state. Relevant personnel must also be notified.
- *SCADA communications, control, or power fails* – The controller should activate backup systems and then determine what caused the original failure. The controller must take whatever actions are appropriate to deal with the situation and return the regular system to operation. Relevant personnel must also be notified.
- *An unexplained change in flow happens* – The controller should determine the cause of the change and correct it. Relevant personnel must also be notified.
- *An unexplained change in pressure happens, or pressure readings cease* – If either of these situations occurs, the controller should determine the cause of the change and correct it. Relevant personnel must also be notified.
- *An unexplained status change in a valve or other unit happens* – The controller should determine the status cause of the change and correct it. Relevant personnel must also be notified.
- *A tank liquid level is not within safe limits* – The controller should halt the current operation, determine the cause of the problem, and correct it. Relevant personnel must also be notified.

2.1.3 Training and Drilling

Given the high stakes associated with the pipeline, control center personnel's response to abnormal operating conditions cannot be left to chance. When an emergency occurs, every person in the control room must be able to be relied upon to carry out the right actions. For this reason, the guidelines for normal procedures differ from those that govern abnormal operating conditions.

Normal procedures may be consulted as required when personnel carry out a particular task. In many control rooms, procedure manuals are simply left out for consultation. Procedures for handling abnormal conditions cannot be handled in this way. Instead, personnel are required to memorize both the entry conditions that lead to the problem as well as the proper responses.

To ensure quick and accurate emergency handling, control room personnel must conduct regular training sessions and drills. According to 49 *CFR* 195.403, the results of these training sessions and drills must be documented.

Control center personnel must also be trained and assessed for skills in performing normal procedures that are less common. These include unusual use of the pipeline, unusual use of pigs, and special switching procedures.

2.2.0 Monitoring and Maintaining the Pipeline

Monitoring and maintaining the pipeline from the control room involves numerous responsibilities. While each of these has definite steps and guidelines, the procedures themselves will probably not be carried out in any particular order. Instead, most activities will be driven by the day's schedule or the needs of the moment, particularly those involving unusual or abnormal conditions, which are not predictable. In some cases, several tasks may have to run simultaneously.

2.2.1 Procedures

Because of the numerous responsibilities involved in monitoring a pipeline, and because personnel must be able to perform any combination of them in any order, the procedures outlined below are not numbered. Their order will depend on the control room situation:

> **CAUTION:** Whenever you are working in the control room, you must be alert to the possibility of abnormal operating conditions and be prepared to initiate the proper responses from memory.

- Continuously monitor the pipeline and confirm that both pressure and flow are in a steady state condition. It may be necessary to monitor the line as it's being packed. If so, confirm that the line eventually reaches the steady state condition. This may take time.
- Adjust pressure and flow limits to appropriate operating values using the SCADA system or other appropriate controls. Confirm that these values are set correctly since deviations could cause AOCs.
- Use the SCADA system to monitor key operating parameters. Monitor pressure, flow, alerts, alarms, trends, tank levels, and the line itself. Also confirm that SCADA communication is working correctly. Performing these steps will require a thorough understanding of the SCADA displays as well as the various controls used to call up the required information.
- Adjust pressure and flow setpoints to appropriate values using the SCADA system or other appropriate controls.
- Using the correct procedures and terminology, communicate with field personnel and/or shippers to coordinate their actions with the control room. Confirm that operations and changes are handled clearly and without confusion.
- Monitor the pipeline for leaks using whatever leak detection systems are available (SCADA, CPM, or direct monitors), and be prepared to take action should a leak occur.
- Monitor alarms through the SCADA system and any other alarm systems present. Know how to acknowledge an alarm and be able to execute the proper response to it.

> **NOTE:** Every facility has its own specific procedures for handling alarms. Confirm that you understand these and are able to carry them out quickly and accurately.

- Be able to document completed procedures, alarm handling, abnormal operating conditions, and any other specified event according to company procedures and standards.

Additional Resources

It is assumed that you will learn to perform this procedure under the direct supervision of your crew leader, foreman, or instructor. For independent study, refer to the recommended Additional Resources at the back of this module.

2.0.0 Section Review

1. To monitor a liquid pipeline, control room personnel must know the pipeline's _____.
   a. standard valve alignment
   b. purge timing
   c. normal range of pressure and flow rates
   d. compatible mixes
2. If an alarm associated with a safety device activates, the controller should first _____.
   a. identify the device involved
   b. properly log the alarm
   c. query SCADA for other problems
   d. alert higher management
3. Key operating parameters, such as pressure, flow, alerts, and tank levels, can be monitored using _____.
   a. automatic controllers
   b. the SCADA system
   c. manual valve actuators
   d. pneumatic systems

Summary

Keeping a pipeline running smoothly is the main task of the control center. But managing a pipeline that may be hundreds or even thousands of miles long requires very sophisticated tools to make it manageable from a single location. The Supervisory Control and Data Acquisition (SCADA) system accomplishes this task by using field devices along the pipeline and in various facilities to collect data, and assembling it using supervisory computers. These computers issue commands from the control room to any location along the pipeline.

Pipeline control operators spend much of their time monitoring the pipeline. Pressure and flow, two crucial pipeline parameters, indicate its basic condition. The computational pipeline management (CPM) system uses this information to create a hydraulic profile and to detect leaks and pipeline breaks. Other monitoring tools watch over pipeline safety, alerting operators to potential and current problems.

Avoiding mixes of incompatible products is another operator task, and requires careful scheduling and a process known as *purging* help achieve this goal. Another purging process can be used to create safe working conditions when personnel must maintain the empty pipeline.

Since most pipeline facilities are unstaffed, remote surveillance technologies are becoming increasingly important. Video systems typically supplement conventional security sensors and monitors. Surveillance tools also help personnel monitor unstaffed facilities and sections of the pipeline that run through empty countryside. Weather-monitoring technologies enable control center personnel to make decisions when dangerous weather threatens.

Pipeline control operators must perform many different procedures in order to maintain the pipeline in a steady state condition. While these tasks have no particular order, they do have specific steps that must be performed in a definite order. Rather, procedures involved in monitoring a pipeline are driven by daily needs and the unpredictable issues that arise. Operators need to know procedures thoroughly, particularly those associated with abnormal operating conditions.

Review Questions

1. SCADA systems provide for pipeline monitoring and _____.
   a. response
   b. alerts
   c. control
   d. emergencies

2. A device that monitors the pipeline and reports to a SCADA supervisory computer is a(n) _____.
   a. RTU
   b. PROFIBUS
   c. PCMCIA
   d. ECC

3. Another name for an HMI is a(n) _____.
   a. HIA
   b. MMI
   c. UI
   d. HCI

4. Abnormal pressure and flow could indicate an impending shutdown or a(n) _____.
   a. product release
   b. communications test cycle
   c. API gravity problem
   d. MOP reset

5. Control room personnel deal with more than alarms and alerts. They also have significant responsibilities involving pipeline _____.
   a. programming
   b. valve maintenance
   c. PLC diagnostics
   d. management

6. CPM can detect line breaks and leaks by monitoring for negative pressure waves or through _____.
   a. RTU inconsistencies
   b. line balance calculations
   c. detecting product in the surrounding soil
   d. alerts from remote stations

7. Gases that are considered "safe" (not toxic to the body), but create a hazardous environment by displacing oxygen are called _____.
   a. asphyxiants
   b. sulfides
   c. combustible gases
   d. vapors

8. A relief device that acts when a controller can't operate a valve for some reason is a(n) _____.
   a. cyclic relief device
   b. re-entrant relief device
   c. bypass relief device
   d. level relief device

9. Gasoline cannot be mixed with fuel oil because it _____.
   a. damages valve seats
   b. reduces product quality
   c. explodes upon contact
   d. lowers the flash point

10. Modern video surveillance systems send their data _____.
    a. in RS-170A format
    b. over a network
    c. in SECAM format
    d. using analog techniques

11. The main tool that a pipeline control operator uses to obtain informational pages and to control the pipeline is the _____.
    a. SCADA system
    b. fieldbus network
    c. main PLC
    d. CPM system

12. If the SCADA system fails, the pipeline control operator must _____.
    a. send a fieldbus reset command
    b. reset the supervisory computer
    c. activate the backup system
    d. open all block valves

13. Control center personnel must memorize abnormal operating condition handling procedures, as well as their _____.
    a. economic consequences
    b. alternate names
    c. API numbers
    d. entry conditions

14. The many different tasks involved in monitoring and maintaining the pipeline _____.
    a. must be carried out in a specific sequence
    b. must be memorized by pipeline control personnel
    c. are driven by the day's schedule and the needs of the moment
    d. are carried out in API- and DOT-specified sequence

15. Pipeline control personnel must monitor the pipeline for leaks using SCADA, direct monitors, or _____.
    a. CPM
    b. RTUs
    c. PLCs
    d. 49 *CFR*

Trade Terms Quiz

Fill in the blank with the correct term that you learned from your study of this module.

1. A pipeline instrument that determines the product's specific gravity is called a(n) _________.
2. A control system used in the pipeline industry that collects and assembles data and also allows operators to control the pipeline remotely is called a(n) _________.
3. The process performed when changing between incompatible products in a pipeline that prevents their mixing is known as _________.
4. The lowest gas/vapor concentration in air that could explode if an ignition source were present is the _________.
5. A display station attached to the SCADA system that allows users to monitor and/or interact with the pipeline is known as a(n) _________.
6. The highest pressure at which a liquid pipeline may be operated is the _________.
7. A large computer in a SCADA system that collects data from and issues commands to field devices such as RTUs and PLCs is a(n) _________.
8. Conditions within a pipeline system that may indicate (1) a malfunction of a component or (2) a condition exceeding design limits or resulting in a hazard to persons, property, or the environment are called _________.
9. A compact industrial computer largely used for machine control and other dedicated tasks requiring intelligence is called a(n) _________.
10. The region between batches of dissimilar product in which both products combine to form a transmix is called the _________.
11. The condition in which pipeline variables are relatively constant over a period of time is called _________.
12. A simple digital communications network with a set of communications standards designed to connect instruments, controllers, and other intelligent devices together is called a(n) _________.
13. A quantity of product that contaminates another incompatible product is a(n) _________.
14. Devices that watch a process parameter and generate informational reports or alarms based on these parameters are _________.
15. An alarm circuit connected to several sensors which will not alarm unless a majority of the sensors activate is called a(n) _________.
16. A computer-like device that connects field devices to the supervisory computer in a SCADA system is called a(n) _________.
17. A pipeline instrument that uses light absorption to detect the interface between batches of dissimilar product is called a(n) _________.
18. The condition in which the pipeline is kept under pressure so its contents do not vaporize is known as _________.
19. An unusable mixture of two different products, usually created by switching from one incompatible product to another, is referred to as _________.

Trade Terms

Abnormal Operating Conditions (AOCs)
Actuator
Fieldbus
Gravitometer
Human-machine interface (HMI)
Interface
Lower explosive limit (LEL)
Maximum operating pressure (MOP)
Monitors
Optical interface detector (OID)
Packed
Programmable logic controller (PLC)
Purging
Remote terminal unit (RTU)
Supervisory Control and Data Acquisition (SCADA)
Slug
Steady state
Supervisory computer
Transmix
Voting circuit

Trade Terms Introduced in This Module

Abnormal operating conditions (AOCs): Conditions within a pipeline system, identified by the operator, that may indicate (1) a malfunction of a component or (2) a deviation from normal operations that may indicate a condition exceeding design limits or result in a hazard to persons, property, or the environment.

Actuator: A powered mechanism that operates a valve.

Fieldbus: A simple digital communications network with a set of communications standards designed to connect instruments, controllers, and other intelligent devices together.

Gravitometer: A pipeline instrument that determines the product's specific gravity.

Human-machine interface (HMI): A display station attached to the SCADA system that allows users to monitor and/or interact with the pipeline. Also known as a *Man-Machine Interface (MMI)*.

Interface: The region in between batches of dissimilar products where both products combine to form a transmix.

Lower explosive limit (LEL): The lowest gas/vapor concentration in air that could explode if an ignition source were present.

Maximum operating pressure (MOP): The pipeline or a component's highest normal operating pressure.

Monitors: Devices that watch a process parameter, such as temperature, flash point, or sediment and water, and generate informational reports or alarms based on these parameters.

Optical interface detector (OID): A pipeline instrument that uses light absorption to detect the interface between batches of dissimilar product.

Packed: A condition in which the pipeline is kept under pressure so its contents do not vaporize.

Programmable logic controller (PLC): A compact industrial computer largely used for machine control and other dedicated tasks requiring intelligence.

Purging: The process performed when changing between incompatible products in a pipeline that prevents their mixing. Also refers to a maintenance procedure in which flammable vapors are forced out of the pipeline by an inert gas.

Remote terminal unit (RTU): A computer-like device that connects field devices to the supervisory computer in a SCADA system; RTU can also stand for remote telemetry unit or remote telecontrol unit.

Supervisory Control and Data Acquisition (SCADA): A control system used in the pipeline industry that collects and assembles data, and also allows operators to control the pipeline remotely.

Slug: Refers to a quantity of product that contaminates another incompatible product. Can also refer to a volume of inert gas injected into a pipeline during a maintenance purge.

Steady state: The condition in which pipeline variables, such as pressure and flow, are relatively constant over a period of time.

Supervisory computer: A large computer in a SCADA system that collects data from and issues commands to field devices such as RTUs and PLCs.

Transmix: An unusable mixture of two different products usually created by switching from one incompatible product to another (stands for "transportation mixture").

Voting circuit: An alarm circuit connected to several sensors. It will not alarm unless a majority of the sensors activate. Also known as a *majority circuit*.

Additional Resources

The following recommended resources can provide additional helpful information related to the requirements for this covered task:

NCCER Module AOCCC-17, *Abnormal Operating Conditions Control Center*
NCCER Module AOCFG-17, *Abnormal Operating Conditions Field and Gas*
NCCER Module 64205-02, *Supervisory Control Systems*
NCCER Module CT55_0-17, *Maintain Fixed Gas Detection Equipment (Gas)*
NCCER Module CT23_1-17, *Maintain/Repair Relief Valves*
NCCER Module CT24_1-17, *Inspect, Test, and Calibrate Pressure Limiting Devices*
NCCER Module CT44_3-16, *Inspect, Test, and Maintain a Liquid Pipeline Leak Detection Flow Computer*
The following websites offer resources for products and training:
- American Petroleum Institute, **www.api.org**
- Automation.com, **www.automation.com**
- National Oceanic and Atmospheric Administration, **www.noaa.gov**
- National Hurricane Center, **www.nhc.noaa.gov**
- OMNI Flow Computers, Inc., **www.omniflow.com**
- Pipeline and Hazardous Materials Safety Administration, **www.phmsa.dot.gov**
- Yokogawa Electric Corporation, **www.yokogawa.com**

Figure Credits

ABB, Module Opener, Figure 6
© Michael Clime/ Dreamstime.com, Module Opener
Photos courtesy of Entergy New Orleans, Inc. © 2016 All Rights Reserved, Figure 1
Yokogawa Corporation of America, Figure 3
Automationdirect.com, Figure 4
Unitronics, Inc., Figure 5
Emerson Process Management, Figure 8
Tim Mills, Figure 9
OMNI Flow Computers, Inc., Figure 10
Courtesy of Honeywell International Inc., Figure 12
General Monitors, Figures 13–16
© Yudesign/Dreamstime.com, Figure 17
© Kittichai Kasemsarn/Dreamstime.com, Figure 18
© Patrik Dietrich/Dreamstime.com, Figure 19
© Sasidis Sasisakulporn/Dreamstime.com, SA01

Section Review Answer Key

Answer	Section Reference	Objective
Section One		
1. b	1.1.5	1a
2. a	1.1.6	1a
3. d	1.2.5	1b
4. b	1.3.1	1c
5. a	1.4.1	1d
6. b	1.5.1	1e
Section Two		
1. c	2.1.1	2a
2. a	2.1.2	2a
3. b	2.2.1	2b

NCCER CURRICULA — USER UPDATE

NCCER makes every effort to keep its textbooks up-to-date and free of technical errors. We appreciate your help in this process. If you find an error, a typographical mistake, or an inaccuracy in NCCER's curricula, please fill out this form (or a photocopy), or complete the online form at **www.nccer.org/olf**. Be sure to include the exact module ID number, page number, a detailed description, and your recommended correction. Your input will be brought to the attention of the Authoring Team. Thank you for your assistance.

Instructors – If you have an idea for improving this textbook, or have found that additional materials were necessary to teach this module effectively, please let us know so that we may present your suggestions to the Authoring Team.

NCCER Product Development and Revision

13614 Progress Blvd., Alachua, FL 32615

Email: curriculum@nccer.org
Online: www.nccer.org/olf

❑ Trainee Guide ❑ Lesson Plans ❑ Exam ❑ PowerPoints Other ______________________

Craft / Level: Copyright Date:

Module ID Number / Title:

Section Number(s):

Description:

Recommended Correction:

Your Name:

Address:

Email: Phone:

CT64_4-17

Remotely Operate Valves on a Liquid Pipeline System

OVERVIEW

The control center's main purpose is moving product through the system efficiently, economically, and safely. Control center personnel must use their training, experience, and judgment to make good decisions that support this goal. This module describes the fundamental task of operating valves by remote control from the control center. While this task seems simple, it carries significant responsibility. Furthermore, modern pipeline control systems are complex. Control room personnel must know how to obtain valve status information quickly so they can make informed decisions.

Trainees with successful module completions may be eligible for credentialing through the NCCER Registry. To learn more, go to **www.nccer.org** or contact us at 1.888.622.3720. Our website has information on the latest product releases and training, as well as online versions of our *Cornerstone* magazine and Pearson's product catalog.

Your feedback is welcome. You may email your comments to **curriculum@nccer.org**, send general comments and inquiries to **info@nccer.org**, or fill in the User Update form at the back of this module.

This information is general in nature and intended for training purposes only. Actual performance of activities described in this manual requires compliance with all applicable operating, service, maintenance, and safety procedures under the direction of qualified personnel. References in this manual to patented or proprietary devices do not constitute a recommendation of their use.

CT64_4-17
Remotely Operate Valves on a Liquid Pipeline System

Objectives

When you have completed this module, you will be able to do the following:

1. Describe key liquid pipeline components and their operations.
   a. Describe manifolds and the valve operations associated with them.
   b. Describe pumps and the operations associated with them.
   c. Describe SCADA technology and its role in valve control.
2. Describe the requirements and procedures for operating valves remotely (CT64_4-17).
   a. Identify the prerequisites, competencies, and abnormal operating conditions associated with remote valve operation.
   b. Describe the procedures used to operate valves remotely.

Performance Task

Under the supervision of your instructor, you should be able to do the following:

1. Remotely operate valves on a liquid pipeline (CT64_4-17).
   - Identify potential abnormal operating conditions that may occur during performance of this CT, and know the appropriate actions to take in response to them.
   - Utilize the appropriate personal protective equipment according to relevant company procedures.
   - Verify that the flows, pressures, and product characteristics are within normal range.
   - Identify the need to open or close a valve or start or stop a unit to control flow or pressure and to maintain safe operation of the pipeline system.
   - Identify the applicable valve, unit, or the appropriate control device.
   - Perform proper operation according to company policy and procedures.
   - Verify that the desired result was achieved.
   - Complete the appropriate documentation as required by operator's procedures.

Trade Terms

Abnormal operating conditions (AOCs)
Actuator
Encoder
Gravitometer
Interface
Line fill
Maximum operating pressure (MOP)
Optical interface detector (OID)
Programmable logic controller (PLC)
Remote terminal unit (RTU)
Supervisory Control and Data Acquisition (SCADA)
Steady state
Transmix
Valve interlock
Variable frequency drive (VFD)

Industry Recognized Credentials

If you are training through an NCCER-accredited sponsor, you may be eligible for credentials from NCCER's Registry. The ID number for this module is CT64_4-17. Note that this module may have been used in other NCCER curricula and may apply to other level completions. Contact NCCER's Registry at 888.622.3720 or go to **www.nccer.org** for more information.

Contents

Figures and Tables

Section One

1.0.0 Liquid Pipeline Components

Objective

Describe key liquid pipeline components and their operations.

a. Describe manifolds and the valve operations associated with them.
b. Describe pumps and the operations associated with them.
c. Describe SCADA technology and its role in valve control.

Trade Terms

Actuator: A powered mechanism that operates a valve.

Encoder: A device that represents a physical quantity, such as position, as a numeric or electrical value that a controller can use for feedback.

Gravitometer: A pipeline instrument that determines the product's specific gravity.

Interface: The region between batches of dissimilar products in which both products combine to form a transmix.

Line fill: The fluid volume required to fill a pipeline section.

Maximum operating pressure (MOP): The highest pressure at which a liquid pipeline may be operated.

Optical interface detector (OID): A pipeline instrument that uses light absorption to detect the interface between batches of dissimilar product.

Programmable logic controller (PLC): A compact industrial computer largely used for machine control and other dedicated tasks requiring intelligence.

Remote terminal unit (RTU): A computer-like device that connects field devices to the supervisory computer in a SCADA system; RTU can also stand for remote telemetry unit or remote telecontrol unit.

Supervisory Control and Data Acquisition (SCADA): A control system used in the pipeline industry that collects and assembles data and also allows operators to control the pipeline remotely.

Steady state: The condition in which pipeline variables, such as pressure and flow, are relatively constant over a period of time.

Transmix: An unusable mixture of two different products usually created by switching from one incompatible product to another (stands for "transportation mixture").

Valve interlock: A control system feature that prevents unsuitable valve alignments and ensures that valve sequencing occurs in the correct order.

Variable frequency drive (VFD): An electric motor paired with an electronic controller that adjusts motor speed by varying the power supply voltage and frequency.

Because pipelines and their associated facilities are now far more automated than they used to be, control room personnel have more responsibilities as a result. Their duties now include some tasks that used to be handled by field personnel. Their actions carry financial consequences as well; the careless contamination of product can be a costly mistake. Control center personnel must be thoroughly familiar with their responsibilities.

Operating a liquid pipeline involves interacting with and controlling many different pieces of equipment. Typical tasks like start-up, shutdown, and valve manipulation all involve many of these devices. For this reason, control center personnel must understand the functions of each piece of equipment, as well as its basic operation.

1.1.0 Manifolds and Valve Operations

Operations personnel control product movement by opening and closing terminal, station, and pipeline valves in the proper sequence. They may do this directly, or remotely through the Supervisory Control and Data Acquisition (SCADA) system. Common tasks associated with these operations include making tank switches, verifying valve alignments, and making pipeline switches.

1.1.1 Valve Types

The pipeline and its associated systems contain numerous types of valves. Many of these perform specialized tasks, and all work together to make the pipeline both efficient and safe. At the most basic level, valves are control devices, opening and closing to start, stop, or regulate flow.

Hand valves (*Figure 1*) must be operated manually. They are common on meter runs and tanks.

Figure 1 Hand-operated valve.

However, almost all valves can be operated by hand if necessary. Field personnel may have to use a portable power source to operate valves manually if they are very large.

Block valves are the main pipeline valves and can isolate pipeline segments as required. They operate in the fully open or fully closed positions only, and offer minimal flow resistance. Block valves play an important role when the control center purges a pump station. The block valve closes while pump suction and discharge remain open. This configuration prevents a slug of the previous product from contaminating the new one.

Pipeline personnel operate block valves either at the local control point or remotely from the control center. While any valve can act as a block valve, gate, plug, butterfly, and ball valves are most commonly used. A gate valve is shown in *Figure 2*.

Control valves (*Figure 3*) function in the fully open, fully closed, or partially open (throttle) positions. They are ideal for precision control, and are often used to manage pressure and flow. For example, control valves manage both the station

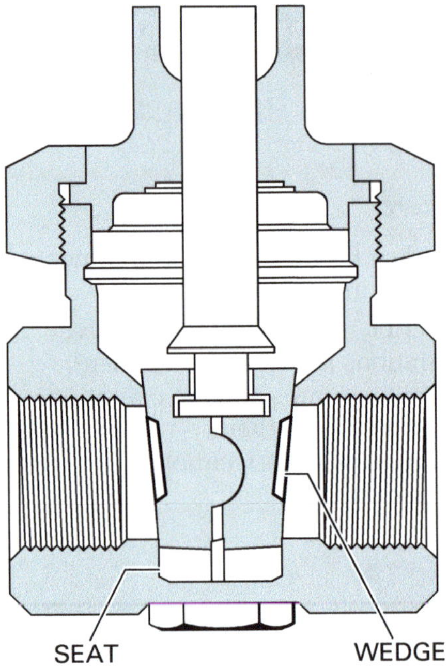

Figure 2 Gate valve.

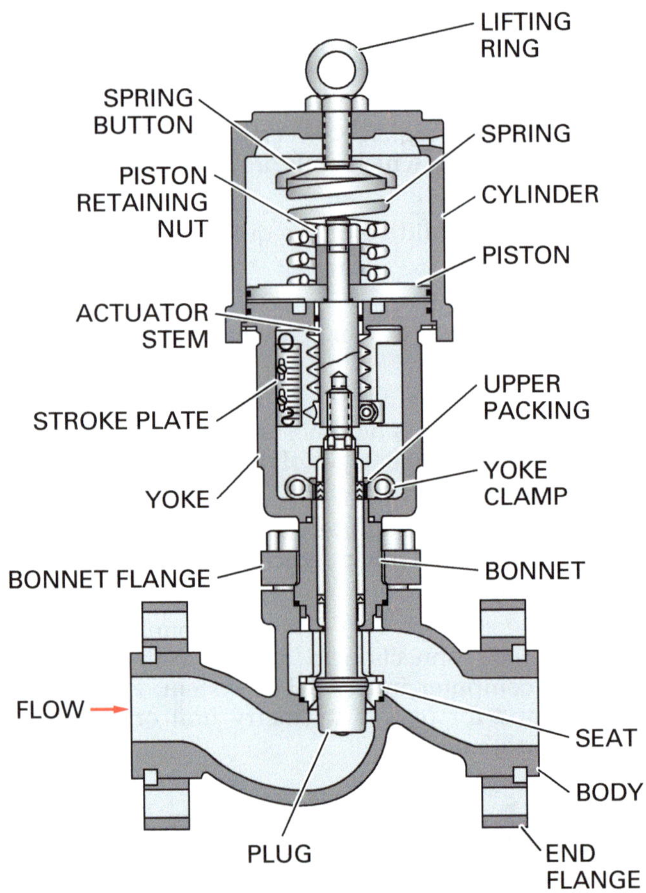

Figure 3 Control valve.

suction and station discharge ends of pumps. Their action protects both the upstream and downstream sides. In delivery facilities, they control the delivery rate. Globe and angle valves are the most common control valves, although butterfly valves are also used. They can be operated by hand, although most are managed remotely.

Control valves are usually smaller than the pipeline itself to provide some built-in throttling action. This reduces wear on the valve seat as well as the stem motion required to control the flow. Because partially open valves reduce pressure downstream, they must be used with care to avoid damaging other components, such as pumps. Constant-speed centrifugal pumps are often used with control valves.

Pipeline personnel use control valves in any situation requiring flow or pressure adjustment. These valves limit flow until conditions normalize in cases of low suction, during pipeline start-up, or after a main-line pump loss.

NOTE

Pipeline controllers may change some valve setpoints through the SCADA system. Other setpoints are dictated by engineering standards and design and may not be changed without authorization. However, pipeline controllers should know all setpoint values, regardless of whether or not they can alter them.

A backpressure valve is a special type of control valve that intentionally creates backpressure through throttling. The valve raises pipeline pressure upstream, preventing a slack line. Backpressure valves reduce pipeline vibration and protect low-pressure equipment, such as metering installations. Normally, these valves come preset to generate a specific amount of backpressure and are not remotely adjustable.

Pressure relief valves protect the pipeline during an overpressure condition by opening and releasing product into a breakout tank. Once the pressure has dropped back to safe levels, the valve closes. Relief valves use a number of different technologies to determine when to open. The most common type uses a spring whose force is equal to the pressure at which the device should open. Relief valves may also be built around electronic or hydraulic technologies.

Safety valves protect the pipeline by activating during abnormal conditions. For example, with electrical control systems, a power loss could cause serious problems. The safety system detects the flow loss and closes the valve to protect the manifold. This kind of valve normally appears in the line below the takeoff valve (the last valve that opens to start the shipment). Safety valves appear in other parts of the pipeline as well. In all cases, however, the safety valve's purpose is to place the pipeline in a safe condition.

1.1.2 Valve Control

Remote-control valves make the pipeline and its associated equipment much more manageable. Ideally, every valve in the system should be both controlled and monitored from the control room. While this is almost never true in practice, most important control valves will be controlled and monitored remotely.

Controlling a valve from another location requires three pieces of technology. The first is a power source for moving the valve. The second is a means of controlling that power source. Third, there must be some form of feedback so the control system will know the valve's current position or status.

An **actuator** is a powered device that operates a valve in response to commands from the control system. Actuators may be electrically, pneumatically, or hydraulically powered. They connect to the valve through a gear drive or other mechanical linkage. Electrical actuators (*Figure 4*) are particularly popular because they can be controlled over great distances through simple and inexpensive technology—wires.

Valves managed by powered actuators usually are controlled by a **programmable logic controller (PLC)** (*Figure 5*) or a **remote terminal unit (RTU)**. These are computer-like devices that communicate with the SCADA system. Control room personnel issue commands to valves through the SCADA system, which talks to the PLC or RTU. This approach allows control decisions to be made hundreds of miles away from the actual valves.

Figure 4 Electrical valve actuator.

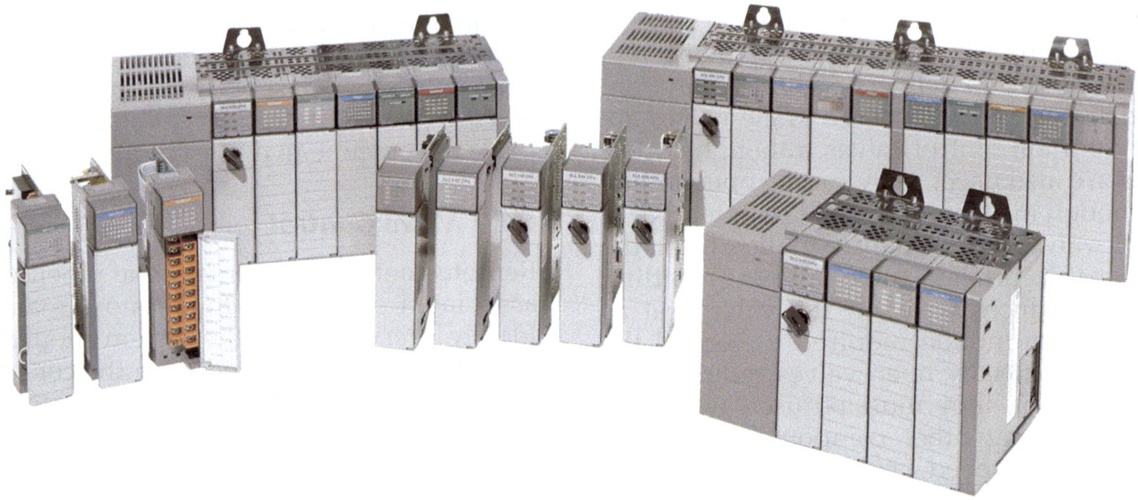

Figure 5 PLCs are often found in control systems.

Many valve actuators include built-in position sensors (*Figure 6*) that monitor the valve's location. For open/closed valves, such as block valves, the position sensor may be nothing more than a pair of switches that detect when the valve has reached its two possible positions. Control valves, which operate across a range of 0 to 100 percent, usually have a more sophisticated **encoder** that provides specific information about the valve's position within the range.

1.1.3 Interlocks

Control systems may include **valve interlock** logic designed to prevent dangerous or problematic valve combinations. For example, the control system's interlock programming might not allow a main-line valve to close until the corresponding pumps are shut down. Such automation prevents many careless accidents, protecting both the station and the pipeline.

Some companies use interlocks on booster stations to prevent the line from becoming blocked.

Figure 6 Actuator with built-in position sensor.

Interlock program logic monitors all major valves at the stations, including the control valves. If a valve begins to close unexpectedly, the control program will send an Open command to reopen the valve. If the control valve closes by more than 65 percent, the program will send the main-line block valve an Open command to ensure that the line is not blocked.

Unstaffed booster stations may have interlock logic that prevents the main-line block valve and the booster pump suction and discharge valves from closing at the same time. They may also have interlock logic designed to open the main-line block valve if the control valve closes by more than 65 percent.

Interlocks can also shut down the station in response to a line block command. This action requires less time than shutting down all units individually.

While the interlock system can fail, human error causes blocked lines more often. These failures occur most frequently with deliveries into staffed facilities. Regardless of the cause, the controller must shut down the line, following company procedures. Interlock failures should be handled by following the specified procedure. In most cases, the time to respond to and resolve overpressure situations is short. Control room personnel must be quick to recognize problems and know the correct response so they can act immediately.

1.1.4 Valve Alignment

Control center personnel must verify valve alignment for every shipment or change in operations. Once the scheduling department issues shipping orders, the control center sets up the procedures to fulfill those orders. This task includes determining the proper valve alignment needed to get the product from point A to point B safely and efficiently. Operators will look at piping layouts to pick the best path and prepare a list of valves to be opened or closed. They will then use the SCADA system to operate the valves in the correct sequence.

Most valve operations occur by remote control. Similarly, SCADA screens show valve status and indicate whether valves are open, closed, or traveling. Control center personnel must keep an eye on status indicators to confirm that everything is working. The operation may not begin until all valve positions have been confirmed. As already mentioned, many control systems include valve interlock logic that prevents inappropriate or unsafe valve combinations or sequencing.

If a valve does not respond correctly, it may be necessary to have field personnel go to the location and check the valve or operate it manually. Standard operating policies require control center personnel to verify that valves are properly set before any product enters the system. Some companies dispatch field personnel to check each valve individually, although this practice is becoming uncommon due to unstaffed facilities. Verification is important, especially when several companies are involved. Valves controlled by other companies, and therefore not visible through the SCADA system, must be confirmed by other means prior to delivery.

1.1.5 Flow Paths and Relief Devices

Before shipping can begin, controllers must confirm through the SCADA system that all valves are properly aligned for the procedure. These alignments provide what is known as the *valid flow path*. Without it in place, product will not move through the pipeline correctly.

Critical flow path, on the other hand, refers to pipes and valve alignments that protect the pipeline and support safe operation (*Figure 7*). These flow paths must remain open when the pipeline is operating. The critical flow path includes the valves and pipes that lead from the pipeline to breakout tanks or the connecting carrier. The pipeline cannot be started without a working critical flow path.

Many components in terminals and stations cannot handle main-line pressure and flow rates. Metering systems and manifolds, for example, can be damaged by excessive pressure. For this reason, pressure relief and pressure control devices must be present and correctly set to protect these vulnerable components.

Pressure relief valves (*Figure 8*) open when pressure becomes excessive, and release product into a breakout tank. Pressure control devices manipulate valves or divert flow to prevent pressure from rising to dangerous levels. Thermal relief devices are similar to pressure relief valves. They relieve pressure caused by thermal expansion—the tendency of the product to expand when it heats up.

Generally, the control center will not be able to change protective device setpoints. In some cases, these may be fixed by the manufacturer and cannot be changed. Other devices may have to be changed in the field either through a mechanical adjustment or by reprogramming a control device. The control center must know protective device setpoints, however, and be alert to pipeline pressure and flow conditions. Controllers must also monitor relief devices to be sure that they're performing properly.

1.1.6 Dealing with Incompatible Products

Many products cannot be mixed. The combination might be dangerous or simply fail to meet standard requirements. For example, gasoline mixing with a shipment of fuel oil is dangerous because it has a lower flash point than fuel oil should have. Mixing regular gasoline with a shipment of premium gasoline isn't dangerous, but the mixture doesn't have the required quality level. If ultra-low-sulfur diesel becomes contaminated with other grades of diesel, it violates the strict quality standards for diesel fuels. For these reasons, switching between incompatible products requires one or more special procedures.

Purging (also known as *line wash* or *line flush*) involves using another product to push out the remaining product still in the line or manifold. Operations personnel must manage the **line fill** between each manifold as well as other volumes of product residue remaining in equipment. In most installations, line fill is known or can be calculated for various valve alignments and delivery points. Control center personnel should use this information when purging. For example, if the line fill is 300 barrels (48 m^3), that amount must be removed and diverted when purging the line.

The line should always be purged with a product that is compatible with the upcoming shipment. Personnel must calculate the correct amount and purge the line, diverting the output into an appropriate tank. Then the second shipment can begin. Knowing or calculating line fills ensures that the line gets purged without using excess product.

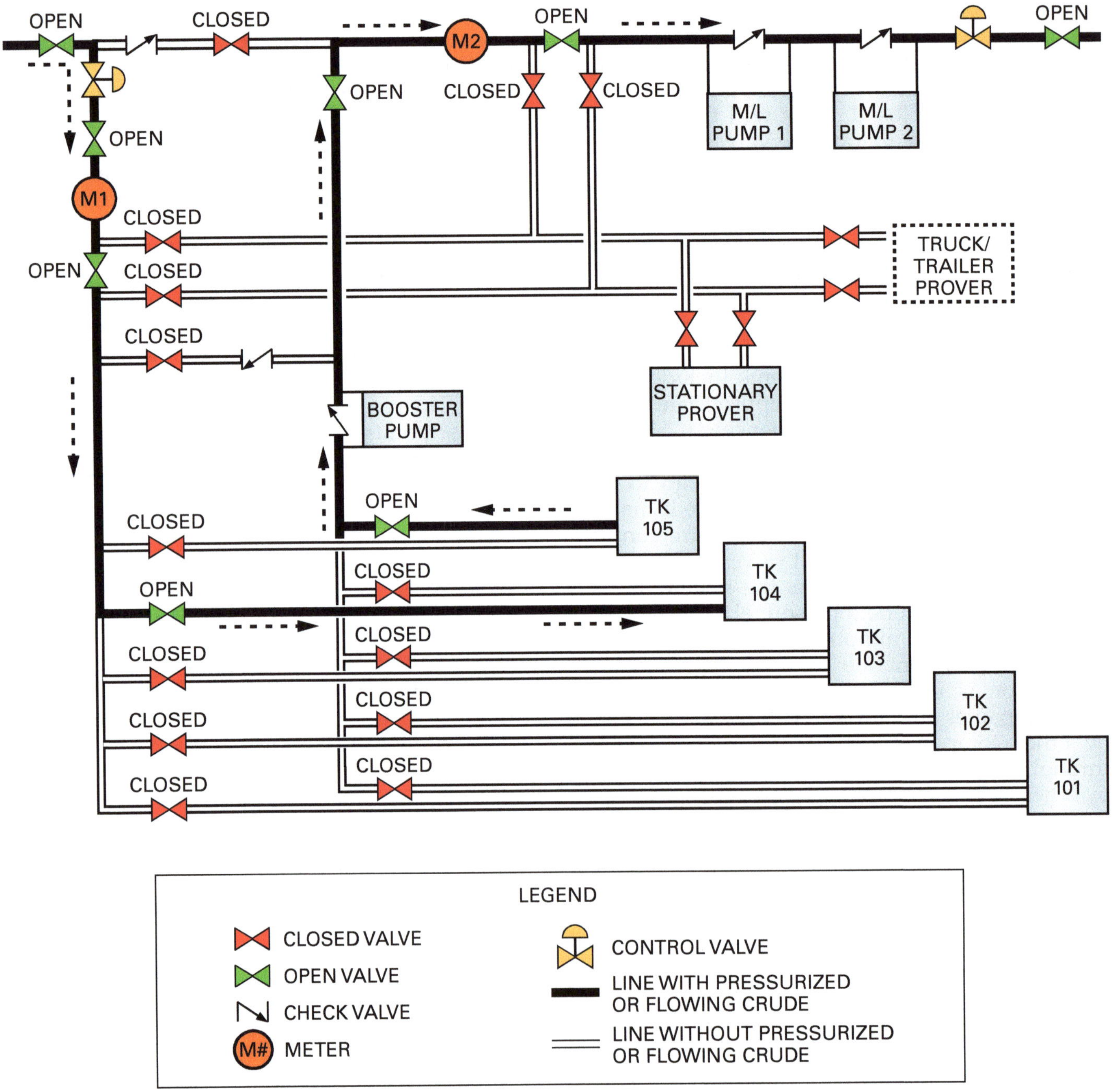

Figure 7 Critical flow path diagram.

Measuring the product's API gravity is one way to identify and track batches as they flow through the pipeline. Gravity testing also reveals changes in the product that may occur in shipment. For this reason, many pipelines have some type of **gravitometer** (*Figure 9*) installed upstream of the terminal. The SCADA system receives data from the unit and relays it to the control room.

NOTE

Gravitometers measure product density or specific gravity. Calculations then convert this information to API gravity. Gravitometers use one of several different methods to measure density. One method involves vibrating an element in the product stream and measuring the oscillation frequency. As the product becomes denser, the element oscillates more slowly. Gravitometers are usually checked periodically against laboratory density measurements.

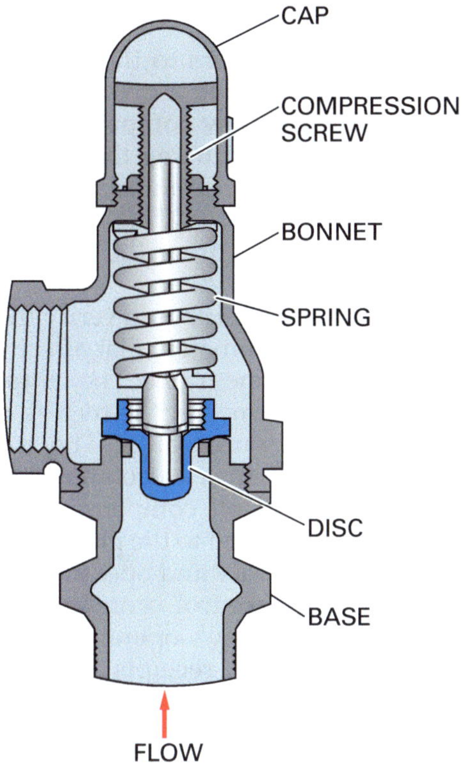

Figure 8 Pressure relief valve.

The SCADA system calculates, records, and displays all gravity information. Control center personnel monitor the API gravity patterns, and should investigate any variations. Significant variations must be documented and addressed according to company specifications.

API gravity measurements alone are not always useful for identifying the **interface** between batches. Some pipelines use a second instrument, called an **optical interface detector (OID)**, to supplement the gravitometer's readings. An OID passes light through the product and reads the absorption pattern. This method is especially useful with products having similar API gravity values.

Before a scheduled change arrives, control center personnel should begin monitoring the mainline stream and documenting the API gravity according to company policy. If the pipeline is not using an OID, the company-specified method for detecting the interface should be used. When the first break occurs, operations personnel should begin constant monitoring and perform the switch according to company policy.

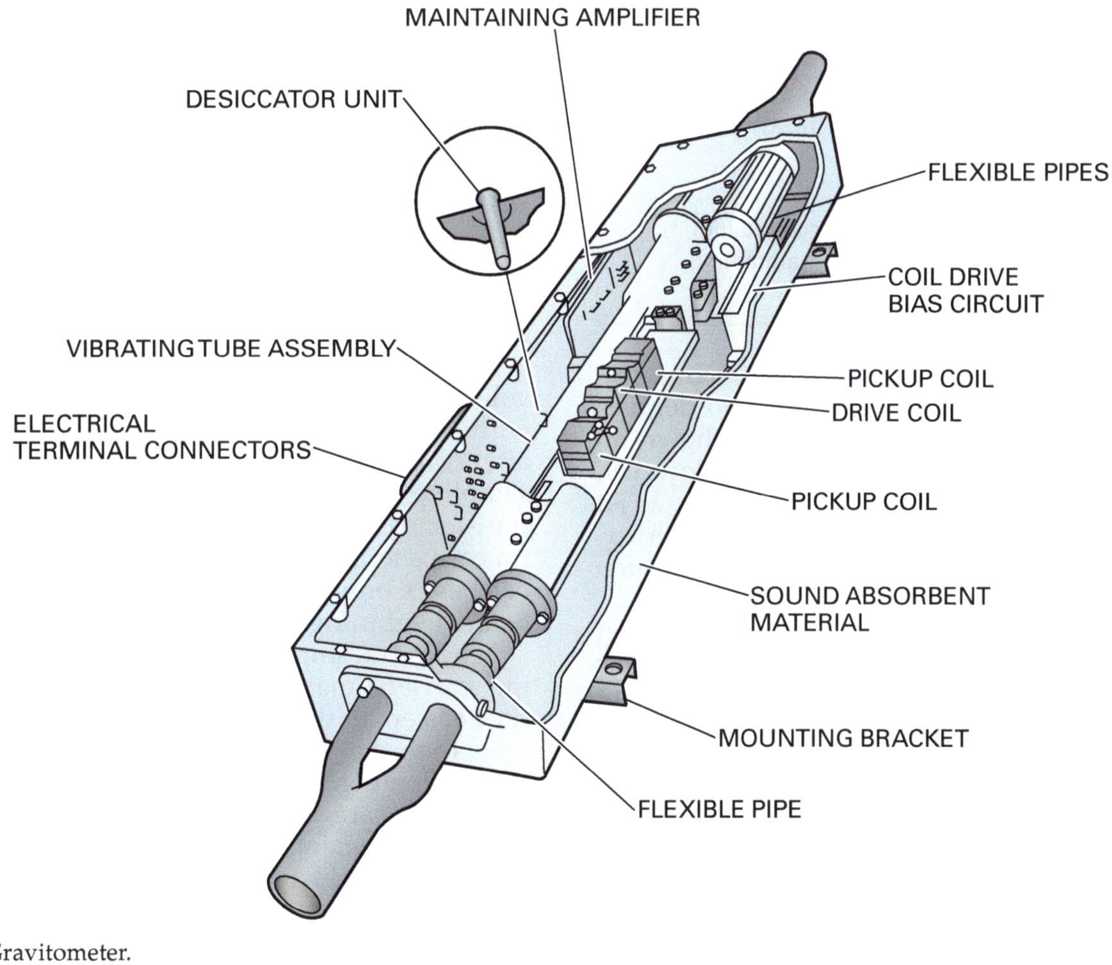

Figure 9 Gravitometer.

Some facilities rely on time calculations to decide when to switch. Using information like pumping rate and the volume to be delivered, they can determine when switches should occur. Some valve and tank switches are made based on the length of time that product flows through the pipe. This method is common in systems with hand valves. Personnel must perform manual switches as precisely as possible.

Because schedules may vary over long distances, operations personnel should supplement timing with measurements, such as API gravity, to confirm interfaces. Switches should occur at the correct gravity point regardless of time.

1.1.7 Batching Operations

Batching operations normally involve a sequence of different products. Control center personnel should be meticulous about documentation and handle every customer's receipts and deliveries accurately. Strict EPA standards, particularly those involving the sulfur content of diesel fuels, have led to many procedural guidelines and rules. Scheduling personnel set up batch sequences to minimize incompatible products being run one after the other. Code numbers and/or letters identify products with unique specifications. Charts may be set up to illustrate which products can be pumped directly in front of and behind other products. *Table 1* is an example of such a chart.

NOTE

In addition to being skillful at scheduling, controllers must be expert at maintaining product quality.

API gravity ranges vary somewhat in similar products. A deviation of one or two degrees API from the maximum or minimum shown on a chart is not considered significant. However, products with unique specifications (such as jet fuel) are handled more carefully in batch arrangements than are variations in products (such as fuel oils).

Testing for the interface and making a correct switch during an interface protects the product integrity. Personnel must sample and document each shipment's product as it arrives at the terminal. Doing so determines the switch point, at which time operations personnel open the pipeline valve, switching the product into the terminal. Transmix goes into separate tankage, and the pure product goes into storage.

1.1.8 Measuring Volume

The term *barrel count* refers to the total number of barrels measured during a given operation or time interval. SCADA flow computers (*Figure 10*) calculate the barrel count using the pipeline metering system. The barrel count may indicate the total volume flowing through the pipeline, the total volume of a shipment, or the volume being moved, blended, or injected.

Control center personnel should verify the totals, calculations, and all parameters that affect volume for each operation or time frame. The temperature and pressure are confirmed at the meter's manual thermometer and pressure gauge. The true total barrel count comes from a calculation involving the meter reading and all compensating factors. This value can be compared to the physical totals in tankage to verify inventory and operations.

Barrel counts give control center personnel a way to account for the day's operations, calculate the over-and-short, verify receipts and deliveries, and track operations. They also can act as a type of leak detection system, since significant discrepancies may indicate leakage.

1.1.9 Pipeline Switching

Various pipeline companies serve different geographical areas. Pipelines often run parallel to each other and sometimes cross (see *Figure 11*). Transferring product from one pipeline to another has numerous advantages to customers, so many companies have developed methods to make it possible. Control center personnel must understand the geography, facilities, and installations of other companies to schedule shipments between pipelines.

During transfers between companies, all involved control centers must closely coordinate with each other. Close communications are vital after confirming schedules. The same testing and sampling methods used for any regular custody transfer still apply. When changing carriers, extra care should be taken to verify and document batches. Confirm readiness with each pipeline before beginning the shipment.

Sending a shipment to a tank before changing carriers makes measurement and verification easier. If the scheduling department specifies that the shipment be tightlined into the other pipeline, control center personnel should take extra care with documenting specifications and measurements. (Tightlining is the process of moving a product from one pipeline to another

Table 1 Sample Batch Arrangement Chart

Grade	Type of Product	60°F API Gravity Range	Normal Pump Sequence Head End	Normal Pump Sequence Tail End
A	Premium Unleaded Gasoline 91/93 Undyed	45-75	N, M	N, M
B	Commercial Aviation Gasoline 100/103 Octane – Dyed Blue	66-72	N, M with HP-7 Buffer	N, M with HP-7 Buffer
C	Refinery Charging Stock Crude Oil	20-55	X5	X5
D	Ultra-Low-Sulfur Diesel Fuel Oil – Normally Undyed 47.5 Cetane Range	36-41	X, X5, Y	X, X5, Y
E	Denatured Fuel Ethanol		Not Pumped	Not pumped
F	#2 Fuel Oil Distillate Cenex Proprietary – Dyed Red	30–37	D, Y, Q, X, X5, N	D, Y, X, X5, N
G	Naptha		Special Delivery	Special Delivery
H	Normal Butane	0.580-0.588	N, L	N, L
I	Isobutane	0.560-0.570	H	H
J	Military Jet Fuel JP8 or J8	41-48	N, V	N, V, D, X
L	Propane	0.500-0.510	H	H
M	Mid-Grade Unleaded Gasoline 89 Octane Range	48-75	A, D, N, X	A, D, N, X
N	Regular Unleaded Gasoline 87 Octane Range	48-75	A, F, J, M, Q, V, X, X5, Y	A, F, J, M, Q, V, X, X5, Y
O	Water	1	Special Delivery	Special Delivery
P	Refinery Blendstock		A, N, X	A, N, X
Q	Commercial Jet Fuel	41-48	N, V	N, V, D, X
R	Leaded Racing Gasoline	60-80	N, V	N, V
T	Transmix Commercial Purchases	35-65	N, X, X5	N, X, X5
V	Sub-Octane Unleaded Gasoline, 84 Octane Range	48-75	F, J, M, N, Q, V, X, X5, Y	F, J, M, N, Q, V, X, X5, Y
W	Natural Gasoline Liquid	79-86	D, X, X5, N, H	D, X, X5, N, H
X5	#2 Fuel Oil High Sulfur (HSFO) Normally Dyed Red 0.50 sulfur maximum (The 5 in X5 may not be present when listed in specific grade)	30-40	D, F, N, V, X, Y	D, F, N, V, X, Y
Y	#1 Fuel Oil GrDistillate	40-45	D, F, N, V, X, X5	D, F, N, V, X, X5

Figure 10 Flow computer.

Figure 11 Abbreviated map of the main US pipelines.

without transferring it into a tank first.) If a question comes up, this information will be needed. Always document valve alignments and verification records so product movement can be easily traced.

1.1.10 Procedures

Each company has standard operating procedures for receipts, deliveries, and custody transfers. In order to manage these transfers efficiently, control center personnel should review scheduled shipments. They should understand the exact shipment type, specifications, time, and flow rate.

> **NOTE**
> Always know and follow your company's procedures for all product transfers.

When possible, prepare and document terminal piping and valves as far in advance of shipments as possible. Make appropriate plans for switching local line fill. Confirm that receiving equipment is operating properly and that there is sufficient tankage, if needed. If tankage is a problem, contact the scheduling and dispatching departments immediately.

Although consecutive batches may be compatible, the characteristics or additives of one batch may be different from others. Because delivery points may be different for each batch, switching will also be different. Operations personnel must plan the switch so each batch reaches the appropriate destination. The following tools may help with this planning:

- Company schedule
- Company product change report form
- API gravity conversion chart
- Software batch cut programs

To signal the change between two batches when there is little or no difference between them, some companies inject a dye flag. If the facility is using an OID, it can detect these batches easily, making switching simple. If the pipeline is using a gravitometer only, the batch is not switched when the dye flag arrives. Rather, the switch occurs at the mid-point of the gravity change. Personnel should verify dye flags or color markers before the shipment begins. Dye flags may be color-coded for various products.

1.2.0 Pump Operations

Pumps move product from one point in the pipeline to another. Centrifugal pumps (*Figure 12*) act as main-line and booster pumps on most pipelines. With this type of pump, fluid enters the case near the axis of a rotating impeller. The impeller accelerates the fluid and flings it outwards at high velocity. The surrounding casing captures the fluid, which then exits the pump at high pressure. Some pipelines also use positive displacement (PD) pumps for auxiliary or special purpose applications.

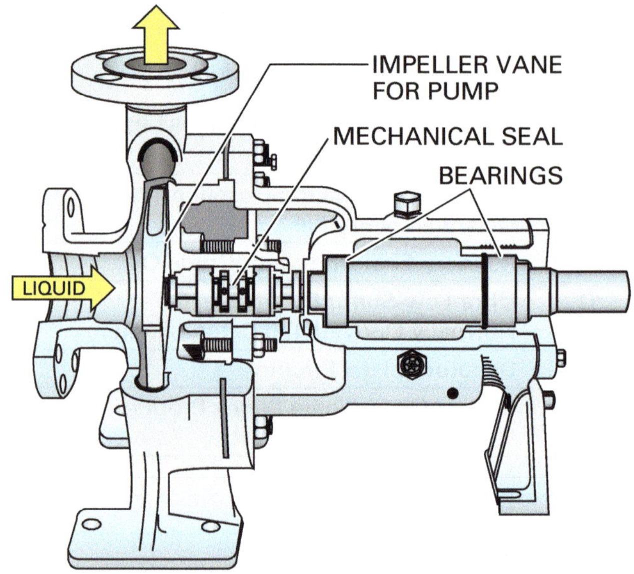

Figure 12 Centrifugal pump.

All pumps use a driver of some kind to supply power. Electric motors, internal combustion engines, and turbines are typical drivers. Power is expressed in horsepower (hp). Moving a heavy product at a high flow rate requires more power than moving a lighter one at a low flow rate. Control center personnel plan and calculate shipping procedures to use the lowest horsepower possible to get the job done.

> **NOTE**
> In countries that use the metric system, power is expressed in watts (W) or kilowatts (kW) rather than horsepower (1 hp = 746 W = 0.746 kW).

Pipeline pumps fall into three main categories:

- Main-line
- Booster
- Sump

1.2.1 Main-Line Pumps

Main-line pumps move the product through the pipeline, generating the pressure and flow required for meeting shipment schedules. Most main-line units are large centrifugal pumps

(*Figure 13*) with high horsepower drivers. Each location will likely have a number of these. Main-line units operate individually, in parallel, or in series. In many installations, main-line pumps receive suction from a common line supplied by a tank manifold and supply flow to the pipeline through a common manifold. The manifold allows several tanks to supply the pumps during an operation.

The control center manages pumps, starting and stopping them as required. It also sets and adjusts suction and discharge setpoints, depending on how the pumps are working. When multiple units work together, control center personnel must ensure compatible flow rates. Check valves prevent the pumps from recirculating.

Pumps do not have to be of the same size or capacity. When operating in parallel, they must be adjusted so their head settings match, because unequal pressure will cause a shutdown. Pumps operating in series complement each other, so it's not necessary to match head settings. Total discharge is the sum of each pump's head.

1.2.2 Booster Pumps

Booster pumps (*Figure 14*) are high-volume, low-head units that pump product from a tank into the suction side of a main-line unit. When operating in series, they also can act as injection pumps. Booster pumps provide the flow rate needed by main-line pumps running at high settings. Product level in a tank may supply sufficient suction pressure for the main-line pump, but if not, booster pumps come on line to supply the needed pressure.

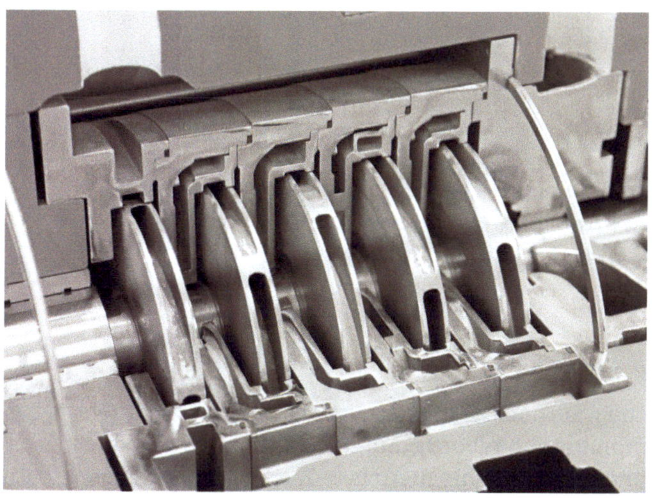

Figure 13 Multi-impeller centrifugal pump.

Figure 14 Booster pumps on a manifold.

1.2.3 Sump Pumps

Sump pumps are low-head centrifugal pumps used to move product from a sump into a tank or the pipeline. Product collects in the sump during normal operations. Automatic sump pumps come on and off in response to a level switch, keeping the sump product level below a certain point.

1.2.4 Managing Pressure and Flow

Safe and efficient pipeline operations depend on control center personnel monitoring and managing pressure. Control valves, installed in the pumping station's incoming and outgoing lines, adjust the pressure and flow. Transducers monitor and report pressure at the pump's suction and discharge sides to the SCADA system. Similarly, SCADA also collects data on line pressure and case pressure. Control center personnel use this information to adjust control valves to sustain **steady state** operating conditions on the outgoing line.

Control valves provide crucial pressure and flow regulation in systems with constant-speed main-line pumps. Valve throttling action at the suction and discharge sides keeps pressure and flow within appropriate ranges. Control center personnel determine the appropriate setpoints and adjust the valves from the control center. Station design may allow personnel to adjust pressure and flow at several different locations (*Figure 15*). Generally, the control center can manage valves at origin and delivery points but not at booster stations.

Backpressure valves maintain line pressure above the product vapor pressure point in case of problems or operational changes. They also protect terminal and station low-pressure piping. These devices are preset and will operate as required.

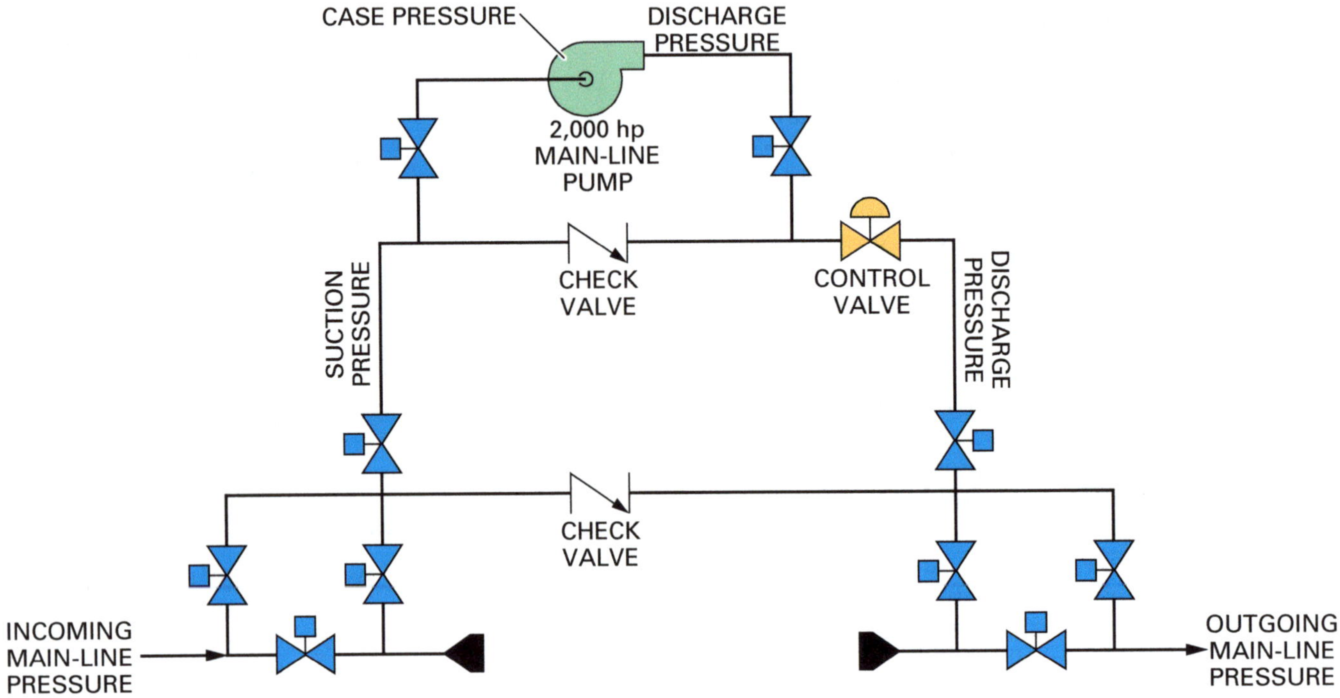

Figure 15 Simplified diagram of a pump station.

1.2.5 Pressure Points

Suction pressure is generated on the pump's upstream side, while discharge pressure is generated on its downstream (pipeline) side. The pump controller, usually a PLC, manages both of these pressures by monitoring them and comparing their current values to programmed setpoints. Control valves throttle the flow to maintain safe and appropriate operating conditions.

Increasing discharge pressure increases flow but decreases suction pressure because of the greater demand on the supply. Suction must be increased or the pump may shut down. Decreasing discharge pressure increases suction pressure, but decreases flow because the pump is ejecting less product.

Product density can significantly affect pressure and flow as well. A pump will experience a density change when an interface passes through it. A density increase will cause discharge pressure to also increase. The pump control PLC may make adjustments to maintain appropriate suction and discharge. The control room may also modify settings through the SCADA system.

In a large pipeline, different segments may operate at different flow rates and pressures. Some sections may supply low-pressure equipment, such as metering and manifolds. On older

Valves and Pigging

Pump stations may include technology designed to handle pipeline pigs. Pigs are cylindrical devices that fit snugly in the pipeline and are forced through it by the flowing product. They are used for one of several reasons:

- Cleaning or decontaminating the pipeline
- Inspecting the pipeline
- Pushing out gases or water
- Separating batches of product

Pigs come in a number of types and styles, depending on their purpose. Some are very simple, while others, such as "smart" pigs, contain sophisticated electronics that monitor and record conditions within the pipeline.

In all cases, however, when pigs are used, appropriate technology to handle them is required. This includes valves for introducing and removing them as well as detectors to sense their presence. The station shown in *Figure 15* includes valves to accommodate pigging.

pipelines, segments may be de-rated to a lower maximum operating pressure (MOP). These must be handled separately. When starting the pipeline, control operators will probably open segments in a specific sequence rather than all at once. They will allow the initially opened segments to build up pressure before opening others.

Control center personnel must note pressure differences from one segment to the next and be careful not to confuse them. Errors could cause an overpressure situation or a slack line condition. If there is a great difference in segment sizes, pressure management may be difficult. One section may require throttling while another may not. Pressure transducers provide full information through the SCADA system so operators can make decisions.

A pump's discharge side setting determines the pressure within the pump case. It will always be higher than the pressure past the discharge side. Pump-case pressure can cause damage over time, but it might not be noticed immediately. As the condition worsens, high pump-case pressure may cause unstable pump controls, piping vibration, and high noise levels. The control center can deal with this problem by making pressure adjustments upstream and/or downstream. Pumps should be inspected periodically for signs of pressure-related damage.

1.2.6 Pumping Applications

Pipeline engineers select pumps based on product type and throughput requirements. Increasing or decreasing horsepower increases or decreases flow rate. Similarly, adding or removing pumps has the same effect. Control center personnel determine the best way to utilize the pumps under their control to make the shipment move on schedule.

If the pipeline climbs to a higher elevation, extra horsepower will be needed beyond that required to handle a level grade. Calculations determine the extra horsepower necessary to do the work. A station may need to start several additional main-line units to supply enough horsepower to go uphill (*Figure 16*).

If pumps come on or go off line unexpectedly, pressure and flow will probably change. The possibility of such occurrences increases with the shipment distance. Pressure and flow rates must be watched closely and adjusted to compensate for these changes.

1.2.7 Pump Timing

Systems made up of multiple stations, terminals, and tank farms require a specific pump start-up

Figure 16 Increased horsepower required for elevation.

sequence. For example, a booster pump may have to be started before a main-line pump in order to provide sufficient suction. The booster pump delivers product to the suction side of the main-line unit. Failing to start the booster pump first could cause damage and force the main-line unit to shut down due to low suction (*Figure 17*).

> **NOTE**
> Raising suction pressure and discharge pressure also raises pump-case pressure.

On the other hand, if discharge or main-line pressure rises too high, the control center may shut down a booster pump or a main-line pump. These actions would reduce suction pressure, causing the pump-case and discharge pressures to drop in response.

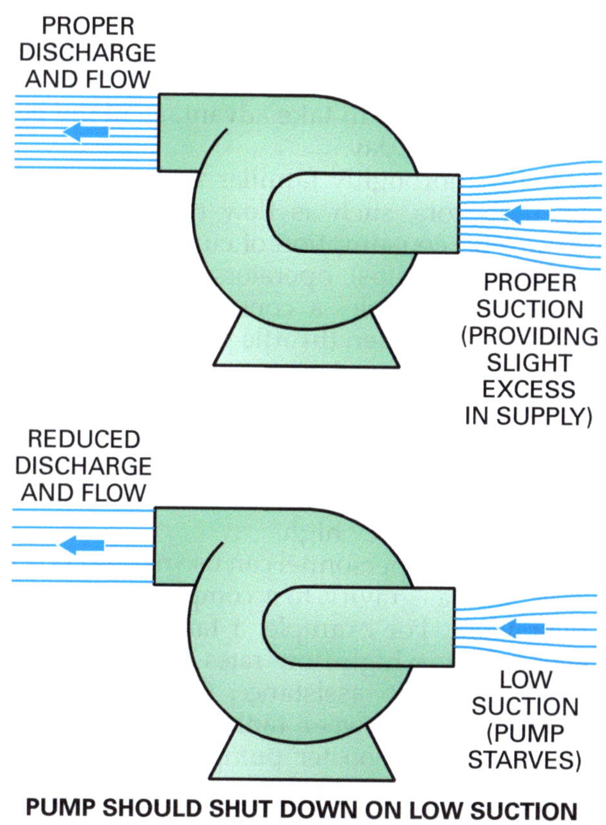

Figure 17 Effects of normal and low suction.

The control center brings up an idle pipeline by starting the pumps at the originating station first and gradually adding downstream stations as the product stream and pressure allow. Prior to start-up, main-line block, terminal, and tank valves must be opened for flow to increase. Booster pumps come on line first. Main-line units then follow. After the initial start-up, each successive pump downstream comes on line. Pressure builds until the system reaches a steady state condition.

1.2.8 Pump Management and Power Optimization

No matter the type, power costs money. Almost every pipeline activity consumes power, so control center personnel work hard to manage power carefully and optimize its use. By using training, experience, and good judgment, personnel set up shipments to make the most economical use of the equipment. Some facilities compare utility rates when planning a delivery. They then schedule pumping routes to take advantage of the best rates and times of day.

No matter the type, power costs money. Almost every pipeline activity consumes power, so control center personnel work hard to manage power carefully and optimize its use. By using training, experience, and good judgment, personnel set up shipments to make the most economical use of the equipment. Some facilities compare utility rates when planning a delivery. They then schedule pumping routes to take advantage of the best rates and times of day.

By being thoroughly familiar with pump condition indicators, such as flow rate, differential pressure, and consumption of current (measured in amperes, or amps), operators can estimate efficiency. For example, a constant-speed pump's efficiency drops when throttle pressure increases. With a variable-speed pump, some speeds are more efficient than others for a given set of conditions. Any unexplained change requires investigation immediately. By staying on top of problems, the control center can reduce costs and keep operational efficiency high.

Control center personnel can use natural power sources, such as gravity, to accomplish some parts of their work. For example, a tank with a high level can deliver high flow rates to pumps without booster pump assistance. Gravity can also transfer product between tanks if they are at different heights. A booster pump may have to be used for part of the transfer, however.

For variable-speed pumps, a **variable frequency drive (VFD)** can control the pump for maximum efficiency. VFDs manage their electric motor's speed by adjusting the voltage and frequency supplied to it. This technique controls pressure and flow without resorting to throttling, which wastes more power. It also offers an efficient way to reduce power when full power is not needed.

Power consumption can be calculated if a few key factors are known. The following formula is one such example based on the pipeline units used in the United States:

$$\text{power} = (\text{flow} \times \text{head}) \div (2{,}450 \times \text{efficiency})$$

Where

power = the power consumed in hp
flow = the pump's flow rate in bph
head = the pump's head in psi
2,450 = a conversion constant
efficiency = the pump's efficiency as a percentage

Another way to manage power is by switching pumps, an option available in installations that have multiple units. Pumps can be alternated based on power requirements in order to meet rate conditions or to perform maintenance, depending on company policy and procedure. One pumping unit can be brought on line and another unit taken off in order to improve pipeline hydraulics or to achieve line balance. Control center personnel manage suction and discharge at the origin station. They must also sequence valves and handle switching. Booster stations have their own control systems.

1.2.9 Drag Reducing Agents

Drag reducing agents (DRAs) are long-chain polymers that enhance flow by reducing pipeline friction. Personnel introduce DRAs into the pipeline through annular ring injectors in order to coat the inner surface (*Figure 18*). Pipelines treated with DRAs experience about 20 percent greater throughput and a corresponding 20 percent reduction in pressure. Consequently, DRA injections greatly reduce the horsepower required to pump certain products, especially crude oil. However, pipeline personnel must calculate the DRA cost and the injection expense to be sure that the energy savings are large enough to justify using a DRA.

NOTE

DRAs are also used during maintenance when a station must be bypassed.

Figure 18 DRA injection control valve.

DRA injections are planned by the scheduling department. By reducing friction, a DRA affects the pumping rate in positive ways. Adding a DRA results in a shorter pumping schedule and creates a negligible change in volume. Since DRAs are expensive and have a strong effect on operations, use care in following schedule orders.

Cautions and guidelines associated with using DRAs include the following:

- DRAs may be used only with compatible products, such as crude oil.
- DRAs may not be used with refined highly volatile liquid (HVL) products.
- DRAs injection must be stopped at least two hours before pumping specialty products, such as jet fuel.

Control center personnel can manage some injections from the control center. Others may require field personnel to be at the site. Control center personnel should consult their company's DRA product compatibility chart and the product schedule to verify that a planned DRA injection is compatible with the product. DRAs should not be injected if any of the following situations exist:

- An incompatible shipment will occur within a specific time period after the injection.
- An incompatible product interface has passed through the injection site less than a specific time before the injection is scheduled to start.
- A residue of product or DRA could cause some type of contamination.

1.3.0 SCADA

The pipeline is a huge system with many different pieces spread over a vast geographic area. Creating a single picture of it all is the job of the SCADA system. SCADA has two main purposes: to collect information (data acquisition) and to enable operators to manage the pipeline (supervisory control). Unlike other industrial control systems, SCADA doesn't make numerous autonomous decisions. Instead, it relies on human operators to make decisions by using the information that it assembles for them.

A typical SCADA system (*Figure 19*) consists of four component categories—field devices, supervisory computers, human-machine interfaces (HMIs), and communications systems. Together, these gather data from equipment in the field, send it to centralized locations for analysis, and provide a means for human operators to examine the data and interact with the pipeline.

1.3.1 Field Devices

Out in the field, sensors gather information reflecting the pipeline's current state. Typical variables include pressure, temperature, and flow. In many cases, this information goes to field control devices (controllers, PLCs, or RTUs) that make local decisions to modify these parameters. Using actuators, the control device opens, closes, or adjusts valves to alter pipeline conditions.

1.3.2 Supervisory Computers

Individual PLCs and RTUs see only a small part of the pipeline, but pulling their data together in a central location makes it possible to create a picture of the entire pipeline and its activities. One or more supervisory computers handle this task. These are large machines that collect data from many different RTUs and PLCs, format it, and supply it in various forms to those who manage the pipeline.

Supervisory computers communicate with RTUs and PLCs over a network of some type. They run sophisticated software that monitors field devices and collects their data. This data becomes the basis for reports. These can be simple screens presented on an HMI, or they can be downloaded in the form of spreadsheets by a standard PC. Pipeline operators use HMI-presented reports to make control decisions. Managers and executives use downloadable reports to make higher-level decisions about the pipeline and to make projections that will help them plan future decisions.

Besides collecting and assembling data to help people make decisions, supervisory computers perform a number of other tasks. A significant one is the historian function, which archives data over long periods of time. Historical data can

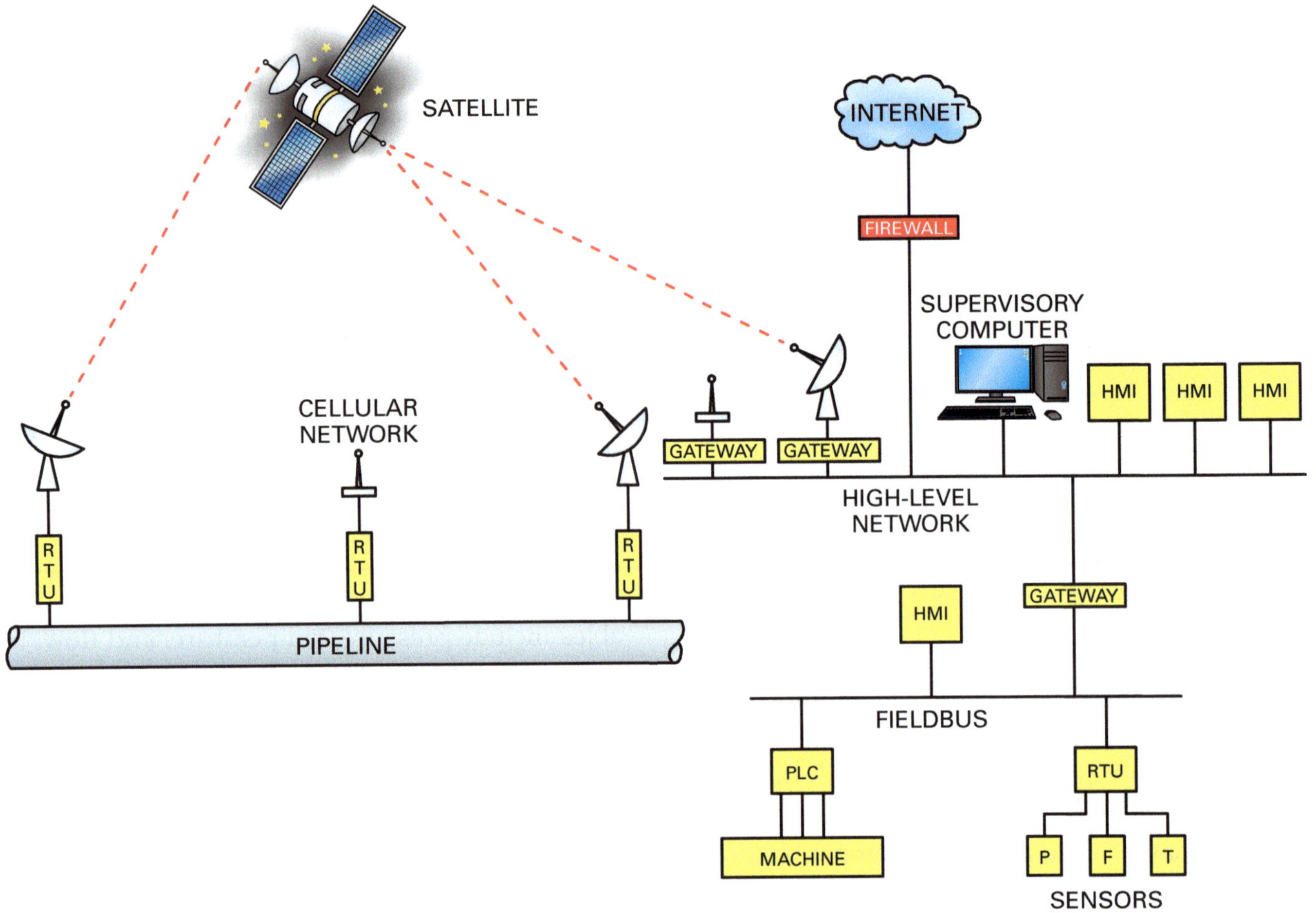

Figure 19 SCADA system.

reveal subtle pipeline trends, facilitating more sophisticated management decisions.

Finally, the supervisory computer centralizes pipeline control. Operators interact with HMIs, which, in turn, interact with the supervisory computer. Based on these interactions, the supervisory computer issues appropriate commands to the correct RTU or PLC, which modifies the pipeline's behavior.

1.3.3 HMIs

Every control system contains one or more HMIs, which allow operators to monitor the process and make control decisions. In some systems, HMIs are called *man-machine interfaces* (MMIs). An HMI can be relatively simple, such as a small, text-only LCD display with a few buttons or perhaps a simple keypad. More sophisticated HMIs have full-color graphical displays with touchscreens for operator input, such as the one shown in *Figure 20*. These are often used by field personnel to interact with more sophisticated field devices, such as PLCs.

Figure 20 A small graphical HMI.

Control rooms usually contain many HMIs, including large wall-mounted ones and smaller panel-mounted units (*Figure 21*). These graphically display key process-control components along with important data, status values, and alarms. HMIs communicate with the supervisory computer through the SCADA network.

Figure 21 Control room with large and small SCADA displays.

When interacting with field personnel, remember that field HMIs are usually smaller and cannot show as much information on a single page as the larger control room units. Field personnel may have to navigate through multiple screens to locate information that appears on a single page in the control room. Be aware of these differences so you can communicate effectively with field personnel.

1.3.4 Communications Technologies

For SCADA components to be effective, they must communicate with each other. For this reason, all SCADA systems have several communications networks that link their components into a unified whole, allowing data to circulate freely.

At the field level, a SCADA system consists of many devices that interact directly with the pipeline. These include sensors, actuators, controllers, RTUs, and PLCs. Field device communication is fairly basic and involves only small amounts of data. There are usually less than a few hundred devices in a given area. For this reason, communications systems linking field devices tend to be small and simple.

A fieldbus is a low-level network (*Figure 22*) designed to function over a relatively limited physical area. It uses a simple cabling system to tie field devices together. A typical fieldbus might link a PLC or RTU and a number of other devices. A local operator HMI might also be connected to the fieldbus. Because fieldbuses cannot operate over large areas, a second class of network, called the high-level network, handles this task.

A high-level network must be fast, robust, and capable of handling large volumes of data efficiently and reliably. There are many different

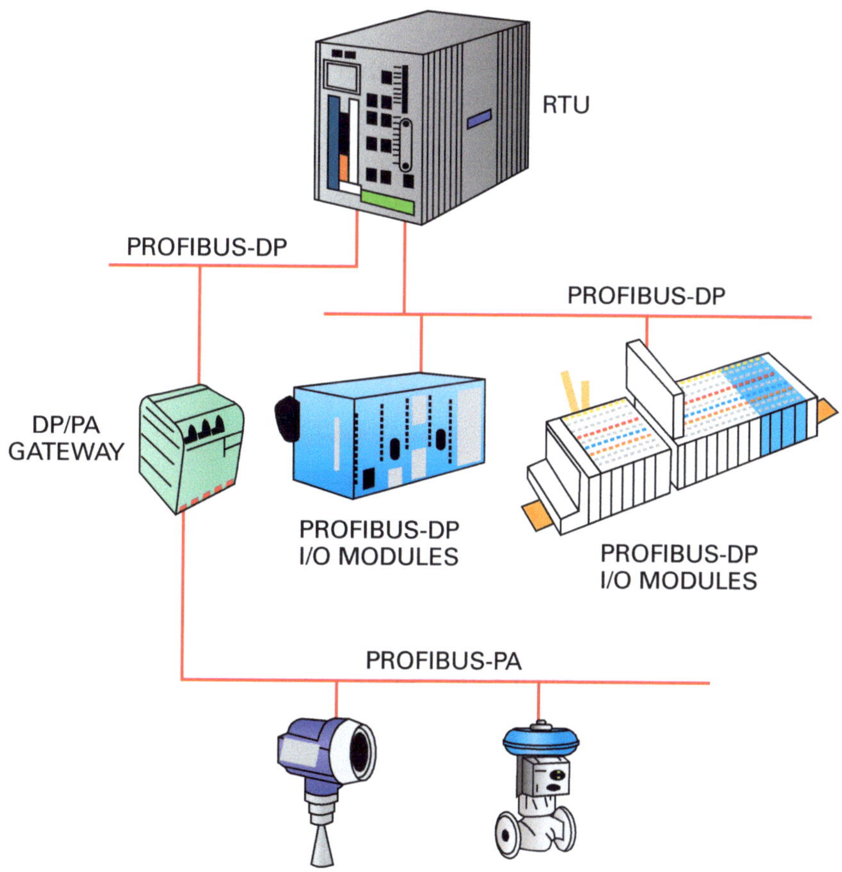

Figure 22 A fieldbus ties local devices together.

technologies available to implement high-level networks. Some of these are used in plants, offices, and on corporate campuses. This kind of technology usually relies on cabled connections (copper or fiber optic).

In some situations, a cabled network just isn't practical. It may not be possible to run cables in certain places, or the distance might be prohibitive. Wireless networking can solve some of these problems by using radio-frequency (RF) technology to carry digital data. Short-range wireless networking, commonly known as Wi-Fi™, is often used in offices and factories as a means of supporting mobile devices. But it does not have the range needed for the pipeline industry. In these situations, satellite radio, cellular phone communications, microwaves, and other types of long-range RF communication are used.

1.3.5 SCADA Screens

Control room operators spend a lot of their time interacting with SCADA through screens—informational pages that appear on HMIs. Screens fall into a number of categories: control, status, trend, and alarm/alert. Operators must become very familiar with the screens available through their particular system in order to call up whatever they require rapidly and confidently.

What appears on the screen depends largely on the age of the particular system. SCADA screens have changed significantly over the years since the technology first appeared. Early SCADA systems used traditional schematic symbols (*Figure 23*), similar to those found on piping and instrumentation diagrams (P&IDs). Most information came in the form of numbers next to the associated components. While this style does work, these screens are often crowded and operators have to keep a mental list of appropriate numeric values.

Later system designs were more colorful and graphical (*Figure 24*). Components were brightly colored, much more three-dimensional, and often animated. But the colors were not standardized and many of the visual effects were distracting rather than useful. Industry experts began to realize that these screens were not much of an improvement, and they often made it difficult for operators to gather information quickly.

The most recent SCADA screens are designed much more thoughtfully. Components are less brightly colored and usually are much simpler. Instead of displaying data simply as numbers, moving pointer indicators and bars help operators know exactly where a variable is within a range. Color coding is more carefully selected since SCADA designers have realized that some operators suffer from color vision deficiency. Instead of relying on color to indicate status, many systems use symbol brightness instead—a light-colored symbol is active, while a dark-colored one is inactive.

Animation appears far less often on SCADA screens today, and when it is used it is much more purposeful. Instead of having operating components moving, spinning, or blinking, an animation might be used on an alarm screen. Flashing alarms effectively catch an operator's attention, because a person's eyes are sensitive to motion. As long as the screen isn't filled with other non-essential motion, a blinking alarm will stand out.

Regardless of the type and age of a particular SCADA system, it's crucial that personnel learn to use it as effectively as possible. They need to know how to call up the desired screens quickly, recognize how status information appears, and what different color or brightness codes signify. These are skills that define a proficient control room operator.

Color Vision Deficiency

Color vision deficiency, inaccurately known as color blindness, affects about 7 percent of males and 0.4 percent of females in the United States. Percentages vary in other countries, depending on their ethnic mix. Most people with color vision deficiency are not truly color blind. Instead, they cannot distinguish between certain colors, such as red and green, but they can perceive others. For this reason, human factors engineers are using means other than color to identify information on control panels and screens. Color may still be used, but alternative methods are utilized as well.

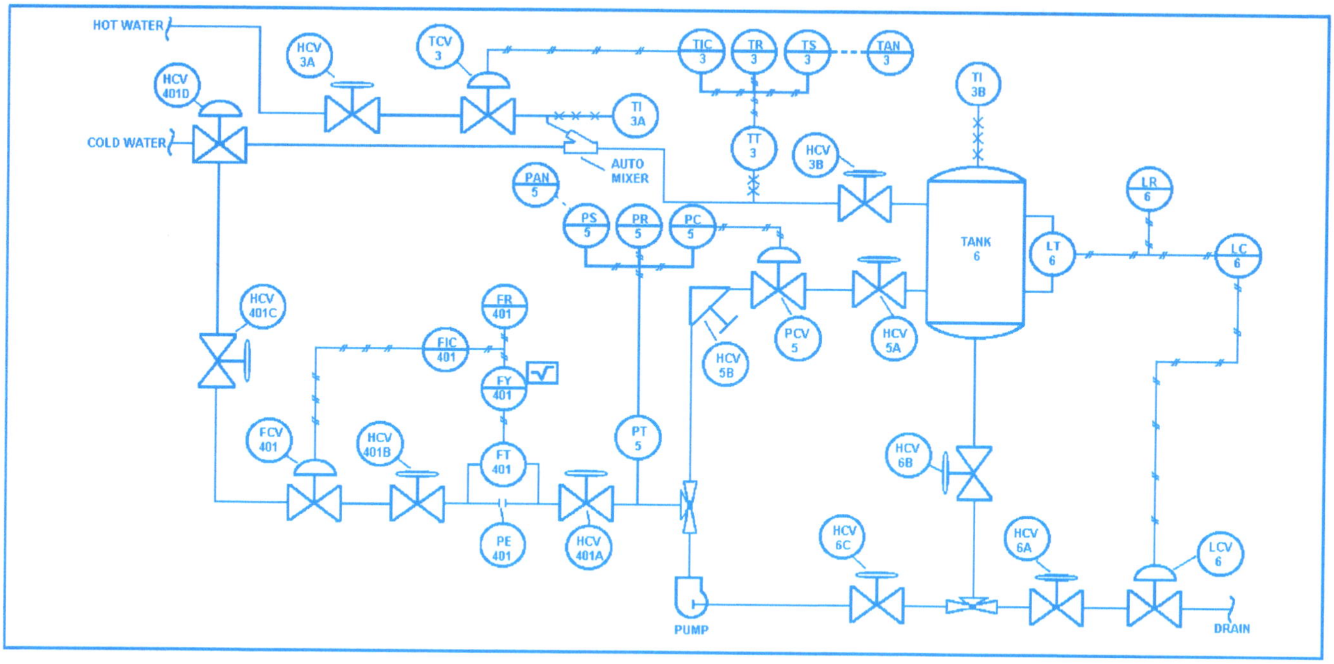

Figure 23 P&ID schematic symbols.

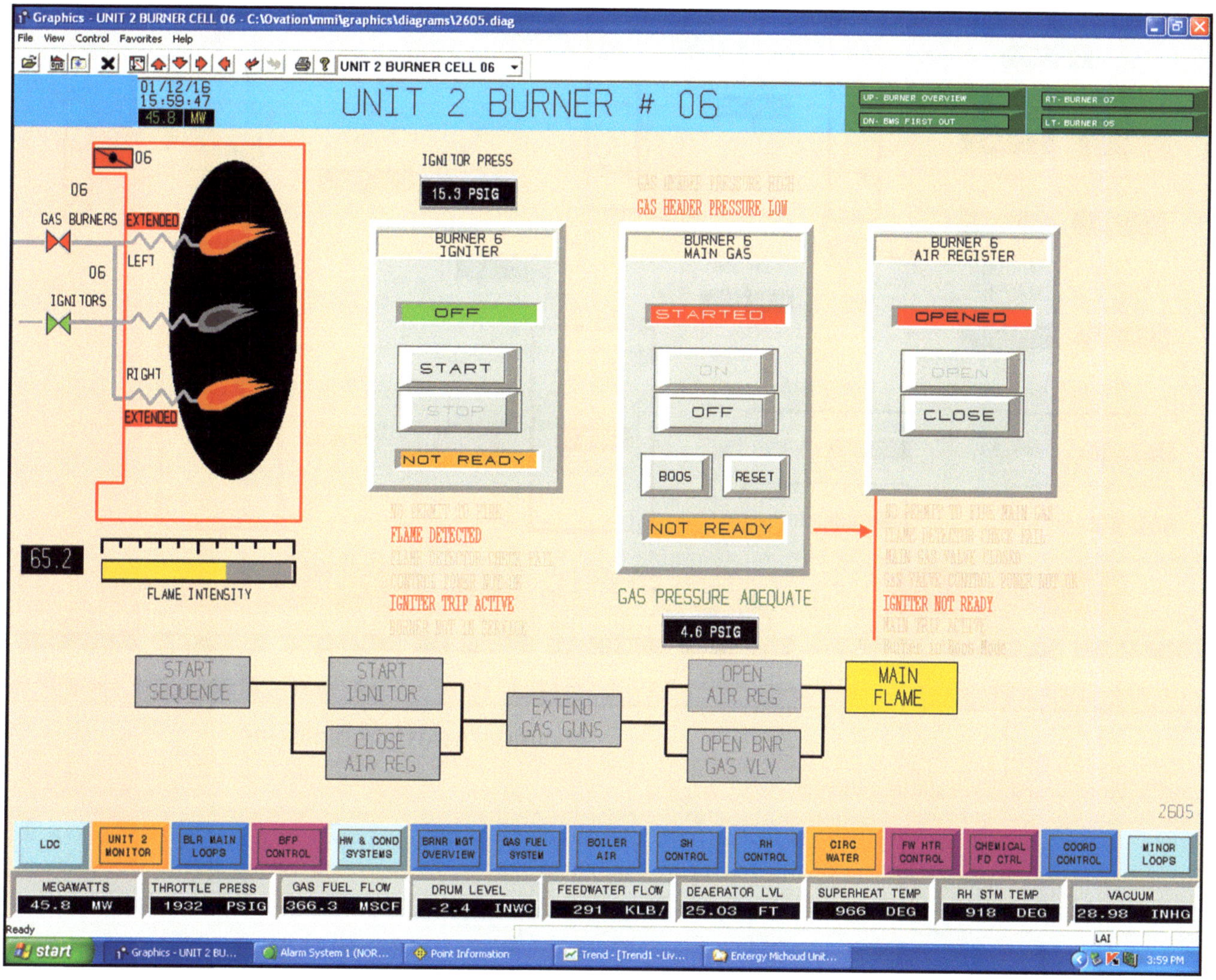

Figure 24 A highly graphical HMI display.

1.3.6 Valve Control through SCADA

Almost all remote valve control and monitoring happen through SCADA screens. These screens provide information on a valve's current status (open, closed, traveling) or throttle position (percentage) Depending on the age and brand of the SCADA system, this information will appear in one or more forms.

Textual displays use words (*Figure 25*) or numbers (*Figure 26*) to indicate the valve's status or position. Symbolic displays (*Figure 27*) change color or brightness to indicate the valve's status or position. Finally, moving bars or pointers (*Figure 27*) visually portray a valve's position. Symbolic and moving indicators usually include a supplementary numeric value as well.

Controlling valves through SCADA also depends on the particular system's age and brand. Open/closed valves may be controlled by buttons that must be clicked or touched (*Figure 28*). Valves that can throttle may be controlled by up/down arrows (*Figure 29*) or sliders. Some systems may allow the operator to key in a specific percentage.

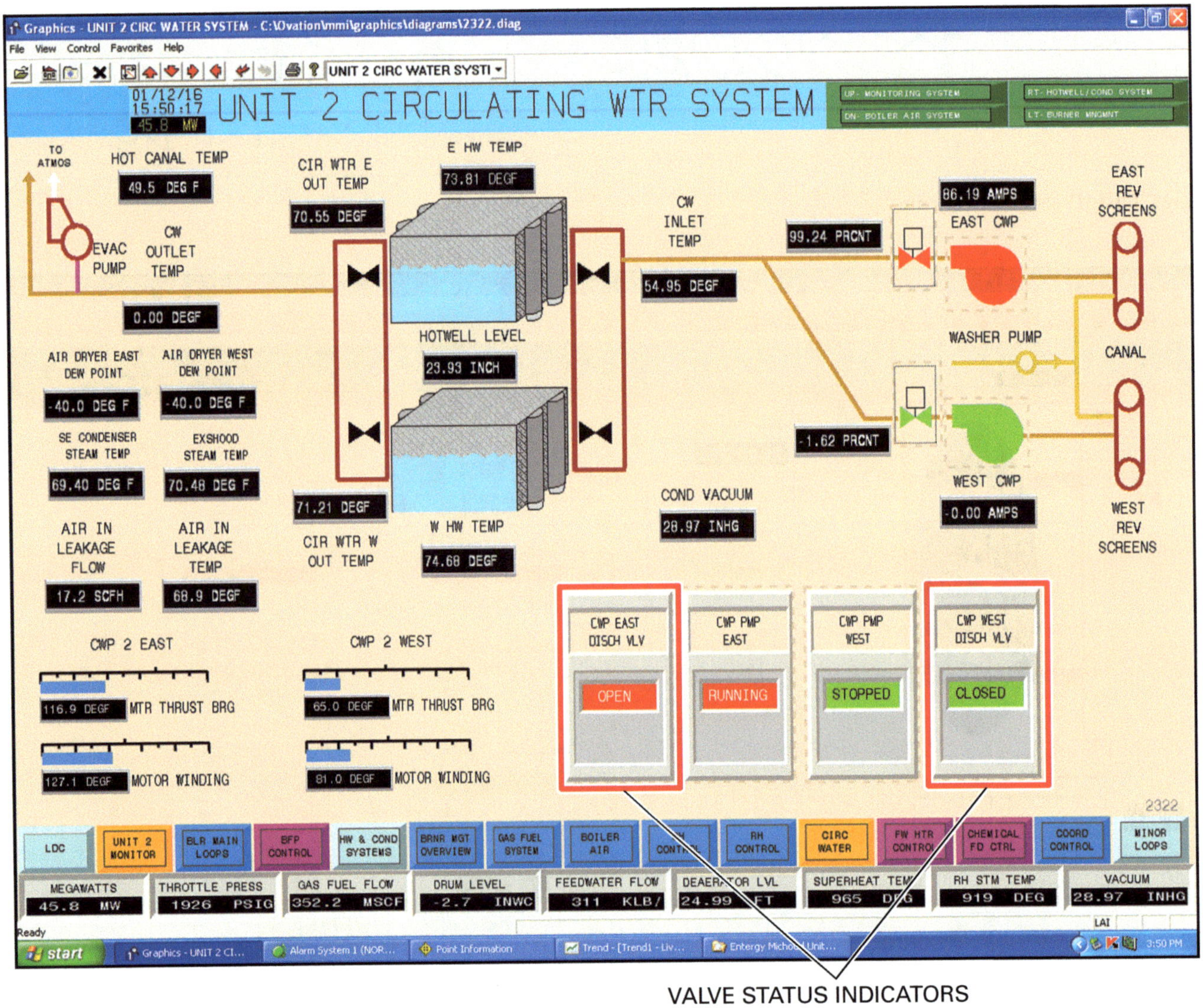

Figure 25 Word-based valve status indicator.

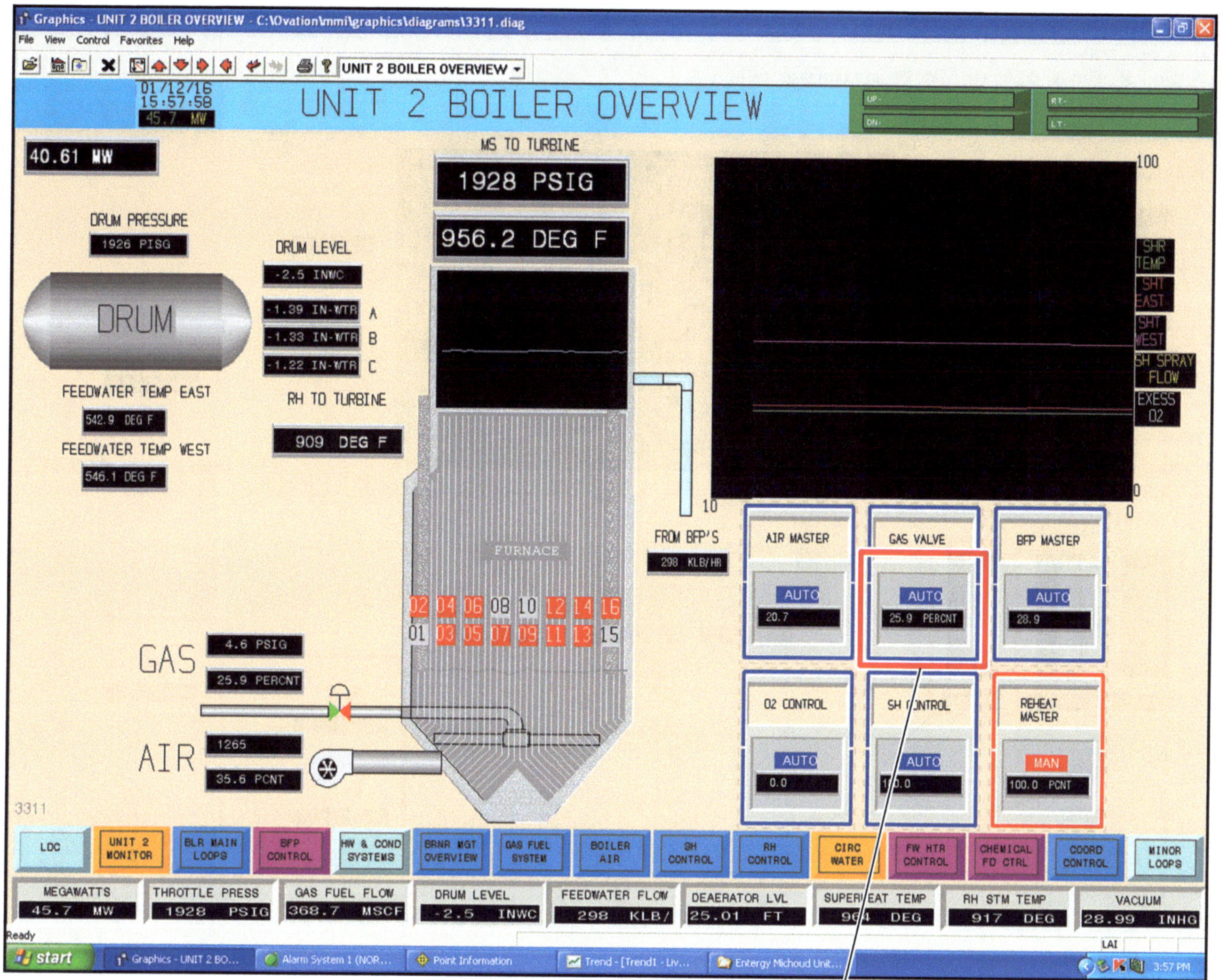

VALVE STATUS INDICATOR

Figure 26 Number-based valve position indicator.

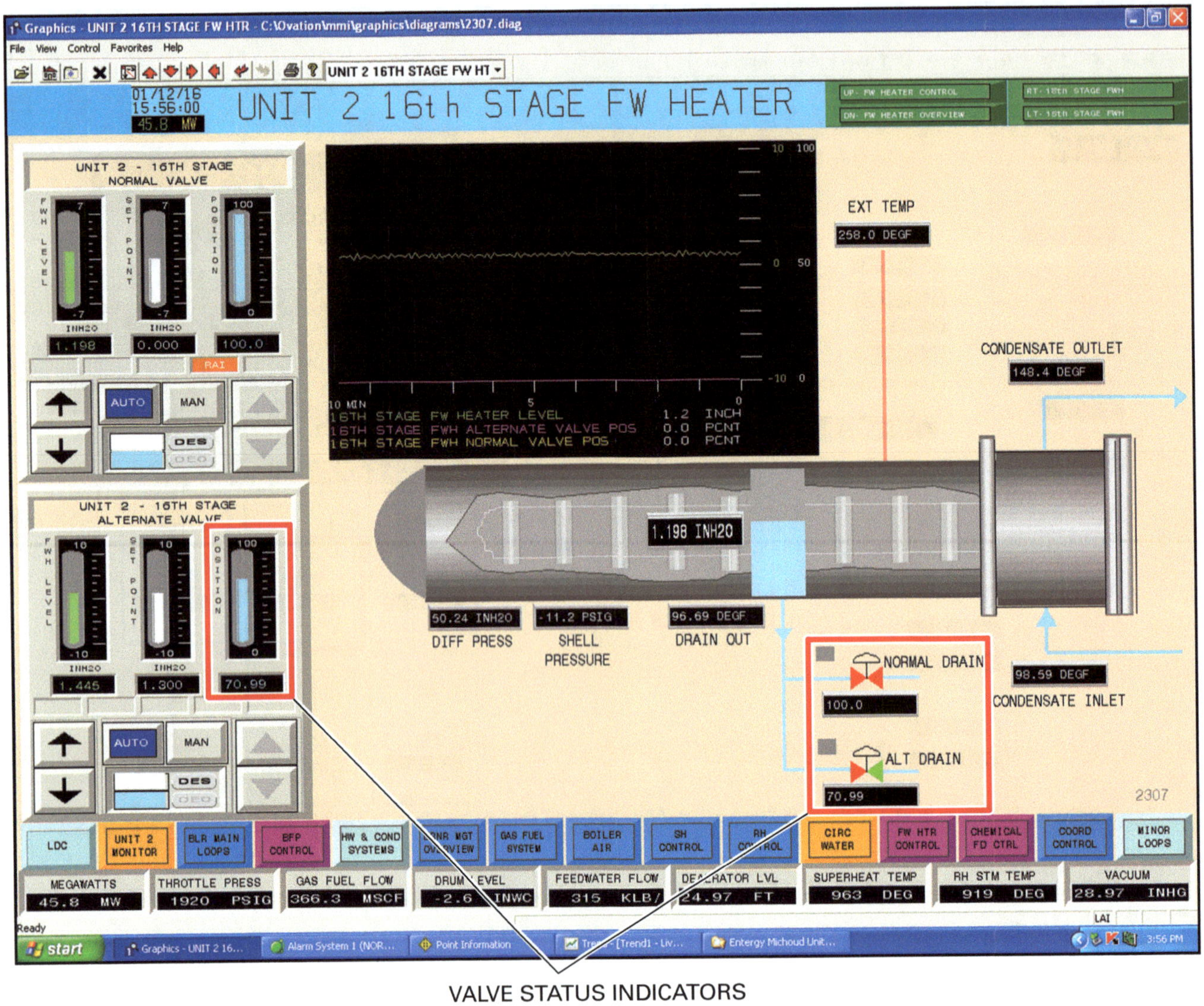

Figure 27 Symbolic and moving bar indicators.

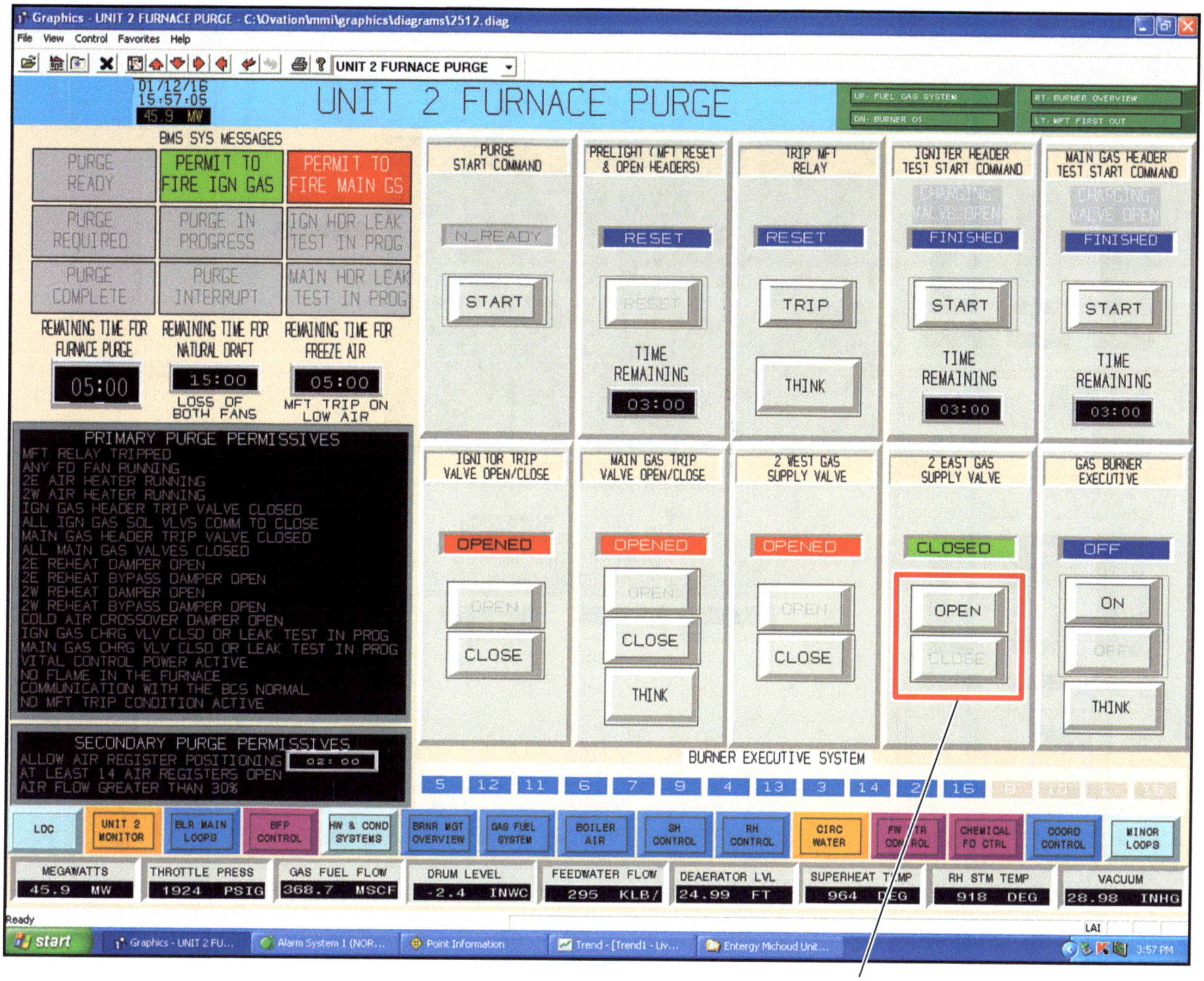

Figure 28 Button-based valve control.

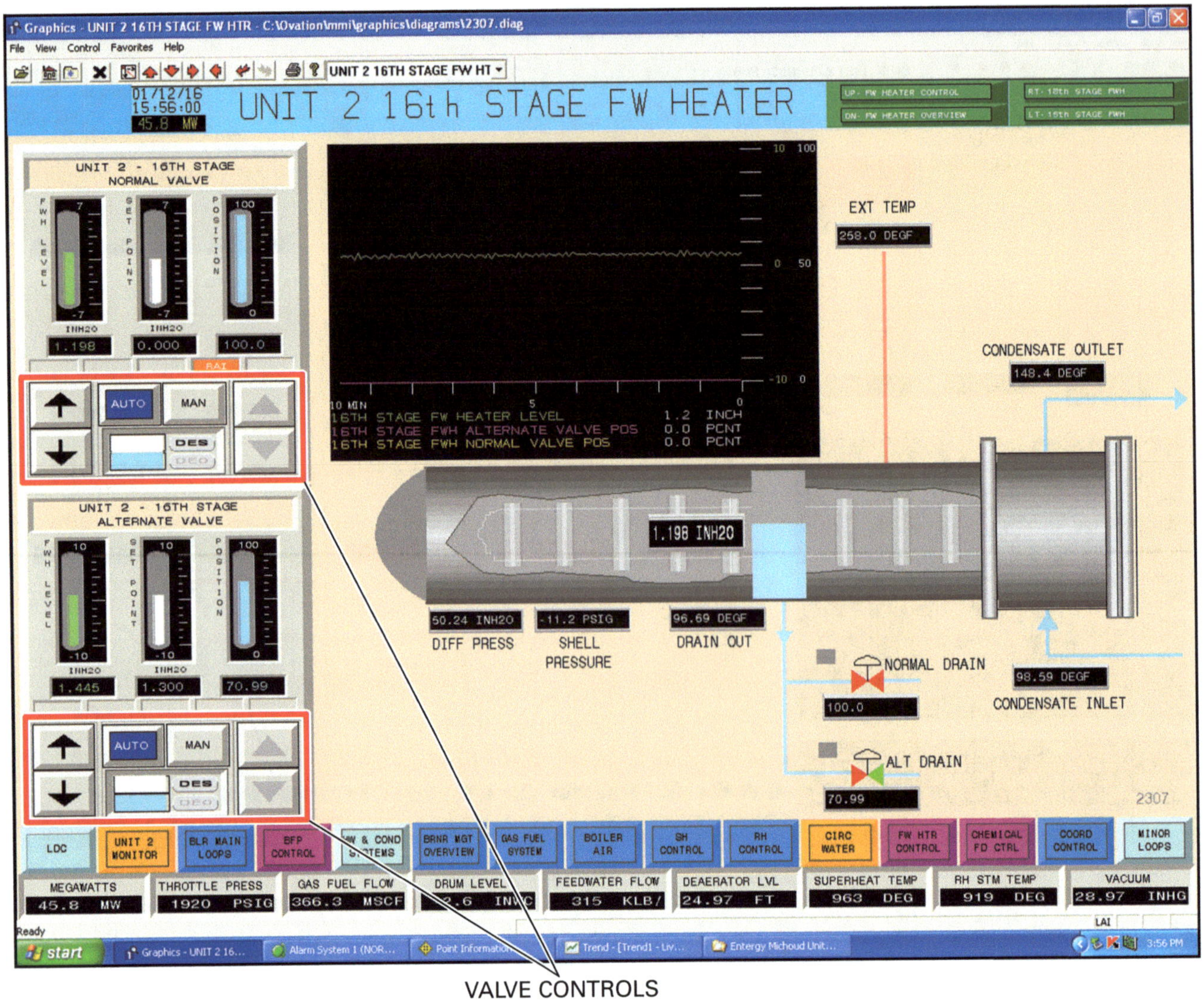

Figure 29 Up/down arrow-based valve control.

Additional Resources

It is assumed that you will learn to perform this procedure under the direct supervision of your crew leader, foreman, or instructor. For independent study, refer to the recommended Additional Resources at the back of this module.

1.0.0 Section Review

1. Which of the following valves can operate in the open, closed, and throttle positions?
   a. Block valves
   b. Safety valves
   c. Gate valves
   d. Control valves

2. When main-line pumps run in parallel, their heads _____.
   a. add together
   b. must match
   c. can be different
   d. subtract from each other

3. PLCs, RTUs, and actuators are all examples of _____.
   a. field devices
   b. HMIs
   c. MMIs
   d. supervisory computers

4. How do the most recent SCADA systems indicate that a component is active or inactive?
   a. They use red and green.
   b. They use brightness levels.
   c. They animate the object.
   d. They cycle through a range of colors.

SECTION TWO

2.0.0 REMOTE VALVE OPERATION

Objective

Describe the requirements and procedures for operating valves remotely (CT64_4-17).

a. Identify the prerequisites, competencies, and abnormal operating conditions associated with remote valve operation.
b. Describe the procedures used to operate valves remotely.

Performance Task

1. Remotely operate valves on a liquid pipeline (CT64_4-17).
   - Identify potential abnormal operating conditions that may occur during performance of this CT, and know the appropriate actions to take in response to them.
   - Utilize the appropriate personal protective equipment according to relevant company procedures.
   - Verify that the flows, pressures, and product characteristics are within normal range.
   - Identify the need to open or close a valve or start or stop a unit to control flow or pressure and to maintain safe operation of the pipeline system.
   - Identify the applicable valve, unit, or the appropriate control device.
   - Perform proper operation according to company policy and procedures.
   - Verify that the desired result was achieved.
   - Complete the appropriate documentation as required by operator's procedures.

Trade Terms

Abnormal operating conditions (AOCs): Conditions within a pipeline system, identified by the operator, that may indicate (1) a malfunction of a component or (2) a deviation from normal operations that may indicate a condition exceeding design limits or result in a hazard to persons, property, or the environment.

Valves are key to managing tanks and manifold operations and help to keep the pipeline safe. Valve alignments configure the pipeline system to deliver or receive product. While some valves require manual control and others activate automatically, many operate in response to remote commands sent from the control center. This section describes some of the issues and procedures associated with controlling these valves.

2.1.0 Prerequisites, Competencies, and Abnormal Operating Conditions

Control room personnel must possess a variety of skills in order to operate valves remotely. They must also be aware of the particular **abnormal operating conditions (AOCs)** that may occur as they carry out these tasks.

2.1.1 Prerequisites/Competencies

Control room personnel must possess the following basic knowledge and skills prior to operating valves by remote control:

- Understand the process and components of the pipeline so valve operations can be made knowledgably.
- Thoroughly understand the consequences of valve operations, particularly those that could lead to unsafe situations or abnormal operating conditions.
- Be proficient with the SCADA system, its displays, and its controls.
- Know how to use SCADA to obtain valve status information and to operate valves.
- Be thoroughly familiar with abnormal operating conditions and emergencies, understanding both their entry conditions and the proper responses to each.
- Understand and be skilled at communicating both within the control room and with field personnel.

2.1.2 Abnormal Operating Conditions

Control room personnel operating valves by remote control must be able to identify the following potential abnormal operating conditions and know the proper response to each, should they occur:

- *A unexpected pressure change appears on a monitor, or a pressure-related alarm activates.* – If either of these situations occurs, the controller should try to return the system to normal and determine if there is any damage or product release. Possible actions could include shutting down the pipeline, correcting a valve setting, or isolating equipment. Relevant personnel should be notified, if required.

- *A unexpected flow change appears on a monitor, or a flow-related alarm activates.* – If either of these situations occurs, the controller should try to return the system to normal and determine if there is any damage or product release. Possible actions could include shutting down the pipeline, correcting a valve setting, or isolating equipment. Relevant personnel should be notified, if required.
- *The SCADA system reports an inappropriate or unexplained valve status or position.* – If this situation occurs, the controller should perform appropriate diagnostics and troubleshooting on the control system and its communications network. Relevant personnel must also be notified.

2.2.0 Controlling Valves

There are two aspects involved in valve control. The first requires understanding the specific SCADA system that oversees the pipeline. Control room personnel must understand how to use this system to monitor and control valves quickly and efficiently. The second aspect requires the control room operator to understand the valves themselves and the roles that they play in pipeline operations. Controllers should not manipulate valves unless they thoroughly understand the consequences of particular actions to the pipeline and its safety.

2.2.1 Procedures

While a pipeline's control system is tailored to that particular installation, valve control tends to reflect a common pattern. Always follow company policy and procedures. Control center personnel must protect life, the environment, and the pipeline's safety. The following general steps summarize the typical procedures involved in controlling valves remotely from the control center:

CAUTION

Whenever you are working in the control room, you must be alert to the possibility of abnormal operating conditions and be prepared to initiate the proper responses from memory.

Step 1 Determine which valve must be controlled. Confirm the correct valve before performing any operations. Company documentation or the SCADA system itself may help with this task.

Step 2 If required, communicate with appropriate personnel before operating the valve. This could include field personnel or those in the control room. Follow company procedures regarding communications.

Step 3 Use the SCADA system to call up the correct screen for checking the valve status and operating the valve. Confirm that you understand the correct procedure for operating the valve. Also confirm that you understand how to interpret the status display.

Step 4 Operate the valve. If the valve is an open/closed valve, command the valve to go to the desired position. If the valve is a control (throttle) valve, command the valve to go to the desired percentage position.

Step 5 Using the SCADA system, confirm that the valve has reached the desired position.

Step 6 If required, communicate with appropriate personnel that the operation is complete.

Additional Resources

It is assumed that you will learn to perform this procedure under the direct supervision of your crew leader, foreman, or instructor. For independent study, refer to the recommended Additional Resources at the back of this module.

2.0.0 Section Review

1. When remotely managing valves, control room personnel must know how to use SCADA to control valves and to _____.
   a. obtain valve status
   b. reset the CPM system
   c. complete required documentation
   d. monitor system security

2. If the SCADA system reports an inappropriate or unexplained valve position, the pipeline controller must first _____.
   a. restart the SCADA supervisory computer
   b. start diagnostics and troubleshooting
   c. manually return the valve to the correct position
   d. send a fieldbus reset command

3. Before calling up the correct SCADA screen to operate a valve, what step might the pipeline operator be required to perform?
   a. Fill out all required documentation.
   b. Send field personnel to observe the valve.
   c. Visually check the valve.
   d. Communicate with any required personnel.

SUMMARY

A liquid pipeline is made up of many different parts, including pumps, valves, tanks, and pipes. Pipelines have sophisticated systems that monitor the equipment and allow personnel to manage almost everything from a central location. It is through this control system that personnel can perform one of the most fundamental pipeline tasks—operating valves remotely from the control room.

In order to manage the pipeline through valve operations, personnel must understand the key components that make up the associated technologies. Valves control product as it moves through tanks, manifolds, and the pipeline itself. Pumps move product from one part of the pipeline to another. Personnel must know how to operate valves and check their status through the SCADA system.

Controlling valves remotely from the control room requires a high level of familiarity with the SCADA system and company procedures in order to operate valves safely and efficiently.

Review Questions

1. Valves that help prevent a slack line condition are called _____.
   a. hand valves
   b. safety valves
   c. block valves
   d. backpressure valves

2. What control system feature prevents inappropriate valve alignments and ensures that valve sequences happen in the correct order?
   a. SCADA
   b. Valve interlocks
   c. Safety valves
   d. Fieldbus networks

3. The pressure that is generated on the downstream side of a pump is called the _____.
   a. discharge pressure
   b. mainstream pressure
   c. pump-case pressure
   d. suction pressure

4. How many hours must elapse after a DRA injection stops before jet fuel can be run through the pipeline?
   a. 1 hour
   b. 2 hours
   c. 5 hours
   d. 10 hours

5. The historian function on a SCADA supervisory computer can be used to _____.
   a. perform diagnostics
   b. control valves
   c. explore trends
   d. monitor the CPM system

6. Another name for an HMI is a(n) _____.
   a. MHI
   b. SSD
   c. UI
   d. MMI

7. A high-level network used for SCADA may be based on copper, fiber, satellite radio, microwaves, or _____.
   a. cellular phone
   b. token bus
   c. ATX
   d. ISO/IP

8. Before performing any control actions, control room personnel must be aware of the possible safety consequences that come from _____.
   a. SCADA errors
   b. valve operations
   c. valve interlock corruption
   d. RTU lockout

9. If a pipeline pressure alarm goes off during valve operation, control room personnel must first _____.
   a. check for damage
   b. return the system to normal
   c. check for product release
   d. notify appropriate personnel

10. After commanding a valve to move to a new position, the pipeline controller must _____.
    a. document the operation in the log
    b. check for SCADA alarms
    c. confirm that the valve arrived at the commanded position
    d. notify appropriate personnel

Trade Terms Quiz

Fill in the blank with the correct term that you learned from your study of this module.

1. A powered mechanism that operates a valve is called a(n) _________.
2. A pipeline instrument that determines the product's specific gravity is called a(n) _________.
3. A control system used in the pipeline industry that collects and assembles data and also allows operators to control the pipeline remotely is called a(n) _________.
4. A device that represents a physical quantity, such as position, as a numeric or electrical value that a controller can use for feedback is called a(n) _________.
5. A control system feature that prevents unsuitable valve alignments and ensures that valve sequencing occurs in the correct order is referred to as _________.
6. The highest pressure at which a liquid pipeline may be operated is referred to as _________.
7. The region between batches of dissimilar product in which both products combine to form a transmix is called the _________.
8. A compact industrial computer largely used for machine control and other dedicated tasks requiring intelligence is called a(n) _________.
9. The condition in which pipeline variables, such as pressure and flow, are relatively constant over a period of time is called _________.
10. A pipeline instrument that uses light absorption to detect the interface between batches of dissimilar product is called a(n) _________.
11. The fluid volume required to fill a pipeline section is called the _________.
12. A computer-like device that connects field devices to the supervisory computer in a SCADA system is called a(n) _________.
13. An electric motor paired with an electronic controller that adjusts motor speed by varying the power supply voltage and frequency is called a(n) _________.
14. An unusable mixture of two different products, usually created by switching from one incompatible product to another, is referred to as _________.
15. Conditions within a pipeline system, identified by the operator, that may indicate (1) a malfunction of a component or (2) a condition exceeding design limits or resulting in a hazard to persons, property, or the environment are called _________.

Trade Terms

Abnormal operating conditions (AOCs)
Actuator
Encoder
Gravitometer
Interface
Line fill
Maximum operating pressure (MOP)
Optical interface detector (OID)
Programmable logic controller (PLC)
Remote terminal unit (RTU)
Supervisory Control and Data Acquisition (SCADA)
Steady state
Transmix
Valve interlock
Variable frequency drive (VFD)

Trade Terms Introduced in This Module

Abnormal operating conditions (AOCs): Conditions within a pipeline system, identified by the operator, that may indicate (1) a malfunction of a component or (2) a deviation from normal operations that may indicate a condition exceeding design limits or result in a hazard to persons, property, or the environment.

Actuator: A powered mechanism that operates a valve.

Encoder: A device that represents a physical quantity, such as position, as a numeric or electrical value that a controller can use for feedback.

Gravitometer: A pipeline instrument that determines the product's specific gravity.

Interface: The region between batches of dissimilar products in which both products combine to form a transmix.

Line fill: The fluid volume required to fill a pipeline section.

Maximum operating pressure (MOP): The highest pressure at which a liquid pipeline may be operated.

Optical interface detector (OID): A pipeline instrument that uses light absorption to detect the interface between batches of dissimilar product.

Programmable logic controller (PLC): A compact industrial computer largely used for machine control and other dedicated tasks requiring intelligence.

Remote terminal unit (RTU): A computer-like device that connects field devices to the supervisory computer in a SCADA system; RTU can also stand for remote telemetry unit, or remote telecontrol unit.

Supervisory Control and Data Acquisition (SCADA): A control system used in the pipeline industry that collects and assembles data and also allows operators to control the pipeline remotely.

Steady state: The condition in which pipeline variables, such as pressure and flow, are relatively constant over a period of time.

Transmix: An unusable mixture of two different products usually created by switching from one incompatible product to another (stands for "transportation mixture").

Valve interlock: A control system feature that prevents unsuitable valve alignments and ensures that valve sequencing occurs in the correct order.

Variable frequency drive (VFD): An electric motor paired with an electronic controller that adjusts motor speed by varying the power supply voltage and frequency.

Additional Resources

The following recommended resources can provide additional helpful information related to the requirements for this covered task:

NCCER Module AOCCC-17, *Abnormal Operating Conditions Control Center*
NCCER Module AOCFG-17, *Abnormal Operating Conditions Field and Gas*
NCCER Module 12406-16, *Programmable Logic Controllers*, from *Instrumentation Level Four*
NCCER Module 64205-02, *Supervisory Control Systems*
NCCER Module CT19_5-17, *Actuator/Operator Adjustment, Electric*
NCCER Module CT19_6-17, *Actuator/Operator Adjustment, Pneumatic*
NCCER Module CT19_7-17, *Actuator/Operator Adjustmen, Hydraulic*
NCCER Module CT23_1-17, *Maintain/Repair Relief Valves*
NCCER Module CT24_1-17, *Inspect, Test, and Calibrate Pressure Limiting Devices*
NCCER Module CT44_3-17, *Inspect, Test, and Maintain a Liquid Pipeline Leak Detection Flow Computer*
The following websites offer resources for products and training:
American Petroleum Institute, **www.api.org**
Pipeline and Hazardous Materials Safety Administration, **www.phmsa.dot.gov**

Figure Credits

ABB, Module Opener, Figure 21
Apollo Valves and Elkhart Products Corporation, Figure 1
Bray Controls USA
Division of Bray International, Inc., Figures 4, 6
Rockwell Automation, Inc., Figure 5
OMNI Flow Computers, Inc., Figure 10
© Stepan Popov/Dreamstime.com, Figure 13
Satam, Figure 14
Tim Mills, Figure 18
Unitronics, Inc., Figure 20
Photos Courtesy of Entergy New Orleans, Inc. ©2016 All Rights Reserved, Figures 24–29

Section Review Answer Key

Answer	Section Reference	Objective
Section One		
1. d	1.1.1	1a
2. b	1.2.1	1b
3. a	1.3.1	1c
4. b	1.3.5	1c
Section Two		
1. a	2.1.1	2a
2. b	2.1.2	2a
3. d	2.2.1	2b

NCCER CURRICULA — USER UPDATE

NCCER makes every effort to keep its textbooks up-to-date and free of technical errors. We appreciate your help in this process. If you find an error, a typographical mistake, or an inaccuracy in NCCER's curricula, please fill out this form (or a photocopy), or complete the online form at **www.nccer.org/olf**. Be sure to include the exact module ID number, page number, a detailed description, and your recommended correction. Your input will be brought to the attention of the Authoring Team. Thank you for your assistance.

Instructors – If you have an idea for improving this textbook, or have found that additional materials were necessary to teach this module effectively, please let us know so that we may present your suggestions to the Authoring Team.

NCCER Product Development and Revision

13614 Progress Blvd., Alachua, FL 32615

Email: curriculum@nccer.org

Online: www.nccer.org/olf

❑ Trainee Guide ❑ Lesson Plans ❑ Exam ❑ PowerPoints Other ____________________

Craft / Level: Copyright Date:

Module ID Number / Title:

Section Number(s):

Description:

Recommended Correction:

Your Name:

Address:

Email: Phone:

CT65_1-17

Start-Up of a Gas Pipeline

Overview

A newly installed natural gas pipeline system must be started up by introducing gas to the line while the main-line compressors at each station are activated in the proper sequence. This module describes the procedure for starting up a gas pipeline.

Trainees with successful module completions may be eligible for credentialing through the NCCER Registry. To learn more, go to **www.nccer.org** or contact us at 1.888.622.3720. Our website has information on the latest product releases and training, as well as online versions of our *Cornerstone* magazine and Pearson's product catalog.

Your feedback is welcome. You may email your comments to **curriculum@nccer.org**, send general comments and inquiries to **info@nccer.org**, or fill in the User Update form at the back of this module.

This information is general in nature and intended for training purposes only. Actual performance of activities described in this manual requires compliance with all applicable operating, service, maintenance, and safety procedures under the direction of qualified personnel. References in this manual to patented or proprietary devices do not constitute a recommendation of their use.

CT65_1-17
Start-Up of a Gas Pipeline

Objective

When you have completed this module, you will be able to do the following:

1. Describe how to start up a gas pipeline (CT65_1-17).
   a. Describe the process for starting up a gas pipeline.
   b. Identify the skills and knowledge required to start up a gas pipeline.

Performance Task

Under the supervision of the instructor, you should be able to do the following:

1. Demonstrate how to start up a gas pipeline (CT65_1-17).
   - Identify potential abnormal operating conditions (AOCs) that may occur during performance of this CT, and know the appropriate actions to take in response to them.
   - Utilize the appropriate personal protective equipment (PPE) according to relevant company procedures.
   - Verify that the flows, pressures, and product characteristics are within normal range.
   - Ensure that upstream and downstream stations are prepared to bring the pipeline up before beginning the start-up procedure.
   - Units shall be activated (started up) beginning at the origination point and proceeding downstream to the delivery point. Units at downstream stations are to be activated upon arrival of a pressure rise from the upstream station.
   - Complete appropriate documentation as required by operator's procedures.

Trade Terms

Abnormal operating conditions (AOCs)
Compressor stations
Line fill
Line pack
Pneumatics
Maximum allowable operating pressure (MAOP)
Steady state

Industry Recognized Credentials

If you are training through an NCCER-accredited sponsor, you may be eligible for credentials from NCCER's Registry. The ID number for this module is CT65_1-17. Note that this module may have been used in other NCCER curricula and may apply to other level completions. Contact NCCER's Registry at 1.888.622.3720 or go to **www.nccer.org** for more information.

Contents

Figures and Tables

Section One

1.0.0 Starting Up a Gas Pipeline

Objective

Describe how to start up a gas pipeline (CT65_1-17).

a. Describe the process for starting up a gas pipeline.
b. Identify the skills and knowledge required to start up a gas pipeline.

Performance Task

Demonstrate how to start up a pipeline (CT65_1-17).

- Identify potential abnormal operating conditions (AOCs) that may occur during performance of this CT, and know the appropriate actions to take in response to them.
- Utilize the appropriate personal protective equipment (PPE) according to relevant company procedures.
- Verify that the flows, pressures, and product characteristics are within normal range.
- Ensure that upstream and downstream stations are prepared to bring the pipeline up before beginning the start-up procedure.
- Units shall be activated (started up) beginning at the origination point and proceeding downstream to the delivery point. Units at downstream stations are to be activated upon arrival of a pressure rise from the upstream station.
- Complete appropriate documentation as required by operator's procedures.

Trade Terms

Abnormal operating conditions (AOCs): Conditions within a pipeline system, identified by the operator, that may indicate (1) a malfunction of a component or (2) a deviation from normal operations that may indicate a condition exceeding design limits or result in a hazard to persons, property, or the environment.

Compressor stations: Facilities along a natural gas pipeline that compress the gas, allowing it to continue traveling along the line to the intended destination.

Line fill: The amount of gas required to fill a new line before deliveries can be made.

Line pack: The introduction of new gas into the pipeline at the receipt point.

Maximum allowable operating pressure (MAOP): The highest level of internal gauge pressure allowed for a particular pipeline or pipeline segment, as set by the US Department of Transportation.

Pneumatics: The branch of physics that deals with the mechanical properties of air and other gases.

Steady state: The point at which pressures and flows are relatively constant over time.

In gas pipeline operations, a full system start-up is performed only on a newly installed system, because an entire pipeline system is almost never shut down. Personnel should be familiar with the procedures involved in starting up a gas pipeline system. They must also understand the abnormal operating conditions (AOCs) that may occur, and the proper response to each.

> **CAUTION**
>
> An individual's actions during the performance of a covered task could cause a safety device to activate. The activation represents a deviation from normal operations, resulting in an abnormal operating condition (AOC).

1.1.0 Starting Up a Gas Pipeline

Personnel who perform a gas pipeline system start-up must possess basic knowledge in the following areas:

- Pipeline pneumatics
- The maximum allowable operating pressure (MAOP) for the pipeline system
- Company general operations policies
- Company policies for the appropriate response to problems or AOCs

Personnel must also be familiar with the pipeline's normal operations. They must know what is normal so that they can quickly spot anything that is not normal.

There must be a clear flow path from the origin point to the delivery point before the pipeline is started up. All equipment must be functioning properly and ready to receive the gas flow, which moves downstream from the origin point.

Compressors at downstream compressor stations are activated by a pressure rise from upstream stations.

> **NOTE**
> In addition to compressors for compressing the gas, compression stations may also contain scrubbers, strainers, or separators to remove water, condensed hydrocarbons, dirt, particles, and other impurities.

Specific steps for a start-up procedure, and the order in which the steps are taken, vary from one company to another. In general, a typical start-up procedure may include some or all of the following steps:

Step 1 Identify potential abnormal operating conditions (AOCS) that may occur, and know the appropriate actions to take in response to them.

Step 2 Utilize the appropriate personal protective equipment (PPE) according to relevant company procedures.

Step 3 Confirm that the proper work orders and work schedules have been issued.

Step 4 Confirm that the gas is ready to enter the pipeline.

Step 5 Verify that the flows, pressures, and product characteristics are within normal range.

Step 6 Confirm that upstream and downstream stations are ready for start-up. This step includes confirming that all necessary equipment is available.

Step 7 Confirm that all main-line station valves are in the normal position and that the flow path is correct.

Step 8 Start the compressors. A main-line compressor is shown in *Figure 1*.

Step 9 Notify the next station downstream that the compressors are running.

Step 10 Start the compressor units at stations on the line that will be affected immediately.

Step 11 Step 11 Notify the next station downstream that the compressor unit at that station is coming on line.

Figure 1 Main-line compressor.

Step 12 Set the compressor unit to the desired pressure. Watch for the rise in pressures downstream.

Step 13 Notify each station downstream that the compressor unit is coming on line.

Step 14 Monitor the system until the pipeline is operating in a steady state.

Step 15 Complete appropriate documentation as required by operator's procedures.

1.2.0 Prerequisites, Competencies, and Abnormal Operating Conditions

Personnel must possess a variety of skills in order to successfully complete a gas pipeline system start-up. The operation itself requires a number of steps, which the personnel conducting the start-up must be competent to perform. Personnel must be aware of the particular abnormal operating conditions (AOCs) that may occur during the procedure and know the proper responses to them. Personnel must also understand the company's safety policies and procedures, including the proper use of personal protective equipment (PPE).

Gas Pipelines

Compressor stations are typically located every 40 to 100 miles along a gas pipeline. Pressures in the pipeline can range anywhere from 200 psi up to 1,500 psi, and the gas flows through the pipeline at speeds of 10 mph to 20 mph.

1.2.1 Prerequisites and Competencies

Personnel who will be conducting a gas pipeline system start-up must possess a basic understanding of pipeline processes, including the pneumatic principles that govern the flow of gas through the line. They must also understand the company policies on operations of the particular pipeline system, including the proper unit setup for the desired rate of flow, as well as company policies for emergency response.

Prior to performing a system start-up, all personnel involved must become familiar with the variations in the general start-up procedure associated with the pipeline to which they are assigned. They must know the MAOP for the line in order to ensure that it is not exceeded during the procedure. They must also be familiar with the normal operating pressure, flow rates, **line pack**, and **line fill** for the pipeline in order to verify that they are within the normal range.

Personnel must be able to verify that a source of gas is available and be able to establish a clear flow path for the gas from origination to delivery before beginning the start-up procedure. This may require making a phone call, opening valves remotely, or having someone physically go to the valves (*Figure 2*) to open them, depending on company procedures and the level of automation within the system.

Personnel must have a clear understanding of the sequence of events that occurs as a pipeline system is being started. Once the proper flow path is established, all stations have confirmed that they are ready, and the next station downstream has been notified that the unit is coming on line, the sequence begins with the starting of the origination compressor. Personnel in the next station downstream set their unit to the desired pressure and watch for a rise in line pressure. When the rise is detected, that unit is brought on line. This sequence continues in an orderly fashion until all stations are on line and the pipeline has reached a steady state. The steady state pressure must be maintained for a certain amount of time as specified in company policy before product delivery can begin.

Figure 2 Pipeline valves.

1.2.2 Compressor Start-Up

Starting compressors properly is one of the primary tasks of a full system start-up. Main-line compressors are shown in *Figure 3*. Operations personnel must understand the posted start-up procedures for the compressors at their facility. The procedures for starting up a compressor and the order to follow vary from one company to another, and each company will have its own notification and reporting requirements. In general, the following steps are taken to start a compressor:

Step 1 Identify potential abnormal operating conditions (AOCS) that may occur, and know the appropriate actions to take in response to them.

Step 2 Utilize the appropriate personal protective equipment (PPE) according to relevant company procedures.

Figure 3 Main-line compressors with electric drivers.

Step 3 Carry out any pre-start-up procedures as recommended by the manufacturer or required by company policy.

Step 4 Review and verify knowledge of the starting unit and start-up procedures.

Step 5 Ensure that the compressor and station valves are properly positioned.

Step 6 Notify all affected personnel that the compressor is about to be started.

Step 7 Confirm that all operating variables are within the safe operating limits of the compressor.

Step 8 Confirm that all safety alarms and shutdown devices are set and are working properly.

Step 9 Initiate the start-up according to the posted start-up procedure for the compressor(s) at your location.

Step 10 Follow the manufacturer's instructions for operating the compressor after it has been started.

Step 11 Complete appropriate documentation as required by operator's procedures.

1.2.3 Abnormal Operating Conditions

A variety of abnormal operating conditions could potentially be encountered while performing a system start-up. Prior to carrying out pipeline start-up procedures, personnel must be aware of this possibility and be prepared to respond appropriately should one or more of these conditions occur.

The following are some potential AOCs, and the appropriate action personnel should take in response to them:

- *An exposed pipe, or missing or damaged pipeline supports occurs* – Personnel should notify the control center and additional personnel as required by company policy to initiate repairs. Inspection of the line should be made prior to start-up to look for this AOC.
- *Activation of a safety device occurs* – Personnel should first verify the activation, and if a hazardous condition exists, act quickly to minimize the situation. Notifications should be made to the control center and all other personnel as specified in company policy. Never ignore activated safety devices.

Figure 4 SCADA display on a smartphone.

- *Failure of communications systems, SCADA, or electrical services occurs* – Personnel should initiate the troubleshooting procedures specified for the component that has failed, and make all contacts as required by company policy. A SCADA display is shown in Figure 4.
- *Fire or explosion occurs* – Personnel should immediately evacuate the area and limit access. Proper notifications, including calling 911, should be made as specified in company policy.
- *Damage to a pipeline system component (such as being broken, bent, crushed, gouged, or burned) occurs* – Personnel should make notifications as specified by company policy. This may include minimizing or eliminating the system pressure or flow rate as appropriate for the severity of the situation, as well as repairing or replacing the damaged component.
- *Unexpected release of hazardous liquid or carbon dioxide occurs* – Personnel should contact the control center to isolate or shut down the line, if the leak cannot be isolated safely at the site. Personnel should evacuate the area.
- *An unexplained flow rate deviation (such as high flow, low flow, or no flow) occurs* – Personnel should first verify equipment readings, and confirm that the proper valves are aligned and open. They should also verify that any units and/or valves operated prior to this deviation are still in the desired position. In any case, personnel must always work to minimize any potentially hazardous situation that may develop as a result of the deviation.

- *Unexplained pressure deviation (such as increase, decrease, or lack of pressure) occurs* – Personnel should assume that the reading is correct and immediately work to identify the cause of the deviation. They should verify that any units and/or valves operated prior to this deviation are still in the desired position. It may be necessary to shut down the line if the cause cannot be immediately determined and corrected. Personnel must always work to minimize any potentially hazardous situation that may develop as a result of the deviation. A pressure transmitter is shown in *Figure 5*.
- *An unexplained status change occurs* – Personnel should act immediately to determine the cause and minimize the situation if necessary.

WARNING!

Under no circumstances should any person attempt to reach a valve when a vapor cloud is observed in the area.

Figure 5 Electronic pressure transmitter.

Additional Resources

It is assumed that you will learn to perform these procedures under the direct supervision of your crew leader, foreman, or instructor. For independent study, refer to the recommended Additional Resources at the back of this module.

1.0.0 Section Review

1. Personnel conducting a gas pipeline system start-up must possess a basic understanding of pipeline processes, including the ______.
   a. dynamic principles governing air and gas
   b. pneumatic principles governing gas flow
   c. hydraulic principles governing water and gas
   d. mechanical principles governing pipeline machinery

2. Before delivery can begin, a steady state pressure must be maintained _____.
   a. for at least one hour
   b. until 70 percent of MAOP is reached
   c. until all compressors are at full capacity
   d. as specified in company policy

Summary

Newly installed pipeline systems must be started up by first ensuring that all work orders and schedules have been issued, all valves are properly aligned to establish the correct flow path, and that gas is ready to enter the line. After notification to all affected personnel that the compressor is coming on line, the compressor at the originating station is activated first. A rise in system pressure activates the next compressor downstream. This sequence continues until all compressors are on line and the system has reached a steady state.

Personnel performing a system start-up must understand basic pipeline processes, pneumatic principles, and how to set up a system for the desired flow rate. They must also know the start-up procedures for their particular system including the MAOP, normal operating pressures, flow rates, line packs, and line fill. Compressors are started and operated in accordance with the manufacturer's instructions and company procedures.

Personnel must be aware of the potential AOCs that could occur during start-up and the proper responses to them. Exposed pipe, missing or damaged pipeline supports, or other structural conditions must be reported and repaired or replaced. Troubleshooting must begin if any failure of communications or electrical supply occurs. If a safety device activates, personnel must first verify the activation and act to minimize any hazardous situation. This may require minimizing or eliminating system pressure or flow rate.

Review Questions

1. Prior to system start-up, _____.
   a. a clear flow path must be established
   b. a steady state condition must be established
   c. confirm that the line pressure is no more than 70 percent of MAOP
   d. ensure that station and compressor valves are closed

2. During system start-up, a pressure rise from upstream stations _____.
   a. activates downstream stations
   b. activates remotely controlled safety devices
   c. signals an abnormal operating condition
   d. must not exceed 89 percent of MAOP

3. Prior to system start-up, confirm that _____.
   a. main-line station valves are closed
   b. main-line station valves are in the normal position
   c. upstream station valves are open and downstream stations valves are closed
   d. upstream station valves are closed and downstream station valves are opened

4. After system start-up, the system should be monitored until _____.
   a. the pipeline pressure reaches MAOP
   b. the pipeline pressure reaches 70 percent of MAOP
   c. the pipeline is operating in a steady state
   d. a rise in downstream pressure is observed

5. Compressor start-up is initiated according to _____.
   a. the target line pressure
   b. 49*CFR*195.402
   c. company policy
   d. the posted start-up procedure

Trade Terms Quiz

Fill in the blank with the correct term that you learned from your study of this module.

1. The introduction of new gas into a pipeline is called ________.
2. The highest allowable internal gauge pressure is called ________.
3. The branch of physics that deals with the mechanical properties of air and other gases is called ________.
4. Conditions within a pipeline system that may indicate (1) a malfunction of a component or (2) a condition exceeding design limits or resulting in a hazard to persons, property, or the environment are called ________.
5. The point at which pressures and flows are relatively constant over time is called ________.
6. The amount of gas needed to fill a new pipeline is called ________.
7. Facilities along a natural gas pipeline that compress the gas, allowing it to continue traveling along the line to the intended destination are known as ________.

Trade Terms

Abnormal operating conditions (AOCs)
Compressor stations
Line fill
Line pack
Pneumatics
Maximum allowable operating pressure (MAOP)
Steady state

Trade Terms Introduced in This Module

Abnormal operating conditions (AOCs): Conditions within a pipeline system, identified by the operator, that may indicate (1) a malfunction of a component or (2) a deviation from normal operations that may indicate a condition exceeding design limits or result in a hazard to persons, property, or the environment.

Compressor stations: Facilities along a natural gas pipeline that compress the gas, allowing it to continue traveling along the line to the intended destination.

Line fill: The amount of gas required to fill a new line before deliveries can be made.

Line pack: The introduction of new gas into the pipeline at the receipt point.

Pneumatics: The branch of physics that deals with the mechanical properties of air and other gases.

Maximum allowable operating pressure (MAOP): The highest level of internal gauge pressure allowed for a particular pipeline or pipeline segment, as set by the US Department of Transportation.

Steady state: The point at which pressures and flows are relatively constant over time.

Additional Resources

The following recommended resources can provide additional helpful information related to the requirements for this covered task:

API Recommended Practice 1161, Third Edition, © 2014 American Petroleum Institute. API Publishing Services, 1220 L Street, NW, Washington, DC 20005. Distributed by Thompson Reuters (Scientific) LLC, **www.techstreet.com**.

49 *CFR* 192, **www.ecfr.gov**.

49 *CFR* 195, **www.ecfr.gov**.

NCCER Module AOCCC-17, *Abnormal Operating Conditions Control Center.*

NCCER Module AOCFG-17, *Abnormal Operating Conditions Field and Gas.*

NCCER Module 62105-14, *Introduction toPipeline Documents.*

NCCER Module 62302-02, *Pipeline Damage Inspection.*

NCCER Module 62304-02, *Pipeline Repair.*

NCCER Module 63104-02, *Piping and Mechanical Blueprint Reading.*

NCCER Module 63208-02, *Introduction to Gas Compressors.*

NCCER Module 67102-15, *Basic Pipeline Pneumatics and Equipment.*

NCCER Module 67103-15, *Pipeline Communications.*

NCCER Module 75210-03, *Work Permit Policies*, from Field Safety.

NCCER Module CT18_1-17, *Utilize Leak Survey Techniques.*

NCCER Module CT21_2-17, *Disassembly/Re-Assembly of a Valve.*

NCCER Module CT21_3-17, *Internal Inspection of Valve and Components.*

NCCER Module CT65_3-17, *Monitor pressures, flows, communications and line integrity and maintain them within allowable limits (gas).*

Figure Credits

arhendrix/Shutterstock.com, Module Opener
GTS Productions/Shutterstock.com, Figure 1
INSAGO/Shutterstock.com, Figure 2
© iStockphoto.com/westphalia, Figure 3
ABB, Figure 4
Tim Mills, Figure 5

Section Review Answer Key

Answer	Section Reference	Objective
Section One		
1. b	1.2.0	1b
2. d	1.2.1	1b

NCCER CURRICULA — USER UPDATE

NCCER makes every effort to keep its textbooks up-to-date and free of technical errors. We appreciate your help in this process. If you find an error, a typographical mistake, or an inaccuracy in NCCER's curricula, please fill out this form (or a photocopy), or complete the online form at **www.nccer.org/olf**. Be sure to include the exact module ID number, page number, a detailed description, and your recommended correction. Your input will be brought to the attention of the Authoring Team. Thank you for your assistance.

Instructors – If you have an idea for improving this textbook, or have found that additional materials were necessary to teach this module effectively, please let us know so that we may present your suggestions to the Authoring Team.

NCCER Product Development and Revision

13614 Progress Blvd., Alachua, FL 32615

Email: curriculum@nccer.org

Online: www.nccer.org/olf

❑ Trainee Guide ❑ Lesson Plans ❑ Exam ❑ PowerPoints Other ______________

Craft / Level: Copyright Date:

Module ID Number / Title:

Section Number(s):

Description:

Recommended Correction:

Your Name:

Address:

Email: Phone:

CT65_2-17

Shutdown of a Gas Pipeline

Overview

A gas pipeline or pipe segment may need to be shut down for maintenance, inspection, or due to an emergency. This module describes general procedures for shutting down a gas pipeline, as well as the prerequisites, competencies, and abnormal operating conditions associated with these procedures.

Trainees with successful module completions may be eligible for credentialing through the NCCER Registry. To learn more, go to **www.nccer.org** or contact us at 1.888.622.3720. Our website has information on the latest product releases and training, as well as online versions of our *Cornerstone* magazine and Pearson's product catalog.

Your feedback is welcome. You may email your comments to **curriculum@nccer.org**, send general comments and inquiries to **info@nccer.org**, or fill in the User Update form at the back of this module.

This information is general in nature and intended for training purposes only. Actual performance of activities described in this manual requires compliance with all applicable operating, service, maintenance, and safety procedures under the direction of qualified personnel. References in this manual to patented or proprietary devices do not constitute a recommendation of their use.

CT65_2-17

Shutdown of a Gas Pipeline

Objective

When you have completed this module, you will be able to do the following:

1. Describe how to shut down a gas pipeline (CT65_2-17).
   a. Describe how to shut down a gas pipeline.
   b. Identify the prerequisites, competencies, and abnormal operating conditions associated with shutting down a gas pipeline.

Performance Task

Under the supervision of the instructor, you should be able to do the following:

1. Shut down a gas pipeline (CT65_2-17).
   - Identify potential abnormal operating conditions (AOCs) that may occur during performance of this CT, and know the appropriate actions to take in response to them.
   - Utilize the appropriate personal protective equipment (PPE) according to relevant company procedures.
   - Understand the desired pressure to keep on a non-running line in accordance with company policies.
   - Understand the desired pressure to keep on a static line in accordance with company policies.
   - Shut down pump units or compressor units beginning at the origination point and proceeding downstream to the destination. Depending on the line profile, sufficient units at the source station and at the other stations may be kept operational. Then shut down in conjunction with valve closure at the delivery terminals or stations to satisfy the policies on maintaining minimum pressures on shutdown of the particular line.
   - The closure of isolation valves at intermediate delivery points should occur in conjunction with the shutdown of that station's units unless otherwise specified in company policy or procedure.
   - Complete appropriate documentation as required by operator's procedures.

Trade Terms

Abnormal operating conditions (AOCs)
Blowdown
Compressor stations
Line fill
Line pack
Maximum allowable operating pressure (MAOP)
Maximum operating pressure (MOP)
Pneumatics
Steady state

Industry Recognized Credentials

If you are training through an NCCER-accredited sponsor, you may be eligible for credentials from NCCER's Registry. The ID number for this module is CT65_2-17. Note that this module may have been used in other NCCER curricula and may apply to other level completions. Contact NCCER's Registry at 1.888.622.3720 or go to **www.nccer.org** for more information.

Contents

Figures

SECTION ONE

1.0.0 SHUTDOWN OF A GAS PIPELINE

Objective

Describe how to shut down a gas pipeline (CT65_2-17).

a. Describe how to shut down a gas pipeline.
b. Identify the prerequisites, competencies, and abnormal operating conditions associated with shutting down a gas pipeline.

Performance Task

Shut down a gas pipeline (CT65_2-17).

- Identify potential abnormal operating conditions (AOCs) that may occur during performance of this CT, and know the appropriate actions to take in response to them.
- Utilize the appropriate personal protective equipment (PPE) according to relevant company procedures.
- Understand the desired pressure to keep on a non-running line in accordance with company policies.
- Understand the desired pressure to keep on a static line in accordance with company policies.
- Shut down pump units or compressor units beginning at the origination point and proceeding downstream to the destination. Depending on the line profile, sufficient units at the source station and at the other stations may be kept operational. Then shut down in conjunction with valve closure at the delivery terminals or stations to satisfy the policies on maintaining minimum pressures on shutdown of the particular line.
- The closure of isolation valves at intermediate delivery points should occur in conjunction with the shutdown of that station's units unless otherwise specified in company policy or procedure.
- Complete appropriate documentation as required by operator's procedures.

Trade Terms

Abnormal operating conditions (AOCs): Conditions within a pipeline system, identified by the operator, that may indicate (1) a malfunction of a component or (2) a deviation from normal operations that may indicate a condition exceeding design limits or result in a hazard to persons, property, or the environment.

Blowdown: The depressurization of gas piping, which is accomplished by releasing the pressure at a safe atmospheric vent location.

Compressor stations: Facilities along a natural gas pipeline that compress the gas, allowing it to continue traveling along the line to the intended destination.

Line fill: The amount of gas required to fill a new line before deliveries can be made.

Line pack: The introduction of new gas into the pipeline at the receipt point.

Maximum allowable operating pressure (MAOP): The highest level of internal gauge pressure allowed for a particular pipeline or pipeline segment, as set by the US Department of Transportation.

Maximum operating pressure (MOP): The maximum internal gauge pressure at which a hazardous liquid or gas pipeline system or segment may be operated.

Pneumatics: The branch of physics that deals with the mechanical properties of air and other gases.

Steady state: The point at which pressures and flows are relatively constant over time.

In gas pipeline operations, the entire pipeline system is almost never shut down. However, segments of the pipeline may require shutdown for maintenance, inspection, or due to an emergency. Personnel should be familiar with the procedures involved in shutdown of a gas pipeline system. They must also understand the **abnormal operating conditions (AOCs)** that may occur, and the proper response to each.

> **CAUTION**
>
> An individual's actions during the performance of a covered task could cause a safety device to activate. The activation represents a deviation from normal operations, resulting in an abnormal operating condition (AOC).

1.1.0 Shutting Down a Gas Pipeline

A gas pipeline shutdown procedure begins by first identifying the portion of the pipeline system to be shut down; confirming the intended shutdown with the control center; ensuring that the necessary valves are aligned; and making notifications to all affected personnel. The pressure and flow rates are monitored during the procedure to ensure that the operation remains within safe limits. The shutdown procedure ends when the portion of the pipeline to be shut down reaches a static or steady state, or when the control center assumes control of the operation.

Specific steps for a shutdown procedure, and the order in which the steps are taken, vary from one company to another. However, a planned shutdown typically involves most or all of the same basic actions and general procedures. The following steps outline the general procedure for shutting down a gas pipeline:

Step 1 Identify potential abnormal operating conditions (AOCs) that may occur during performance of this CT, and know the appropriate actions to take in response to them.

Step 2 Utilize the appropriate personal protective equipment (PPE) according to relevant company procedures.

Step 3 Confirm that the proper work orders and work schedules have been issued.

Step 4 Confirm that all proper notifications have been made, including which compressor stations will remain on line and which will be shutting down their compressor units. A main-line compressor is shown in *Figure 1*.

Step 5 Confirm that upstream and downstream compressor stations are ready for shutdown. This step includes confirming that all necessary equipment is available.

Step 6 Shut down designated compressor units beginning at the origination point and proceeding downstream to the delivery terminal. Depending on the line profile, sufficient units at the source station and at other stations may be kept operational in order to maintain the desired end pressure.

Step 7 As each unit is shut down, close the associated delivery valves in accordance with company policy in order to maintain the minimum pressure on the line. Pipeline valves are shown in *Figure 2*. Isolating the pipeline from compressor stations and delivery equipment will minimize the potential for pressures to bleed down over time.

Step 8 Monitor the system until the pipeline is operating in a static or steady state.

Step 9 After shutdown, maintain the appropriate end pressure on the line to minimize contamination of the product and to leave the pipeline in the proper condition for start-up.

Step 10 Ensure that all the required documentation has been completed.

1.2.0 Prerequisites, Competencies, and Abnormal Operating Conditions

Personnel must possess a variety of skills in order to successfully complete a system shutdown. The operation itself requires a number of steps that

Figure 1 Main-line compressor.

Figure 2 Pipeline valves.

the personnel conducting the shutdown must be competent to perform. Personnel must be aware of the particular abnormal operating conditions that may occur during the procedure and know the proper responses to them. Personnel must also understand the company's safety policies and procedures, including the proper use of personal protective equipment (PPE), and policies for emergency response.

1.2.1 Prerequisites and Competencies

Personnel who will be conducting a system shutdown must possess a basic understanding of pipeline processes. They must understand the **pneumatics** of the pipeline, the principles that govern the flow of gas through the line. They must also understand the company policies on operations of the particular pipeline system, including the proper unit setup for the desired pressure and rate of flow.

Before performing a system shutdown, all personnel involved must become familiar with the variations in the general shutdown procedure associated with the pipeline to which they are assigned. They must know the **maximum operating pressure (MOP)** and the **maximum allowable operating pressure (MAOP)** for the line in order to ensure that they are not exceeded during the procedure. They must also be familiar with the normal operating pressures, flow rates, **line pack**, and **line fill** for the pipeline in order to verify that they are within the normal range.

Personnel must be familiar with the desired end pressure to maintain on a static or nonrunning line in accordance with company policy. They must know how to shut down compressor units following the posted shutdown procedure, and know to close the delivery valves after unit shutdown in order to prevent the pipeline pressure from bleeding down.

Personnel must have a clear understanding of the sequence of events that occurs as a pipeline system is being shut down. The sequence begins by shutting down the designated compressor units at the origin point and continuing downstream to the terminal point. As each unit is shut down, the associated valves are closed to isolate the line from the units or other equipment. This sequence continues in an orderly fashion until all designated compressor units are offline and the pipeline has reached a static or steady state. The end pressure must be maintained to prevent product contamination and to leave the line in the proper condition for start-up.

Did You Know?

Gas Pipelines

Compressor stations are typically located every 40 to 100 miles along a gas pipeline. Pressures in the pipeline can range between 200 psi and 1,500 psi, and the gas flows through the pipeline at speeds from 10 mph to 20 mph.

1.2.2 Compressor Shutdown

Shutting down the compressors properly is one of the primary tasks of a system shutdown. A main-line compressor is shown in *Figure 3*. Operations personnel must understand and be able to follow the posted shutdown procedures for the compressors at their facility. The procedures for shutting down a compressor and the order to follow vary from one company to another, and each company will have its own notification and reporting requirements. In general, the following steps are performed to shut down a compressor:

Step 1 Identify potential abnormal operating conditions (AOCs) that may occur during performance of this CT, and know the appropriate actions to take in response to them.

Step 2 Utilize the appropriate personal protective equipment (PPE) according to relevant company procedures.

Figure 3 Main-line compressors with electric drivers.

Step 3 Notify all affected personnel of a pending compressor shutdown per company policy.

Step 4 Slow down the compressor to minimum speed.

Step 5 Unload the compressor and open the bypass valve.

Step 6 Allow the compressor to cool down for the length of time specified in the manufacturer's instructions.

Step 7 Shut off the fuel or electrical supply to the unit.

Step 8 Close the suction and discharge valves.

Step 9 Initiate compressor **blowdown** to depressurize the system.

Step 10 Complete appropriate documentation according to operator's procedures.

The amount of time that the compressor may be left depressurized depends on a number of factors, including the condition of the compressor, the need to make repairs, and the availability of backup compressors. Generally, the maintenance supervisor decides how long the compressor should remain out of service.

> **NOTE**
> In addition to compressors for compressing the gas, compression stations may also contain scrubbers, strainers, or separators to remove water, condensed hydrocarbons, dirt, particles, or other impurities.

1.2.3 Abnormal Operating Conditions

A variety of abnormal operating conditions could be encountered while performing a system shutdown. Prior to carrying out pipeline start-up procedures, personnel must be aware of this possibility and be prepared to respond appropriately should one or more occur.

The following are some potential AOCs, and the appropriate action personnel should take in response to them:

- *An exposed pipe, or missing or damaged pipeline supports occurs* – Personnel should notify the control center and additional personnel as required by company policy to initiate repairs. Inspection of the line should be made prior to start-up to look for this AOC.
- *Activation of a safety device occurs* – Personnel should first verify the activation, and if a hazardous condition exists, act quickly to minimize the situation. Notifications should be made to the control center and all other personnel as specified in company policy. Never ignore activated safety devices.
- *Failure of communications systems, SCADA, or electrical services occurs* – Personnel should initiate the troubleshooting procedures specified for the component that has failed, and make all contacts as required by company policy. A SCADA display is shown in *Figure 4*.
- *Fire or explosion occurs* – Personnel should immediately evacuate the area and limit access. Proper notifications, including calling 911, should be made as specified in company policy.
- *Damage to a pipeline system component (such as being broken, bent, crushed, gouged, or burned) occurs* – Personnel should make notifications as specified by company policy. This may include minimizing or eliminating the system pressure or flow rate as appropriate for the severity of the situation, as well as repairing or replacing the damaged component.
- *Unexpected hazardous liquid or carbon dioxide occurs* – Personnel should contact the control center to isolate or shut down the line, if the leak cannot be isolated safely at the site. Personnel should evacuate the area.
- *An unexplained flow rate deviation (such as high flow, low flow, or no flow) occurs* – Personnel should first verify equipment readings, and confirm that the proper valves are aligned and open. They should also verify that any units and/or valves operated prior to this deviation are still in the desired position. In any case, personnel must always work to minimize any

Figure 4 SCADA display on a smartphone.

potentially hazardous situation that may develop as a result of the deviation.

- *Unexplained pressure deviation (such as increase, decrease, or lack of pressure) occurs* – Personnel should assume that the reading is correct and immediately work to identify the cause of the deviation. They should verify that any units and/or valves operated prior to this deviation are still in the desired position. It may be necessary to shut down the line if the cause cannot be immediately determined and corrected. Personnel must always work to minimize any potentially hazardous situation that may develop as a result of the deviation. A pressure transmitter is shown in *Figure 5*.
- *An unexplained status change occurs* – Personnel should act immediately to determine the cause and minimize the situation if necessary.

WARNING!

Under no circumstances should any person attempt to reach a valve when a vapor cloud is observed in the area.

Figure 5 Electronic pressure transmitter.

Additional Resources

It is assumed that you will learn to perform these procedures under the direct supervision of your crew leader, foreman, or instructor. For independent study, refer to the recommended Additional Resources at the back of this module.

1.0.0 Section Review

1. During a system shutdown, what must be done as each designated unit is being shut down?
   a. Blow down the main-line.
   b. Bleed the delivery line.
   c. Close the associated delivery valves.
   d. Ensure that the system is in a static state.

2. Personnel conducting a gas pipeline system shutdown must possess a basic understanding of pipeline processes, including the ______.
   a. dynamic principles governing air and gas
   b. pneumatic principles governing the flow of gas
   c. hydraulic principles of water and air
   d. mechanical principles governing pipeline machinery

SUMMARY

Pipeline systems or segments may need to be shut down for maintenance, inspection, or due to an emergency. The shutdown begins when a portion of the line to be shut down is identified and confirmed, valves are aligned, and notifications are made. The shutdown ends when the line reaches a static or steady state, or when the control center assumes control. The steps for shutdown include confirming that work orders, schedules, and notifications are complete, and that all stations are ready.

Compressor shutdown procedures vary from company to company. Typical steps for compressor shutdown include making notifications that the compressor is to be shut down; slowing the unit to minimum speed; unloading the compressor; and opening the bypass valve.

Personnel must be aware of the potential AOCs that could occur during shutdown and the proper responses to them. Exposed pipe, missing or damaged pipeline supports, or other structural conditions must be reported and repaired or replaced. This may require minimizing or eliminating system pressure or flow rate. Troubleshooting must begin if any failure of communications or electrical supply occurs. If a safety device activates, personnel must first verify the activation and act to minimize any hazardous situation.

Review Questions

1. Prior to system shutdown, confirm that the compressor stations that will be shutting down their compressor units _____.
   a. are at idle
   b. have been notified
   c. have closed their delivery valves
   d. have depressurized their system

2. During system shutdown, the system should be monitored until the pipeline _____.
   a. pressure reaches MOP
   b. pressure reaches 50 percent of MOP
   c. is operating in a static or steady state
   d. pressure reaches zero

3. Personnel conducting a system shutdown must be familiar with the proper unit setup for _____.
   a. the desired pressure and flow rate
   b. achieving system blowdown
   c. maintaining a system static state prior to shutdown
   d. achieving the desired MOP and MAOP

4. The steps to follow for compressor shutdown are dictated by _____.
   a. the target line pressure
   b. 49*CFR*195.402
   c. the gas flow rate
   d. the posted shutdown procedures

5. If an unexplained status change occurs, personnel should _____.
   a. contact the control center for assistance
   b. determine the cause and minimize the situation
   c. shut down the system until the cause of the change is determined
   d. reduce the pressure and/or flow rates to minimize the situation

Trade Terms Quiz

Fill in the blank with the correct term that you learned from your study of this module.

1. The introduction of new gas into a pipeline is called _________.
2. The highest allowable internal gauge pressure is called _________.
3. The branch of physics that deals with the mechanical properties of air and other gases is called _________.
4. The point at which pressures and flows are relatively constant over time is called _________.
5. Conditions within a pipeline system that may indicate (1) a malfunction of a component or (2) a condition exceeding design limits or resulting in a hazard to persons, property, or the environment are called _________.
6. The amount of gas needed to fill a new line is called _________.
7. The maximum internal gauge pressure at which a pipeline system or segment may be operated is called the _________.
8. The depressurization of gas piping by releasing pressure through a vent is called _________.
9. Facilities along a pipeline that compress the gas are called _________.

Trade Terms

Abnormal operating conditions (AOCs)
Blowdown
Compressor stations
Line fill
Line pack
Maximum allowable operating pressure (MAOP)
Maximum operating pressure (MOP)
Pneumatics
Steady state

Trade Terms Introduced in This Module

Abnormal operating conditions (AOCs): Conditions within a pipeline system, identified by the operator, that may indicate (1) a malfunction of a component or (2) a deviation from normal operations that may indicate a condition exceeding design limits or result in a hazard to persons, property, or the environment.

Blowdown: The depressurization of gas piping, which is accomplished by releasing the pressure at a safe atmospheric vent location.

Compressor stations: Facilities along a natural gas pipeline that compress the gas, allowing it to continue traveling along the line to the intended destination.

Line fill: The amount of gas required to fill a new line before deliveries can be made.

Line pack: The introduction of new gas into the pipeline at the receipt point.

Maximum allowable operating pressure (MAOP): The highest level of internal gauge pressure allowed for a particular pipeline or pipeline segment, as set by the US Department of Transportation.

Maximum operating pressure (MOP): The maximum internal gauge pressure at which a hazardous liquid or gas pipeline system or segment may be operated.

Pneumatics: The branch of physics that deals with the mechanical properties of air and other gases.

Steady state: The point at which pressures and flows are relatively constant over time.

Additional Resources

The following recommended resources can provide additional helpful information related to the requirements for this covered task:

API Recommended Practice 1161, Third Edition, © 2014 American Petroleum Institute. API Publishing Services, 1220 L Street, NW, Washington, DC 20005. Distributed by Thompson Reuters (Scientific) LLC, **www.techstreet.com**.

49 *CFR* 192, **www.ecfr.gov**.

49 *CFR* 195, **www.ecfr.gov**.

NCCER Module AOCCC-17, *Abnormal Operating Conditions Control Center*

NCCER Module AOCFG-17, *Abnormal Operating Conditions Field and Gas*

NCCER Module 62105-14, *Introduction to Pipeline Documents*

NCCER Module 62302-02, *Pipeline Damage Inspection*

NCCER Module 62304-02, *Pipeline Repair*

NCCER Module 63104-02, *Piping and Mechanical Blueprint Reading*

NCCER Module 63208-02, *Introduction to Gas Compressors*

NCCER Module 67102-15, *Basic Pipeline Pneumatics and Equipment*

NCCER Module 67103-15, *Pipeline Communications*

NCCER Module 75210-03, *Work Permit Policies*, from Field Safety

NCCER Module CT18_1-17, *Utilize Leak Survey Techniques*

NCCER Module CT21_2-17, *Disassembly/Re-Assembly of a Valve*

NCCER Module CT21_3-17, *Internal Inspection of Valve and Components*

NCCER Module CT65_3-17, *Monitor pressures, flows, communications and line integrity and maintain them within allowable limits (gas)*

Figure Credits

arhendrix/Shutterstock.com, Module Opener

GTS Productions/Shutterstock.com, Figure 1

INSAGO/Shutterstock.com, Figure 2

© iStockphoto.com/westphalia, Figure 3

ABB, Figure 4

Tim Mills, Figure 5

Section Review Answer Key

Answer	Section Reference	Objective
Section One		
1. c	1.1.1	1a
2. b	1.2.1	1b

NCCER CURRICULA — USER UPDATE

NCCER makes every effort to keep its textbooks up-to-date and free of technical errors. We appreciate your help in this process. If you find an error, a typographical mistake, or an inaccuracy in NCCER's curricula, please fill out this form (or a photocopy), or complete the online form at **www.nccer.org/olf**. Be sure to include the exact module ID number, page number, a detailed description, and your recommended correction. Your input will be brought to the attention of the Authoring Team. Thank you for your assistance.

Instructors – If you have an idea for improving this textbook, or have found that additional materials were necessary to teach this module effectively, please let us know so that we may present your suggestions to the Authoring Team.

NCCER Product Development and Revision

13614 Progress Blvd., Alachua, FL 32615

Email: curriculum@nccer.org

Online: www.nccer.org/olf

❑ Trainee Guide ❑ Lesson Plans ❑ Exam ❑ PowerPoints Other ____________________

Craft / Level: Copyright Date:

Module ID Number / Title:

Section Number(s):

Description:

Recommended Correction:

Your Name:

Address:

Email: Phone:

CT65_3-17

Monitor Pressures, Flows, Communications, and Line Integrity and Maintain Them Within Allowable Limits (Gas)

Overview

This module describes the methods and procedures for monitoring pipeline parameters that include pipeline pressure, flow rate, and status changes. It describes the knowledge and skills required to safely monitor and maintain pipeline pressure, flow, communications, and line integrity within allowable limits according to regulations and company policy. Potential abnormal operating conditions associated with pipeline monitoring are also discussed, with the appropriate responses to each.

Trainees with successful module completions may be eligible for credentialing through the NCCER Registry. To learn more, go to **www.nccer.org** or contact us at 1.888.622.3720. Our website has information on the latest product releases and training, as well as online versions of our *Cornerstone* magazine and Pearson's product catalog.

Your feedback is welcome. You may email your comments to **curriculum@nccer.org**, send general comments and inquiries to **info@nccer.org**, or fill in the User Update form at the back of this module.

This information is general in nature and intended for training purposes only. Actual performance of activities described in this manual requires compliance with all applicable operating, service, maintenance, and safety procedures under the direction of qualified personnel. References in this manual to patented or proprietary devices do not constitute a recommendation of their use.

CT65_3-17

Monitor Pressures, Flows, Communications, and Line Integrity and Maintain Them Within Allowable Limits (Gas)

Objectives

When you have completed this module, you will be able to do the following:

1. Describe how to monitor pipeline parameters (CT65_3-17).
   a. Describe how to monitor and maintain pipeline pressure.
   b. Describe how to monitor compressor stations.
   c. Describe how to monitor pipeline right-of-way (ROW).
   d. Describe how to monitor weather conditions.
   e. Describe how to document pipeline activities.
2. Identify the knowledge and skills required to monitor pipeline parameters.
   a. Identify prerequisites, competencies, and abnormal operating conditions associated with monitoring pipeline parameters.
   b. Describe the procedure for monitoring pipeline parameters.

Performance Task

Under the supervision of the instructor, you should be able to do the following:

1. Monitor pipeline parameters (CT65_3-17).
   - Identify potential abnormal operating conditions (AOCS) that may occur during performance of this CT, and know the appropriate actions to take in response to them.
   - Utilize the appropriate personal protective equipment (PPE) according to relevant company procedures.
   - Verify that the flows, pressures, and product characteristics are within normal range.
   - Continuously monitor, by whatever means available (SCADA, pressure gauge, meter), the pressures and flow rates of the pipeline to look for any changes. Monitor the line balance for changes by monitoring line pack in gas pipelines. Monitor storage facilities. Verify continuous communications and that data is accurate and updated.
   - Determine whether the pressures, flows, and line balance are within normal range and whether you have communication capabilities.
   - If necessary, determine the cause of variation from normal. Correct the situation if possible. For pressure or flow variations, use corrective methods such as starting and/or stopping pumps/units or using adjustable control devices such as control valves, back pressure valves, and variable speed drives for electric motors or using throttle controls for combustion engines. For communications variations, maintain monitored operations by using alternative communication methods. If necessary, seek assistance from and make notifications to proper personnel.
   - Complete appropriate documentation as required by operator's procedures.

Trade Terms

Abnormal operating conditions (AOCs)
Compressor stations
Derating
Discharge pressure
Line fill
Line pack
Maximum allowable operating pressure (MAOP)
Maximum operating pressure (MOP)
Parameters
Pneumatics
Suction pressure
Supervisory Control and Data Acquisition (SCADA)

Industry Recognized Credentials

If you are training through an NCCER-accredited sponsor, you may be eligible for credentials from NCCER's Registry. The ID number for this module is CT65_3-17. Note that this module may have been used in other NCCER curricula and may apply to other level completions. Contact NCCER's Registry at 1.888.622.3720 or go to **www.nccer.org** for more information.

Contents

Figures and Table

Section One

1.0.0 Pipeline Parameters, Conditions, and Documentation

Objective

Describe how monitor pipeline parameters (CT65_3-17).

a. Describe how to monitor and maintain pipeline pressure.
b. Describe how to monitor compressor stations.
c. Describe how to monitor pipeline right-of-way (ROW).
d. Describe how to monitor weather conditions.
e. Describe how to document pipeline activities.

Trade Terms

Abnormal operating conditions (AOCs): Conditions within a pipeline system, identified by the operator, that may indicate (1) a malfunction of a component or (2) a deviation from normal operations that may indicate a condition exceeding design limits or result in a hazard to persons, property, or the environment.

Compressor stations: Facilities installed along pipelines approximately every 100 miles that compress natural gas to reduce its volume and force it through the pipeline.

Derating: Reducing the rating of a pipeline or other device to permit operation at parameters other than those originally established.

Discharge pressure: The amount of pressure generated on the downstream side of a compressor.

Maximum allowable operating pressure (MAOP): The highest level of internal gauge pressure allowed for a particular pipeline or pipeline segment, as set by the US Department of Transportation.

Parameters: Measures that set the conditions for the operation of a system.

Suction pressure: The amount of pressure generated on the upstream side of a compressor.

Supervisory Control and Data Acquisition (SCADA): A computer-based system or systems used by a controller in a control room that displays information about a pipeline facility, and which may also have the ability to send commands back to the facility.

In order to effectively monitor a pipeline or pipeline facility, personnel must be familiar with the normal operating pressure, flow, product characteristics, and other parameters for the pipeline. With this knowledge of what is normal, they must continually monitor the system for indications of deviations.

Overall pipeline parameters are monitored by the control center (*Figure 1*). When an alarm sounds, indicating a problem, control center personnel notify field personnel. In most cases, field personnel investigate and evaluate the problem and report back to the control center. On the basis of this information, the field maintenance supervisor notifies all affected personnel and schedules the necessary repairs.

Control center personnel monitor critical factors such as pressure, flow rate, and status changes on the pipeline and at facility stations. The Supervisory Control and Data Acquisition (SCADA) system (*Figure 2*) reports and displays deviations quickly. To identify and respond quickly to deviations, control center personnel should be familiar with the physical pipeline system, including operating conditions; equipment being operated upstream and downstream; conditions of the facilities and right-of-ways; and weather conditions. Any abnormal operating conditions (AOCs) should be dealt with appropriately. All pipeline activities must be documented.

CAUTION

An individual's actions during the performance of a covered task could cause a safety device to activate. The activation represents a deviation from normal operations, resulting in an abnormal operating condition (AOC).

Figure 1 Typical pipeline control center.

1.1.0 Monitoring and Maintaining Pipeline Pressure

Pressure is the force that a gas exerts uniformly in all directions within a vessel, pipeline, hole in the ground, or other containment area. Pressure is measured as force exerted per unit of area, such as pounds per square inch. Virtually every action on the pipeline depends on keeping the pressure properly controlled. If the pressure goes too high, the pipeline can rupture. Certain equipment, metering devices, and terminal installations often must be protected from main-line pressure. If the pressure falls too low, service to customers will be affected.

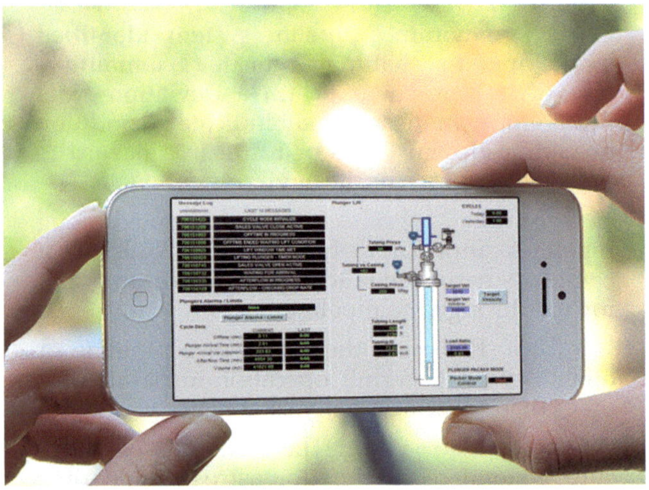

Figure 2 SCADA display on a smartphone.

1.1.1 Monitoring Gas Pipeline Pressure

For safety and efficiency, pipeline facilities and systems are designed to operate within certain limits or parameters. The **maximum allowable operating pressure (MAOP)** is established for each pipeline system or segment. Safety devices and related alarms alert control center personnel when the system strays from within the established operating pressure limits.

The MAOP for each pipeline system is established during the design phase of the system. The MAOP will not change unless there is a physical change in the pipeline system or the pipeline is derated because of the effects of corrosion. (**Derating** is reducing the rating of a pipeline to permit operation parameters different than those originally established.) Normally, pressure readings vary between 1 and 2 psi (6.89 to 13.79 kPa). Unusual variations in pressure could indicate a serious problem.

Qualified control center personnel must be completely familiar with the entire pipeline system and its parts. They should be able to look at the pressure meters, gauges, and rate of flow and correctly identify conditions on the line at all times. Control center personnel must be completely familiar with the company's policies and procedures and be able to quickly and correctly respond when any problem occurs.

If pressure begins to change at some point in the line, the SCADA system shows it within seconds. Control center personnel see the pressure fluctuation by observing the analog or digital meters and monitor screens. A rise in pressure at some point could indicate an unintended valve closure. Control center personnel must investigate every possible cause and quickly return the pipeline to normal operating limits.

Many companies require control center personnel to begin their shift by checking and verifying pipeline pressure, compressor pressure, delivery pressure, and relief setpoints. Personnel observe the readings on the meters and on the appropriate monitor screen. Control center personnel should be familiar with the normal operating pressures for the pipeline under similar conditions and verify that current readings are within normal range. Knowing what is normal helps personnel recognize what is not normal. Personnel should continuously monitor pressures using the SCADA system, pressure gauges, or meters, looking for any changes. Computer screens and analog readings provide an ongoing display of pressure in several formats. Personnel should note changes in pressure before alarms sound.

1.1.2 Maintaining Gas Pipeline Pressure

Deviations in pressure can occur at any time, and control center personnel should identify them quickly so that appropriate actions may be taken. Pressure deviations may not be serious enough to require anything more than close monitoring. More serious deviations may require that control center personnel take action in accordance with company policy.

Pressure detection devices, located at all critical locations on the pipeline, send signals that update the computers in the control center every few seconds. An alarm sounds only when the line pressure exceeds the setpoint. If an alarm sounds indicating that the pressure is too high or too low, control center personnel must contact the appropriate personnel to determine the cause. Every possible cause should be investigated.

Changes in pressure may be caused by compressors that have been started or stopped at other stations, or by valves unexpectedly opening or closing. Control center personnel should determine the cause of the alarm and confirm it before initiating any action. Generally, if the cause cannot be identified by the control center, field personnel may be instructed to investigate.

1.2.0 Monitoring Compressor Stations

In order to work efficiently, compressors must have an ample supply of gas. Personnel monitor all **compressor stations** and related equipment from the control center. The SCADA system (*Figure 3*) receives all operating information from individual stations and shows conditions in real time.

In certain situations, control center personnel must switch control from the control center to an individual compressor station. This is usually done for routine maintenance, but may also be done in response to an emergency. The SCADA system continues to receive information from the station so that there is no break in data recording. To ensure safety, control center personnel must follow company procedures when switching control to stations.

Control center personnel are responsible for monitoring pressure at all critical points in each station. Pressure points that should be monitored include the **suction pressure** and **discharge pressure** at the compressor, line pressure throughout the pipeline, and the differential pressure between any two measured points.

Suction pressure is the amount of pressure generated on the upstream side of the compressor. Discharge pressure is the amount of pressure generated on the downstream side of the compressor and supplied to the pipeline. Pressure transmitters help regulate the proper amount of suction and discharge and the proper flow of gas.

The SCADA system compares suction pressure against discharge pressure, and the control valve opens or closes as necessary to maintain the proper amount of suction. The suction pressure is constantly updated on meters in the control center and on the SCADA monitor screens. When the compressor starts, the discharge pressure and flow rate increase while the suction pressure decreases. Adjustments may be made to keep suction and discharge pressure within an acceptable range.

Line pressure (the pressure on the main pipeline) is an indicator of system integrity. Line pressure transmitters (*Figure 4*) take pressure measurements at various points and relay them to the control center.

Differential pressure is the difference between the pressure at two points in a closed system, such as the upstream and downstream sides of a gate valve. Differential pressure is monitored at various locations so that the pressure change does not damage piping and equipment. The SCADA system uses the transmitted pressure readings to calculate the flow rate of the gas through the pipeline, and compares the rate to a

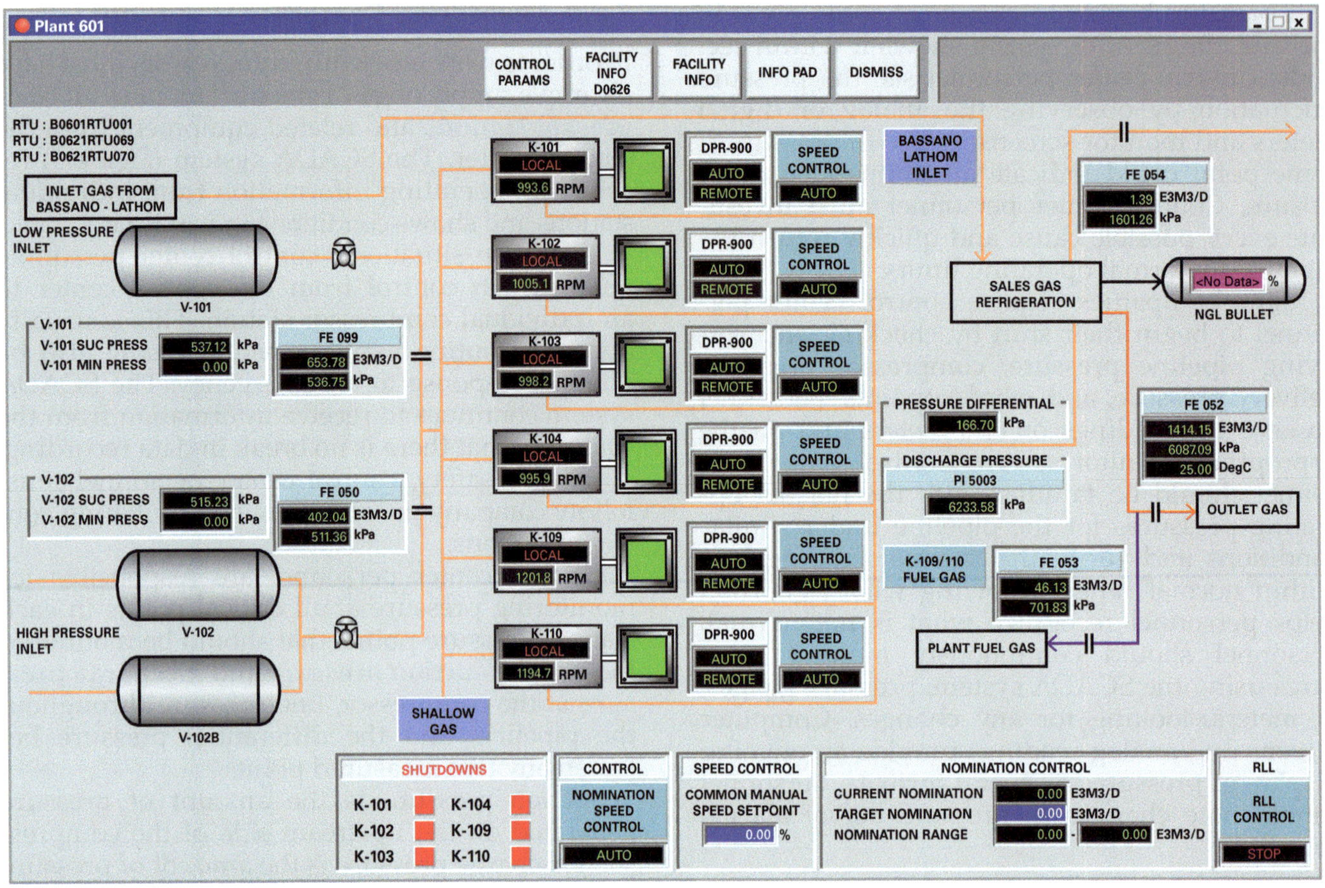

Figure 3 Typical SCADA display of compressor station parameters.

predetermined ideal flow rate. Differences in the two measurements could indicate deposits or corrosion, which would require a pigging operation to clean the line.

1.3.0 Monitoring the Pipeline Right-of-Way (ROW)

Because pipelines cover a vast area, companies must use several methods to survey them (49 *CFR* 192.705). These methods include walking, driving, and flying. Video equipment is often used to allow personnel to monitor certain areas remotely. Aerial patrols, which are the most common form of ROW monitoring, are performed on a schedule set by the pipeline company. Using current maps of the pipeline, pilots fly slowly at a safe altitude to allow for a good view. They radio their reports to the control center. Most companies require aerial patrols to keep a written record of their observations. In some cases, field personnel are sent to investigate an aerial report and complete a follow-up report. Pilots may be required to review any follow-up reports and compare them with the original patrol log.

Although aerial patrols are more commonly used to inspect the pipeline, experienced ground

Figure 4 Electronic pressure transmitter.

patrols may be required to inspect areas where

low-level flying is prohibited. Ground patrols may also be required when visibility is low or when bad weather grounds aircraft. In some cases, the patrols are sent to investigate an aerial report more closely on foot, or they may drive along areas that are accessible by vehicle (*Figure 5*).

Some facilities provide 24-hour security, as well as video cameras and monitors. Being able to continuously see all parts of a facility can reveal equipment failure, as well as break-ins and other security problems. Surveillance cameras are also able to focus on critical equipment and locations so that security and field personnel can monitor these areas for problems.

Inspection personnel check the pipeline right-of-way (ROW) for any conditions affecting the integrity of the pipeline. Gas pipelines must be patrolled regularly and the patrols documented (refer to *Table 1*).

Personnel check for a variety of conditions along the pipeline ROW, including the following:

- Construction work
- Soil erosion or growth of vegetation
- Missing or damaged pipeline markers
- Encroachment
- Signs of vandalism

> **NOTE**
> Inspection of surface conditions along ROW is covered in more detail in NCCER Module CT15_1-17, *Visually Inspect Surface Conditions of Right of Way.*

1.4.0 Monitoring Weather Conditions

Many pipeline operations depend on moderate weather. A tornado, hurricane, or blizzard can threaten personnel and the pipeline itself. High winds, snow, ice, and cold temperatures may also hinder operations. Sometimes the approach of an electrical storm can bring potentially dangerous lightning and wind.

The weather must be monitored to protect the safety of employees working on the pipeline. Company procedures must be followed for properly recording all weather changes. The sooner that personnel become aware of a breaking weather situation, the more likely it is that they can take action before problems become critical. To monitor weather changes effectively, personnel sometimes rely on weather-reporting websites. Many stations use small weather monitors to gauge local conditions, including the following:

- Temperature
- Wind direction and speed (with an alarm for high-wind speeds)
- Wind chill factor (with an alarm for low wind chill)
- Daily rainfall
- Barometric pressure

1.5.0 Documenting Pipeline Activities

Everything that happens on the pipeline is documented for company records. The SCADA system keeps records of all pressure, temperature, flow, and any other pertinent readings for personnel to retrieve at any time. Archived files are kept permanently.

Companies also investigate and document abnormal operating conditions (AOCs) for their own records and to provide documentation required by state and federal regulations. Required documentation for AOCs includes the following:

- Incident reports
- Event notification forms
- Incident investigation and analysis reports
- Control center logs
- Station logs
- Records of recommendations for corrective actions and training, if applicable

Figure 5 Pipeline right-of-way with path for inspection.

Table 1 Maximum Interval Between Patrols

Class Location of Line	Maximum Interval Between Patrols	
	At Highway and Railroad Crossings	**At All Other Places**
1 and 2	7 1/2 months, but at least twice each calendar year	15 months, but at least once each calendar year
3	4 1/2 months, but at least four times each calendar year	7 1/2 months, but at least twice each calendar year
4	4 1/2 months, but at least four times each calendar year	4 1/2 months, but at least four times each calendar year

Additional Resources

It is assumed that you will learn to perform these procedures under the direct supervision of your crew leader, foreman, or instructor. For independent study, refer to the recommended Additional Resources at the back of this module.

1.0.0 Section Review

1. Virtually every action on the pipeline depends on _____
   a. keeping the line at the MOP
   b. keeping the flow rate consistent
   c. maintaining the proper line balance
   d. keeping the pressure properly controlled

Section Two

2.0.0 Monitoring and Maintaining a Gas Pipeline System

Objective

Identify the knowledge and skills required to monitor pipeline parameters.

Identify the prerequisites, competencies, and abnormal operating conditions associated with monitoring pipeline parameters.

Describe the procedure for monitoring pipeline parameters.

Performance Task

1. Monitor pipeline parameters (CT65_3-17).
   - Identify potential abnormal operating conditions (AOCS) that may occur during performance of this CT, and know the appropriate actions to take in response to them.
   - Utilize the appropriate personal protective equipment (PPE) according to relevant company procedures.
   - Verify that the flows, pressures, and product characteristics are within normal range.
   - Continuously monitor, by whatever means available (SCADA, pressure gauge, meter), the pressures and flow rates of the pipeline to look for any changes. Monitor the line balance for changes by monitoring line pack in gas pipelines. Monitor storage facilities. Verify continuous communications and that data is accurate and updated.
   - Determine whether the pressures, flows, and line balance are within normal range and whether you have communication capabilities.
   - If necessary, determine the cause of variation from normal. Correct the situation if possible. For pressure or flow variations, use corrective methods such as starting and/or stopping pumps/units or using adjustable control devices such as control valves, back pressure valves, and variable speed drives for electric motors or using throttle controls for combustion engines. For communications variations, maintain monitored operations by using alternative communication methods. If necessary, seek assistance from and make notifications to proper personnel.
   - Complete appropriate documentation as required by operator's procedures.

Trade Terms

Line fill: The amount of product required to fill a new line before deliveries can be made.

Line pack: The introduction of new gas at the receipt point.

Maximum operating pressure (MOP): The maximum pressure at which a pipeline or segment of a pipeline may be normally operated.

Pneumatics: The branch of physics that deals with the mechanical properties of air and other gases.

Pipeline personnel must possess a range of knowledge and skills in order to safely and efficiently monitor pipeline parameters. Abnormal operating conditions (AOCs) must also be taken into consideration when monitoring a gas pipeline. Time is of the essence when AOCs occur and field personnel are expected to take decisive action. Personnel must always follow established company-specific policies and procedures.

2.1.0 Prerequisites, Competencies, and Abnormal Operating Conditions

Personnel must have a variety of skills to accurately locate pipelines, including the ability to understand specific pipeline system characteristics such as normal operating pressures, and familiarity with SCADA. They must also be familiar with safety practices, and emergency policies and procedures. Personnel must be aware of the particular AOCs that may occur when monitoring a gas pipeline and know the proper responses to them.

2.1.1 Prerequisites and Competencies

Personnel must understand the pipeline system, processes, and product characteristics for their particular pipeline. They must understand what constitutes normal pressure and flow, and the **pneumatics** of different product ranges for their particular line. Personnel assigned to gas pipelines must understand pipeline pneumatics, and the concept of **line pack** and **line fill**. All personnel must know the **maximum operating pressure (MOP)** and MAOP requirements for their pipeline, and they must understand the effect of unit start-up, shutdown, or throttling on the line pressures and flow rates.

Pipeline personnel must be able to perform the activities associated with monitoring and maintaining the pressures, flows, communications, and line integrity within allowable limits according to regulation and company policy. Variables can change at any time during the operation of a pipeline. These changes may be an indication of a pipeline leak, an overfill, or an as yet unidentified AOC. Personnel must continuously monitor, by whatever means available (SCADA, pressure gauge, meter), the pressures and flow rates of the pipeline to look for any changes. They should monitor the line pack.

Pipeline personnel should be familiar with the normal operating pressures, flows, line pack and line fill for their pipeline. By knowing what is normal and what is not normal, personnel in the control room and in the field are able to verify that the flows, pressures, and product characteristics are within the normal range, and are quickly able to identify any that are not. If these parameters are not in the normal range, they must determine the cause of the variation and take action to correct the situation if possible. Personnel should be familiar with company emergency response procedures.

If high or low pressure or flow variations are discovered, there are several corrective methods that personnel can apply to mitigate the situation. Pumps (*Figure 6*) or compressor units (*Figure 7*) may need to be started or stopped, depending on the situation. The use of adjustable control devices such as control valves, back pressure valves, and variable speed drives for electric motors, or the use of throttle controls for combustion engines may also be necessary.

Personnel must verify that they have continuous communication capabilities, and that the data they are receiving is accurate and current. For communications variations, they must maintain monitored operations by using alternative communication methods. If necessary, they must seek assistance from, and make notifications to, the appropriate personnel.

Figure 6 Liquid pipeline booster pump.

Figure 7 Gas pipeline compressors with electric drivers.

2.1.2 Abnormal Operating Conditions

Pipeline personnel must respond to AOCs such as a gas leak, equipment malfunction, or facility damage quickly, but not hastily. It is important to follow established company procedures so that each problem is confirmed, investigated, and addressed safely and correctly. Personnel must confirm that an alarm is not a false alarm. They need to investigate problems to determine what is wrong and what needs to be done to correct it. Depending on the situation, steps may have to be taken immediately to secure the area and ensure the safety of personnel. For other situations, repairs may need to be scheduled and necessary work permits issued. In general, personnel should follow these steps when responding to an AOC:

Step 1 Evaluate the situation.

Step 2 Notify affected personnel.

Step 3 Carry out company safety procedures to protect personnel, property, and the environment.

Step 4 Make necessary repairs or report that repairs are required.

Step 5 Submit the required reports.

Personnel maintaining pipeline parameters must be able to identify potential abnormal operating conditions (AOCs) and know the proper responses to each, should they occur. The following are some potential AOCs, and the appropriate action personnel should take in response to them:

- *An exposed pipe, or missing or damaged pipeline supports occurs* – Personnel should contact the control center and any additional persons as required by company policy, if the line is operated from a central control center. It is likely that these conditions would be noticed if the operator is physically at the location, but not if the operator is in a remote control center. This damage may require immediate attention to secure the integrity of the system, or it may only require submitting a work order for later repair.
- *Activation of a safety device occurs* – Personnel should first verify the activation, and if a hazardous condition exists, act quickly to minimize the situation. Notifications should be made to the control center and all other personnel as specified in company policy. Never ignore activated safety devices.
- *Failure of communications systems, SCADA, or electrical services occurs* – Personnel should initiate the troubleshooting procedures specified for the component that has failed, and make all contacts as required by company policy.
- *Fire or explosion occurs* – Personnel should immediately evacuate the area and limit access. Proper notifications, including calling 911, should be made as specified in company policy.

WARNING!

Proper PPE must be worn and proper safety precautions must be taken when dealing with an unexpected release of hazardous liquids or vapors.

- *Damage to a pipeline system component (such as being broken, bent, crushed, gouged, or burned) occurs* – Personnel should minimize or eliminate pressures and/or flows as appropriate for the severity of the situation and make necessary notifications according to company policy.
- *Unexpected hazardous liquid or carbon dioxide occurs* – Personnel should contact the control center to isolate or shut down the line, if the leak cannot be isolated safely at the site. Personnel should evacuate the area.
- *An unexplained flow rate deviation (such as high flow, low flow, or no flow) occurs* – Personnel should first verify equipment readings, and confirm that the proper valves are aligned and open. They should also verify that any units and/or valves operated prior to this deviation are still in the desired position. Personnel must check equipment readings and verify that they are accurate. In any case, personnel must always work to minimize any potentially hazardous situation that may develop as a result of the deviation.

- *Unexplained pressure deviation (such as increase, decrease, or lack of pressure) occurs* – Personnel should assume that the reading is correct and immediately work to identify the cause of the deviation. They should verify that any units and/or valves operated prior to this deviation are still in the desired position. It may be necessary to shut down the line if the cause cannot be immediately determined and corrected. Personnel must always work to minimize any potentially hazardous situation that may develop as a result of the deviation.
- *An unexplained status change occurs* – Personnel should act immediately to determine the cause and minimize the situation if necessary.

2.2.0 Maintaining Pipeline Parameters

The following steps outline the general procedure for monitoring pressures, flows, communications, and line integrity and maintaining them within allowable limits on a gas pipeline system:

Step 1 Identify potential abnormal operating conditions (AOCs) that may occur during performance of this CT, and know the appropriate actions to take in response to them.

Step 2 Utilize the appropriate personal protective equipment (PPE) according to relevant company procedures.

Step 3 Verify that the flows, pressures, and product characteristics are within normal range.

Step 4 Continuously monitor, by whatever means available (SCADA, pressure gauge, meter), the pressures and flow rates of the pipeline to look for any changes. Monitor the line balance for changes by monitoring line pack in gas pipelines. Monitor storage facilities. Verify continuous communications and that data is accurate and updated.

Step 5 Determine whether the pressures, flows, and line balance are within normal range and whether you have communication capabilities. If not, proceed to the next step.

Step 6 Determine the cause of variation from normal. Correct the situation if possible. For pressure or flow variations, use corrective methods such as starting and/or stopping pumps/units or using adjustable control devices such as control valves, back pressure valves, and variable speed drives for electric motors or using throttle controls for combustion engines. For communications variations, maintain monitored operations by using alternative communication methods. If necessary, seek assistance from and make notifications to proper personnel.

Step 7 Complete appropriate documentation as required by operator's procedures.

Additional Resources

It is assumed that you will learn to perform these procedures under the direct supervision of your crew leader, foreman, or instructor. For independent study, refer to the recommended Additional Resources at the back of this module.

2.0.0 Section Review

1. Personnel monitoring pipeline parameters must have an understanding of the effects of unit start-up and shutdown on _____.
   a. pressure and product temperature
   b. flow rates and line packs
   c. pressure and flow rate
   d. line packs and product characteristics

SUMMARY

Personnel must be familiar with and monitor normal operating pressure, flow, product characteristics, pipeline system processes, MOP, and MAOP in order to maintain the gas pipeline system within set parameters. They must understand pneumatic principles and the concepts of line balance, line pack, and line fill. Personnel must be able to recognize what is normal and what is not normal.

Personnel must monitor compressor station parameters, the pipeline right-of-way (ROW), and the weather. They must document all that happens on a pipeline, both for company records and as required by regulation.

Personnel must be able to recognize and respond appropriately to abnormal operating conditions (AOCs) including damage to the pipeline or equipment, pressure and flow deviations, safety device activation, communication failures, unexplained status changes, and an unexpected release of hazardous liquid or gas.

Review Questions

1. A pressure variation of 1 to 2 psi (6.89 to 13.79 kPa) _____.
   a. indicates a reduced flow rate
   b. is a normal condition
   c. requires immediate attention
   d. will signal an alarm

2. The SCADA system compares _____.
   a. suction pressure against discharge pressure
   b. line pack against line fill
   c. the MOP against the MAOP
   d. line pressure against flow rate

3. For operations on their particular pipeline, personnel must have an understanding of the product _____.
   a. pneumatic principles
   b. packing rate
   c. characteristics
   d. hydraulic principles

4. If there is a failure of communications, electrical power, or the SCADA system, personnel must _____.
   a. alert all affected personnel
   b. follow troubleshooting procedures to determine the cause
   c. notify the control center
   d. notify the maintenance team

Trade Terms Quiz

Fill in the blank with the correct term that you learned from your study of this module.

1. The highest level of internal gauge pressure allowed is called the ________.
2. Conditions within a pipeline system that exceed operating design limits are known as ________.
3. The amount of product required to fill a new line is called ________.
4. The maximum pressure at which a pipeline may be normally operated is called the ________.
5. The introduction of new gas at the receipt point is called ________.
6. A computer-based system that displays information about a pipeline facility is called ________.
7. The branch of physics that deals with the mechanical properties of gases is called ________.
8. Measures that set the conditions for the operation of a system are called ________.
9. The pressure generated on the upstream side of a compressor is called the ________.
10. Reducing the rating of a pipeline or other device to permit operation at parameters other than those originally established is called ________.
11. The pressure generated on the downstream side of a compressor is called the ________.
12. Facilities installed along a pipeline that compress natural gas are called ________.

Trade Terms

Abnormal operating conditions (AOCs)
Compressor stations
Derating
Discharge pressure
Line fill
Line pack
Maximum operating pressure (MOP)
Maximum allowable operating pressure (MAOP)
Parameters
Pneumatics
Suction pressure
Supervisory Control and Data Acquisition (SCADA)

Trade Terms Introduced in This Module

Abnormal operating conditions (AOCs): Conditions within a pipeline system, identified by the operator, that may indicate (1) a malfunction of a component or (2) a deviation from normal operations that may indicate a condition exceeding design limits or result in a hazard to persons, property, or the environment.

Compressor stations: Facilities installed along pipelines approximately every 100 miles that compress natural gas to reduce its volume and force it through the pipeline.

Derating: Reducing the rating of a pipeline or other device to permit operation at parameters other than those originally established.

Discharge pressure: The amount of pressure generated on the downstream side of a compressor.

Line fill: The amount of product required to fill a new line before deliveries can be made.

Line pack: The introduction of new gas at the receipt point.

Maximum allowable operating pressure (MAOP): The highest level of internal gauge pressure allowed for a particular pipeline or pipeline segment, as set by the US Department of Transportation.

Maximum operating pressure (MOP): The maximum pressure at which a pipeline or segment of a pipeline may be normally operated.

Parameters: Measures that set the conditions for the operation of a system.

Pneumatics: The branch of physics that deals with the mechanical properties of air and other gases.

Suction pressure: The amount of pressure generated on the upstream side of a compressor.

Supervisory Control and Data Acquisition (SCADA): A computer-based system or systems used by a controller in a control room that displays information about a pipeline facility, and which may also have the ability to send commands back to the facility.

Additional Resources

The following recommended resources can provide additional helpful information related to the requirements for this covered task:

API Recommended Practice 1161, Third Edition, © 2014 American Petroleum Institute. API Publishing Services, 1220 L Street, NW, Washington, DC 20005. Distributed by Thompson Reuters (Scientific) LLC, **www.techstreet.com**.

49 *CFR* 192, **www.ecfr.gov**.

49 *CFR* 195, **www.ecfr.gov**.

NCCER Module AOCCC-17, *Abnormal Operating Conditions Control Center*

NCCER Module AOCFG-17, *Abnormal Operating Conditions Field and Gas*

NCCER Module 60102-15, *Basic Pipeline Hydraulics and Equipment*

NCCER Module 62105-14, *Introduction toPipeline Documents*

NCCER Module 64106-02, *Pipeline Operations*

NCCER Module 64307-02, *SCADA*

NCCER Module 67102-15, *Basic Pipeline Pneumatics and Equipment*

NCCER Module 67103-15, *Pipeline Communications*

Figure Credits

arhendrix/Shutterstock.com, Module Opener
Courtesy of Honeywell International Inc., Figure 1
ABB, Figures 2 and 3
Tim Mills, Figures 4 and 6
leungchopan/Shutterstock.com, Figure 5
© iStockphoto.com/westphalia, Figure 7

Section Review Answer Key

Answer	Section Reference	Objective
Section One		
1. d	1.1.0	1a
Section Two		
1. c	2.1.1	1a

NCCER CURRICULA — USER UPDATE

NCCER makes every effort to keep its textbooks up-to-date and free of technical errors. We appreciate your help in this process. If you find an error, a typographical mistake, or an inaccuracy in NCCER's curricula, please fill out this form (or a photocopy), or complete the online form at **www.nccer.org/olf**. Be sure to include the exact module ID number, page number, a detailed description, and your recommended correction. Your input will be brought to the attention of the Authoring Team. Thank you for your assistance.

Instructors – If you have an idea for improving this textbook, or have found that additional materials were necessary to teach this module effectively, please let us know so that we may present your suggestions to the Authoring Team.

NCCER Product Development and Revision

13614 Progress Blvd., Alachua, FL 32615

Email: curriculum@nccer.org

Online: www.nccer.org/olf

❑ Trainee Guide ❑ Lesson Plans ❑ Exam ❑ PowerPoints Other ______________

Craft / Level: Copyright Date:

Module ID Number / Title:

Section Number(s):

Description:

Recommended Correction:

Your Name:

Address:

Email: Phone:

CT65_4-17

Manually or Remotely Open or Close Valves or Other Equipment

Overview

Compressors are used to compress natural gas to a specified pressure, allowing it to flow through the pipeline. Valves direct the flow of the gas as it moves through the pipeline. This module describes the procedures for manually or remotely opening and closing valves, starting and stopping compressors, and monitoring the compressor drivers.

Trainees with successful module completions may be eligible for credentialing through the NCCER Registry. To learn more, go to **www.nccer.org** or contact us at 1.888.622.3720. Our website has information on the latest product releases and training, as well as online versions of our *Cornerstone* magazine and Pearson's product catalog.

Your feedback is welcome. You may email your comments to **curriculum@nccer.org**, send general comments and inquiries to **info@nccer.org**, or fill in the User Update form at the back of this module.

This information is general in nature and intended for training purposes only. Actual performance of activities described in this manual requires compliance with all applicable operating, service, maintenance, and safety procedures under the direction of qualified personnel. References in this manual to patented or proprietary devices do not constitute a recommendation of their use.

CT65_4-17

Manually or Remotely Open or Close Valves or Other Equipment

Objective

When you have completed this module, you will be able to do the following:

1. Describe how to manually or remotely open or close valves or other equipment (CT65_4-17).
   a. Describe how to operate valves.
   b. Describe how to start and stop compressors.
   c. Identify the knowledge and skills required to manually or remotely open or close valves or other equipment.

Performance Task

Under the supervision of the instructor, you should be able to do the following:

1. Manually or remotely open or close valves or other equipment (CT65_4-17).
   - Identify potential abnormal operating conditions (AOCs) that may occur during performance of this CT, and know the appropriate actions to take in response to them.
   - Utilize the appropriate personal protective equipment (PPE) according to relevant company procedures.
   - Verify that the flows, pressures, and product characteristics are within normal range.
   - Identify the need to open or close a valve or start or stop a unit to control flow or pressure and to maintain safe operation of the pipeline system.
   - Identify the applicable valve, unit, or the appropriate control device.
   - Perform proper operation according to company policy and procedures.
   - Verify that the desired result was achieved.
   - Complete appropriate documentation as required by operator's procedures.

Trade Terms

Abnormal operating conditions (AOCs)
Actuator
Ball valve
Blowdown
Check valve
Compressor stations
Driver
Gate valve
Line fill
Line pack
Manifold
Maximum allowable operating pressure (MAOP)
Maximum operating pressure (MOP)
Plug valve
Portable valve operator
Pressure-relief valve
Valve alignment

Industry Recognized Credentials

If you are training through an NCCER-accredited sponsor, you may be eligible for credentials from NCCER's Registry. The ID number for this module is CT65_4-17. Note that this module may have been used in other NCCER curricula and may apply to other level completions. Contact NCCER's Registry at 1.888.622.3720 or go to **www.nccer.org** for more information.

Contents

Figures

SECTION ONE

1.0.0 MANUALLY OR REMOTELY OPEN OR CLOSE VALVES OR OTHER EQUIPMENT

Objective

Describe how to manually or remotely open or close valves or other equipment (CT65_4-17).

a. Describe how to operate valves.
b. Describe how to start and stop compressors.
c. Identify the knowledge and skills required to manually or remotely open or close valves or other equipment.

Performance Task

Manually or remotely open or close valves or other equipment (CT65_4-17).

- Identify potential abnormal operating conditions (AOCs) that may occur during performance of this CT, and know the appropriate actions to take in response to them.
- Utilize the appropriate personal protective equipment (PPE) according to relevant company procedures.
- Verify that the flows, pressures, and product characteristics are within normal range.
- Identify the need to open or close a valve or start or stop a unit to control flow or pressure and to maintain safe operation of the pipeline system.
- Identify the applicable valve, unit, or the appropriate control device.
- Perform proper operation according to company policy and procedures.
- Verify that the desired result was achieved.
- Complete appropriate documentation as required by operator's procedures.

Trade Terms

Abnormal operating conditions (AOCs): Conditions within a pipeline system, identified by the operator, that may indicate (1) a malfunction of a component or (2) a deviation from normal operations that may indicate a condition exceeding design limits or result in a hazard to persons, property, or the environment.

Actuator: A device used to operate a valve using electrical, pneumatic, or hydraulic means. Actuators are often used for the remote control or sequencing of pipeline valves.

Ball valve: A valve controlling flow by means of a ball pierced with an opening in one direction and fixed to rotate on a spindle at right angles to the opening.

Blowdown: The depressurization of gas piping, which is accomplished by releasing the pressure at a safe atmospheric vent location.

Check valve: A valve held open by the fluid flow in one direction, which closes when the flow is stopped or reversed.

Compressor station: Facilities installed along pipelines approximately every 100 miles that compress natural gas to reduce its volume and force it through the pipeline.

Driver: A mechanical component that supplies power to move another mechanical component.

Gate valve: A valve with a closing element consisting of a wedge or slab positioned between two seating surfaces that can be moved into or out of the flow stream perpendicular to the direction of the flow.

Line fill: The amount of gas needed to fill a new line before deliveries can be made.

Line pack: The introduction of new gas at the receipt point.

Manifold: A pipeline fitting with several side outlets to connect it with other piping; also known as a header.

Maximum allowable operating pressure (MAOP): The highest level of internal gauge pressure allowed for a particular pipeline or pipeline segment, as set by the US Department of Transportation.

Maximum operating pressure (MOP): The maximum internal gauge pressure at which a hazardous liquid or gas pipeline system or segment may be operated.

Plug valve: A valve whose closure element is a tapered plug.

Portable valve operator: A device that uses manual, hydraulic, electrical, or pneumatic energy to open or close valves.

Pressure-relief valve: A valve that relieves pressure on a system by allowing the pressurized liquid or gas to flow from an auxiliary passage out of the system.

Valve alignment: The exact sequence of valves that must be opened and closed at the same time so that gas can be received at, and delivered to, specific locations.

Compressors force the gas through the pipeline. The flow must be started and stopped following specific procedures in order to ensure that the pipeline system operates continuously and efficiently. Valves are installed at specific locations along pipelines to provide a variety of functions. These valves can be opened, partially opened, or closed to control and direct the flow. The failure of a valve to operate, or the improper operation of a valve, can be both costly and dangerous. Personnel must be aware of the **abnormal operating conditions (AOCs)** that may occur, and how to properly respond to them.

> **CAUTION**
>
> An individual's actions during the performance of a covered task could cause a safety device to activate. The activation represents a deviation from normal operation, resulting in an abnormal operating condition (AOC).

1.1.0 Operating Valves

Valves are mechanical units that function to control the flow through a pipeline. They are used to open or completely shut off a pipeline. There are many kinds of valves. Among the most common types used on a pipeline are the **ball valve**, **check valve**, **gate valve**, **plug valve**, and **pressure-relief valve**. Most valves have position indicators, which are shafts, rods, or pointers that show whether the valves are open or closed. Some valves have automatic or semi-automatic safety devices.

Most valves are operated from the control center. However, field personnel must know how to open and close valves manually. These operations must be performed carefully to safely control flow and pressure and maintain normal operations. *Figure 1* shows an example of a pipeline valve that can be opened manually as well as remotely.

The following steps outline the general procedure for manually or remotely opening or closing valves or other equipment:

Step 1 Identify potential abnormal operating conditions (AOCs) that may occur, and know the appropriate actions to take in response to them.

Step 2 Utilize the appropriate personal protective equipment (PPE) according to relevant company procedures.

Step 3 Verify that the flows, pressures, and product characteristics are within normal range.

Figure 1 Valve with both manual and remote controls.

Step 4 Identify the need to open or close a valve or start or stop a unit to control flow or pressure and to maintain safe operation of the pipeline system.

Step 5 Identify the applicable valve, unit, or the appropriate control device.

Step 6 Perform proper operation according to company policy and procedures.

Step 7 Verify that the desired result was achieved.

Step 8 Complete appropriate documentation as required by operator's procedures.

1.1.1 Valve Placement and Alignment

Valves are located at particular points on pipelines for specific reasons, and they must be opened and closed in the right order. Opening and closing valves in the correct sequence is called **valve alignment**. Pipeline companies have specific valve alignments for every operation. These alignments are laid out for employees on charts or lists so that costly and potentially dangerous errors can be avoided.

Many valves are operated electrically by remote control (*Figure 2*). The remote signal activates an electrical, pneumatic, or hydraulic **actuator** that opens, closes, or positions the valve. Multiple valves are sometimes grouped into a **manifold**, also known as a *header*. Backup generator units or battery-operated units can be used to control the valves during a power failure, but the valves can also be operated manually. Some valves can only be operated by hand.

Field personnel must know where the valves in their work area are located and how to operate them manually, if necessary. Manual operation may be required when valves do not respond to

Figure 2 Valve control panel.

a signal from the control center. In cases where there is no electrical power, a **portable valve operator** may be used. These are handheld devices that engage and turn the shaft of the valve, thus speeding the process and reducing the physical strain involved. Portable valve operators may be powered pneumatically or hydraulically, or by electricity from a small generator.

Although hand-cranking or hand-turning operations might be possible on some valves, the number of turns required to open other valves would make an attempt useless, if not prohibitive, in an emergency. The requirement for a portable valve operator in such a case can readily be seen and is a practical necessity.

1.1.2 Failure to Operate

Field personnel must know how to repair malfunctioning valves. Operations personnel may notice that a valve indicator is not moving up and down as it should, or an alarm in the control center may signal that a valve is not operating. When there is a problem with a valve, field personnel may be asked to investigate and evaluate the situation. They report their findings to the maintenance supervisor or the control center, and a decision about how to address the problem is made based on the report. A typical control center is shown in *Figure 3*.

1.1.3 Leaking Valves

If a valve is leaking, it is not safe to manually operate it until the leak is stopped and the air is cleared. Hazardous vapors can build up around a leak and catch fire or explode. Personnel must isolate the leak if it is safe to do so. The air must be monitored with a combustible gas indicator. After the hazardous vapors have been cleared, repairs can be made. The air must be continuously monitored during repairs. Once repairs have been made, personnel must submit any required reports.

WARNING!

Under no circumstances should any person attempt to reach a valve when a vapor cloud is observed in the area.

1.1.4 Valve Access

Valves are used to control the flow and to isolate segments of pipe or piping systems for maintenance or emergency response activities. In order to perform these important functions, ready access to the valves is critical, particularly if they must be manually operated by field personnel in response to a release. In an emergency, mitigation of the hazards may be hampered by valves that are obscured by overgrown vegetation, or are difficult to access because of marginal roads or flooding.

The location and function of all valves should be clearly indicated on facility maps or pipeline alignment sheets, or associated with latitude-longitude coordinates so that they can be quickly and positively identified by field personnel. A

Portable Valve Operators

To illustrate the practical use of a portable valve operator, consider the example of a large gate valve fitted with a hand wheel and gear box. In low gear, this particular valve requires 10,000 turns of the wheel to fully open or close manually. If it were not for the geared drive mechanism, it might be impossible to turn it at all. In high speed, the same valve still requires 2,500 turns to fully operate the valve. The time consumed making hundreds or thousands of turns can seriously hamper operations.

Figure 3 Typical pipeline control center.

potentially serious incident such as an inadvertent release can be minimized if location personnel are able to quickly locate and close the necessary valve.

Although valves must be accessible for qualified field personnel, they must also be locked or fenced to prevent unauthorized access by the general public. By regulation, pipeline operators are required to provide protection for each valve from unauthorized operation and from vandalism.

1.2.0 Operating Compressors

Natural gas is sent through a pipeline at high pressures. As the gas moves through the pipeline, friction reduces its speed and the pressure drops. To maintain the necessary pressure, the gas must be recompressed periodically as it moves through the pipeline. That is why compressor stations are installed approximately every 100 miles along the pipeline. Often, several compressors are located at one station.

Some compressors use a small part of the gas that flows through the pipeline to power the engine that drives the compressor. Other compressors are electrically powered by turbine engines. Because different types of compressors are available for use on the gas pipeline, it is important to read and follow the manufacturer's recommendations for installation, start-up, shutdown, and maintenance procedures for the specific compressor in use. These procedures should be available at the pipeline facility for easy reference if needed.

1.2.1 Compressor Start-Up

The procedures for starting up a compressor and the order of the steps to follow vary from one company to another. Each company will also have its own notification and reporting requirements. In general, the following steps are followed to start a compressor (*Figure 4*):

Figure 4 Large main-line compressor.

Step 1 Carry out any pre-start-up procedures as recommended by the manufacturer or required by company policy.

Step 2 Review and verify knowledge of the starting unit and start-up procedures.

Step 3 Ensure that the compressor and station valves are properly aligned.

Step 4 Notify all affected personnel that the compressor is about to be started.

Step 5 Confirm that all operating variables are within the safe operating limits of the compressor.

Step 6 Confirm that all safety alarms and shutdown devices are set and are working properly.

Step 7 Initiate the start-up according to the posted start-up procedure for the compressor(s) at the local facility.

Step 8 Follow the manufacturer's instructions for operating the compressor after it has been started.

1.2.2 Compressor Shutdown

Compressors may have to be shut down for regular maintenance or in response to a problem. This procedure must be performed carefully, safely, and in accordance with the manufacturer's specifications. In general, the following steps should be followed to shut down a compressor:

Step 1 Notify all affected personnel of a pending compressor shutdown per company policy.

Step 2 Slow down the compressor to minimum speed.

Step 3 Unload the compressor and open the bypass valve.

Step 4 Allow the compressor to cool down for the length of time specified in the manufacturer's instructions.

Step 5 Shut off the fuel supply.

Step 6 Close the suction and discharge valves.

Step 7 Initiate compressor **blowdown** to depressurize the system.

The amount of time that the compressor may be left depressurized depends on a number of factors, including the condition of the compressor, the need to make repairs, and the availability of backup compressors. Generally, the maintenance supervisor decides how long the compressor should remain out of service.

1.2.3 Monitoring Drivers

A **driver** is a mechanical component that supplies power to move another mechanical component. Drivers are the engines that operate pipeline equipment, including compressors. Some drivers are electrical (*Figure 5*), while others may be turbine or reciprocating engines. Drivers are in constant motion, which results in significant wear and tear on the parts. Drivers must be monitored periodically to ensure that equipment does not fail. Two common problems that occur on drivers are excess vibration and driver surge.

Every driver vibrates as it operates, but too much vibration can cause damage. Causes of harmful vibration include bent shafts or gears, damaged bearings, loose bolts, and misaligned couplings or bearings. In compressors, a common cause of excess vibration is an imbalance in the rotating parts.

Periodically checking drivers for such problems saves time, money, and energy. Recalibrating devices per manufacturer's recommendations, lubricating on schedule, and checking gauges also help to ensure the most efficient operation. Most manufacturers provide a recommended maintenance schedule. It is best to follow the manufacturer's recommendations for maintaining drivers at peak operating efficiency.

Figure 5 Large main-line compressors with electric drivers.

Some drivers are powered by a small portion of the natural gas flowing through the line. Driver surge may occur in drivers fueled by natural gas as the result of an improper fuel/air mixture. It is important to recognize the signs of a surge. Drivers can surge at both low and high speeds. A surge at a low speed may show up in the driver's failure to accelerate, an increase in exhaust temperatures, and the sound of fluttering air. A surge at a high speed results in a loud roar or popping noises, and the driver may fail to accelerate to its rated speed.

If a surge occurs, the driver must be shut down immediately to prevent a rapid increase in temperature. After the driver has slowed to a stop, investigate the problem and notify the appropriate personnel.

1.3.0 Knowledge and Skills

Personnel must possess a variety of skills to operate valves, and to start and stop compressors in a safe and efficient manner. These operations require a number of steps, which personnel must be competent to perform. Personnel must also be aware of the particular abnormal operating conditions (AOCs) that may occur during the procedures and must know the proper responses to them.

1.3.1 Prerequisites and Competencies

Personnel must be able to verify that the flows, pressures, and product characteristics for their pipeline are within normal range. Personnel should have a working understanding of pipeline hydraulics or pneumatics and pipeline elevations, if applicable. They should be familiar with the normal operating pressure, flow, **line pack**, and **line fill**, as well as the **maximum operating pressure (MOP)** and the **maximum allowable operating pressure (MAOP)** for their line. By recognizing what is normal and what is not normal, personnel are able to quickly identify potential problems, thus preventing them from developing into a more serious AOC.

Personnel must be able to recognize a developing situation in which the need to open or close a valve, start or stop a unit, or operate a control device is necessary to control the flow and pressure and maintain safe operation of the pipeline system. To do so, they must be able to quickly identify the flow path, as well as locate and identify the applicable valve, unit, or appropriate control device before executing the sequence. They must be able to perform this operation according to company policy and procedures, and to verify that the desired result was achieved.

Personnel operating pipeline valves must know how to determine the valve's position. Each valve, other than a check valve, has some means of indicating the valve's position. Examples of indicators include the following:

- Rising stem indicator
- Arrow
- Handle position
- Open/close flag or display

Before operating a valve, personnel must first consider how the operation will affect the pipeline system as a whole. The following factors must be considered prior to valve operation:

- Opening or closing a valve may affect the pressure and flow of the pipeline system.
- The operation of an incorrect valve could cause an unsafe condition.
- A thermal trap could be created by shutting in segments of the pipeline system where it could be over-pressured because of an increase in product temperature.

Operations personnel must be able to use the appropriate references to correctly identify the valve to be operated. They must understand how to locally operate a valve, either manually or by the use of a motor-operated actuator. They must also know how to determine that the valve is in the intended position, and be able to communicate this information to field personnel as well as to the control center, which has ultimate responsibility and authorization for actions that affect the safe operation of the pipeline. Personnel must understand what communications are required between local operators and/or the control center in association with valve operation. Personnel should be familiar with company emergency response procedures.

1.3.2 Abnormal Operating Conditions

Prior to opening or closing valves or other equipment, personnel must be able to identify abnormal operating conditions (AOCs) and know the proper response to each, should they occur. The following are some potential AOCs, and the appropriate action personnel should take in response to them:

- *Unexplained pressure deviation occurs* – Personnel should make the condition safe to the extent possible according to the operator's procedures. Assess the situation for safety, and be aware of any environmental or physical damage.

Appropriate responses to an unexplained pressure deviation could include the following:

- Shutting down the system (if qualified)
- Returning the valve to its original position
- Operating another valve to relieve the situation
- Isolating damaged equipment
- Making the proper notifications
- An unexplained pressure deviation may be indicated by pressure display(s), a sound or vibration, or an alarm or other pressure indicator showing the unexplained pressure deviation. *Figure 6* shows a pressure transmitter.

• *Unexplained flow rate deviation occurs* – Personnel should make the condition safe to the extent possible according to the operator's procedures. Assess the situation for safety, and be aware of any environmental or physical damage. Appropriate responses to an unexplained flow rate deviation could include the following:

- Returning the valve to its original position
- Operating another valve to relieve the situation
- Isolating damaged equipment
- Shutting down the system (if qualified)
- Making the proper notifications

Figure 6 Electronic pressure transmitter.

• An unexplained flow rate deviation may be indicated by flow gauge readings, alarms, improper tank levels, or other flow indicators. *Unexplained valve position indication occurs* – Personnel should first confirm the actual valve position. If the valve is not in the indicated position, investigate the cause and resolve the source of discrepancy between the actual valve position and the position indicator. Proper valve position is required before safe pipeline operation resumes. Make the proper notifications.

• *Inoperable valve occurs* – Personnel should make the condition safe to the extent possible according to the operator's procedures. Assess the situation for safety, and be aware of any environmental or physical damage. Appropriate responses to a valve that is inoperable could include the following:

- Trying again to operate the valve to see if the situation can be corrected
- Relieving excessive differential pressure across the valve
- Shutting down the system (if qualified)
- Making all appropriate notifications
- A valve may not operate as intended, or may not open or close. An inoperable valve may be detected by observing that the valve position indicator does not show the intended position, or that valve operation results in unexpected flow or pressure outcomes. A valve may be stuck and unable to be operated either manually or by a valve operator or actuator. Failure to operate may also be due to excessive differential pressure across the valve.

• *Unexpected presence of hazardous liquid or vapor occurs* – Personnel should minimize the situation to the extent possible if it can be safely done, make all appropriate notifications and evacuate the area. Incorrect valve operation could lead to an unintended release of product that may be observed by sight, smell, sound, or alarms.

WARNING!

Proper PPE must be worn and proper safety precautions must be taken when dealing with an unexpected release of hazardous liquids or vapors.

- *An exposed pipe, or missing or damaged pipeline supports occurs* – Personnel should notify the control center and additional personnel as required by company policy to initiate repairs.
- *Activation of a safety device occurs* – Personnel should first verify the activation, and if a hazardous condition exists, act quickly to minimize the situation. Notifications should be made to the control center and all other personnel as specified in company policy. Never ignore activated safety devices.
- *Failure of communications systems, SCADA, or electrical services occurs* – Personnel should follow troubleshooting procedures and make contacts as required by company policy.
- *Fire or explosion occurs* – Personnel should immediately evacuate the area and limit access. Proper notifications, including calling 911, should be made as specified in company policy.
- *Damage to a pipeline system component (such as being broken, bent, crushed, gouged, or burned) occurs* – Personnel should make notifications as specified by company policy and minimizing or eliminating the system pressure or flow rate as appropriate for the severity of the situation.
- *An unexplained status change occurs* – Personnel should act immediately to determine the cause and minimize the situation if necessary.

Several AOCs are associated with the start-up and shutdown of a compressor. When AOCs occur, personnel must follow company policy for identifying the problem and notifying the appropriate personnel. In addition, personnel must conduct any investigations and/or submit reports required by law and company policy.

Gas leaks and excessive vibration or noise are two common AOCs that may occur in conjunction with compressor operations. A hazardous gas alarm will sound in the event of a gas leak. Personnel may be required to leave the area until the leak is repaired and the air is checked with a combustible gas indicator to ensure safety. Too much vibration or noise may indicate that equipment is malfunctioning or about to fail. The compressor may have to be shut down while personnel investigate the cause.

Additional Resources

It is assumed that you will learn to perform these procedures under the direct supervision of your crew leader, foreman, or instructor. For independent study, refer to the recommended Additional Resources at the back of this module.

1.0.0 Section Review

1. When valves do not respond to a signal from the control center, _____.
   a. a work order is issued
   b. manual operation may be required
   c. flow rate must be reduced until the issue is resolved
   d. a backup generator may be used to restore power

2. Before starting a compressor, personnel must _____.
   a. allow the compressor to warm up per manufacturer's instructions
   b. ensure that all of the associated valves are in the open position
   c. review and verify knowledge of the starting unit and starting procedures
   d. ensure that compressor and station valves are properly aligned

3. Before executing a sequence to control flow or pressure in order to maintain safe operation of the line, personnel must be able to quickly _____.
   a. determine the MOP
   b. stop a compressor
   c. start a compressor
   d. identify the flow path

SUMMARY

Valves are used to control and direct the flow of gas. Opening and closing valves in a specific sequence is called valve alignment. Inoperative valves must be investigated and evaluated by field personnel. Leaking valves must be monitored for hazardous vapors until repairs are made. Valves must be kept free of obstructions for quick access in case of an emergency. They must also be protected from unauthorized operation or vandalism.

Compressors compress the gas to reduce volume and force it through the pipeline. Compressors may be powered by electricity, natural gas, a turbine, or other fuels. Compressor start-up, shutdown, and maintenance must follow company policy and the manufacturer's recommendations. Drivers must be monitored to ensure they do not fail. Two common problems are excessive vibration and driver surge. If a unit experiences excessive noise or vibration it may need to be shut down until the cause is determined. If a surge occurs, shut down the driver and investigate the cause.

Personnel must be familiar with the pressures, flows, product characteristics, line pack, line fill, MOP, and MAOP for their line. They must be able to recognize what is normal and what is not normal. An unexplained pressure or flow rate deviation may be corrected by returning a valve to its original position, operating another valve, isolating damaged equipment, or shutting down the system. If an unexpected release occurs, minimize the situation to the extent possible. If a hazardous gas alarm occurs, leave the area and monitor with a combustible gas indicator until clear.

Review Questions

1. Most valves are operated _____.
   a. with a portable valve operator
   b. remotely from the control center
   c. remotely in the field
   d. with a control wheel by hand

2. Compressor start-up is initiated according to _____.
   a. company policy
   b. federal regulations
   c. posted procedures
   d. pipeline regulations

3. A driver surge at low speeds may show up as _____.
   a. a loud roar or popping noise
   b. failure to decelerate when required
   c. a sudden decrease in line pressure
   d. an increase in exhaust gas temperature

4. If a hazardous gas leak occurs during compressor start-up, personnel may be required to _____.
   a. leave the area
   b. slow or stop the compressor
   c. reposition a valve
   d. activate a safety device

Trade Terms Quiz

Fill in the blank with the correct term that you learned from your study of this module.

1. A mechanical component that supplies power to move another mechanical component is called a(n) _________.
2. A pipeline fitting with several side outlets to connect it with other piping is called a(n) _________.
3. Facilities installed along pipelines that compress natural gas to reduce its volume and force it through the pipeline are called _________.
4. The depressurization of gas piping by releasing the pressure through a vent is called _________.
5. The amount of gas needed to fill a new line before deliveries can be made is called _________.
6. A valve whose closure unit is a tapered plug is called a(n) _________.
7. The maximum internal gauge pressure at which a pipeline may be operated is called the _________.
8. A valve that is held open by the fluid flow in one direction and closes when the flow is stopped or reversed is call a(n) _________.
9. The introduction of new gas at the receipt point is called _________.
10. The exact sequence of valves that must be opened and closed at the same time is called _________.
11. A device used to operate a valve using electrical, pneumatic, or hydraulic means is called a(n) _________.
12. The highest level of internal gauge pressure allowed for a pipeline is called the _________.
13. Conditions within a pipeline system, identified by the operator, that may indicate (1) a malfunction of a component or (2) a deviation from normal operations that may indicate a condition exceeding design limits or result in a hazard to persons, property, or the environment are called _________.
14. A valve that is used to relieve pressure on a system is called a(n) _________.
15. A valve with a closing element consisting of a wedge or slab positioned between two seating surfaces is called a(n) _________.
16. A device that uses manual, hydraulic, pneumatic, or electrical energy to open or close valves is called a(n) _________.
17. A valve that controls flow by means of a ball pierced in one direction that rotates at right angles to the opening is called a(n) _________.

Trade Terms

Abnormal operating conditions (AOCs)
Actuator
Ball valve
Blowdown
Check valve
Compressor stations
Driver
Gate valve
Line fill
Line pack
Manifold
Maximum allowable operating pressure (MAOP)
Maximum operating pressure (MOP)
Plug valve
Portable valve operator
Pressure-relief valve
Valve alignment

Trade Terms Introduced in This Module

Abnormal operating conditions (AOCs): Conditions within a pipeline system, identified by the operator, that may indicate (1) a malfunction of a component or (2) a deviation from normal operations that may indicate a condition exceeding design limits or result in a hazard to persons, property, or the environment.

Actuator: A device used to operate a valve using electrical, pneumatic, or hydraulic means. Actuators are often used for the remote control or sequencing of pipeline valves.

Ball valve: A valve controlling flow by means of a ball pierced with an opening in one direction and fixed to rotate on a spindle at right angles to the opening.

Blowdown: The depressurization of gas piping, which is accomplished by releasing the pressure at a safe atmospheric vent location.

Check valve: A valve held open by the fluid flow in one direction, which closes when the flow is stopped or reversed.

Compressor stations: Facilities installed along pipelines approximately every 100 miles that compress natural gas to reduce its volume and force it through the pipeline.

Driver: A mechanical component that supplies power to move another mechanical component.

Gate valve: A valve with a closing element consisting of a wedge or slab positioned between two seating surfaces that can be moved into or out of the flow stream perpendicular to the direction of the flow.

Line fill: The amount of gas needed to fill a new line before deliveries can be made.

Line pack: The introduction of new gas at the receipt point.

Manifold: A pipeline fitting with several side outlets to connect it with other piping. It is also known as a header.

Maximum allowable operating pressure (MAOP): The highest level of internal gauge pressure allowed for a particular pipeline or pipeline segment, as set by the US Department of Transportation.

Maximum operating pressure (MOP): The maximum internal gauge pressure at which a hazardous liquid or gas pipeline system or segment may be operated.

Plug valve: A valve whose closure element is a tapered plug.

Portable valve operator: A device that uses manual, hydraulic, electrical, or pneumatic energy to open or close valves.

Pressure-relief valve: A valve that relieves pressure on a system by allowing the pressurized liquid or gas to flow from an auxiliary passage out of the system.

Valve alignment: The exact sequence of valves that must be opened and closed at the same time so that gas can be received at, and delivered to, specific locations.

Additional Resources

The following recommended resources can provide additional helpful information related to the requirements for this covered task:

API Recommended Practice 1161, Third Edition, © 2014 American Petroleum Institute. API Publishing Services, 1220 L Street, NW, Washington, DC 20005. Distributed by Thompson Reuters (Scientific) LLC, **www.techstreet.com**.

NCCER Module AOCCC-17, *Abnormal Operating Conditions Control Center*

NCCER Module AOCFG-17, *Abnormal Operating Conditions Field and Gas*

NCCER Module 62105-02, *Pipeline Documentation*

NCCER Module 63104-02, *Piping and Mechanical Blueprint Reading*

NCCER Module 63208-02, *Introduction to Gas Compressors*

NCCER Module 67102-15, *Basic Pipeline Pneumatics and Equipment*

NCCER Module 67103-15, *Pipeline Communications*

NCCER Module CT18_1-17, *Utilize Leak Survey Techniques*

NCCER Module CT65_3-17, *Monitor pressures, flows, communications and line integrity and maintain them within allowable limits (gas)*

49 *CFR* 192, **www.ecfr.gov**.

49 *CFR* 195, **www.ecfr.gov**.

Figure Credits

arhendrix/Shutterstock.com, Module Opener

© iStockphoto.com/ekipaj, Figure 1

Tim Mills, Figures 2, 6

Courtesy of Honeywell International Inc., Figure 3

GTS Productions/Shutterstock.com, Figure 4

© iStockphoto.com/westphalia, Figure 5

Section Review Answer Key

Answer	Section Reference	Objective
Section One		
1. b	1.1.1	1a
2. d	1.2.1	1b
3. d	1.3.1	1c

NCCER CURRICULA — USER UPDATE

NCCER makes every effort to keep its textbooks up-to-date and free of technical errors. We appreciate your help in this process. If you find an error, a typographical mistake, or an inaccuracy in NCCER's curricula, please fill out this form (or a photocopy), or complete the online form at **www.nccer.org/olf**. Be sure to include the exact module ID number, page number, a detailed description, and your recommended correction. Your input will be brought to the attention of the Authoring Team. Thank you for your assistance.

Instructors – If you have an idea for improving this textbook, or have found that additional materials were necessary to teach this module effectively, please let us know so that we may present your suggestions to the Authoring Team.

NCCER Product Development and Revision

13614 Progress Blvd., Alachua, FL 32615

Email: curriculum@nccer.org

Online: www.nccer.org/olf

❑ Trainee Guide ❑ Lesson Plans ❑ Exam ❑ PowerPoints Other ____________________

Craft / Level: Copyright Date:

Module ID Number / Title:

Section Number(s):

Description:

Recommended Correction:

Your Name:

Address:

Email: Phone: